FISH IN RESEARCH

A SYMPOSIUM ON THE USE OF FISH AS AN EXPERIMENTAL ANIMAL IN BASIC RESEARCH

SPONSORED BY THE UNIVERSITY OF SOUTH DAKOTA
VERMILLION, SOUTH DAKOTA

NOVEMBER 15-16, 1968

FISH IN RESEARCH

EDITED BY

OTTO W. NEUHAUS
DEPARTMENT OF BIOCHEMISTRY
THE UNIVERSITY OF SOUTH DAKOTA
VERMILLION, SOUTH DAKOTA

JOHN E. HALVER
WESTERN FISH NUTRITION LABORATORY
U.S. BUREAU OF SPORT FISHERIES AND WILDLIFE
COOK, WASHINGTON

ACADEMIC PRESS

NEW YORK | LONDON | 1969

ACADEMIC PRESS, INC.
111 Fifth Avenue, New York, New York 10003

United Kingdom Edition published by
ACADEMIC PRESS, INC. (LONDON) LTD.
Berkeley Square House, London W1X 6BA

LIBRARY OF CONGRESS CATALOG CARD NUMBER: 74-107020

PRINTED IN THE UNITED STATES OF AMERICA

INVITED SPEAKERS

Ashley, Laurence M., Western Fish Nutrition Laboratory, U. S. Fish and Wildlife Service, Cook, Washington.

Bilinski, E., Vancouver Laboratory, Fisheries Research Board of Canada, Vancouver, British Columbia, Canada.

Conte, Frank P., Department of Zoology, Oregon State University, Corvallis, Oregon.

Gall, Graham A. E., Agriculture Experiment Station, College of Agriculture, Department of Animal Science, University of California, Davis, California.

Goldberg, Erwin, Department of Biological Sciences, Northwestern University, Evanston, Illinois.

Halver, John E., Western Fish Nutrition Laboratory, U. S. Bureau of Sport Fisheries and Wildlife, Cook, Washington.

Hastings, W. H., Fish Farming Experimental Station, U. S. Fish and Wildlife Service, Stuttgart, Arkansas.

Idler, D. R., Halifax Laboratory, Fisheries Research Board of Canada, Halifax, Nova Scotia, Canada.

Mertz, Edwin T., Department of Biochemistry, Purdue University, Lafayette, Indiana.

Ridgway, George J., Biological Laboratory, U. S. Fish and Wildlife Service, West Boothbay Harbor, Maine.

Scarpelli, D. G., Department of Pathology and Oncology, University of Kansas Medical Center, Kansas City, Kansas.

Sinnhuber, R. O., Department of Food Science, Oregon State University, Corvallis, Oregon.

Siperstein, M. D., Department of Internal Medicine, Southwestern Medical School, University of Texas, Dallas, Texas.

INVITED SPEAKERS

Tarr, H. L. A., Vancouver Laboratory, Fisheries Research Board of Canada, Vancouver, British Columbia, Canada.

Wellings, S. R. Department of Pathology, University of Oregon Medical School, Portland, Oregon.

Zaugg, W. S., Western Fish Nutrition Laboratory, U. S. Fish and Wildlife Service, Cook, Washington.

PREFACE

This volume is comprised of papers presented at a symposium entitled "Fish in Research" sponsored by the University of South Dakota in Vermillion. The purpose of the symposium was to ask those directly involved in research on fish, "What unique information of biochemical and physiological processes can be gained by using fish as experimental animals?" Discussions at the meeting indicated this approach to be of great interest, but never adequately considered. We sincerely hope that the unique question explored will stimulate future meetings and other volumes of proceedings dealing with the use of fish in research.

We wish to express thanks to the Bureau of Sport Fisheries and Wildlife of the U. S. Department of the Interior for its interest and participation. Special thanks are due Mr. John S. Gottschalk, Director, Bureau of Sport Fisheries and Wildlife, for addressing the guests of the symposium at a special banquet. His subject was "Basic Research Programs in the Bureau of Sport Fisheries and Wildlife Using Fish as Experimental Animals." Also we thank Clarence Johnson, Western Fish Nutrition Laboratory, Cook, Washington, Bruno von Limbach, Fish Genetics Laboratory, Beulah, Wyoming, and Dr. Norman Benson, North Central Reservoir Investigations, Yankton, South Dakota, for their participation in the displays of fish.

Our sincere appreciation is extended to the president of the University of South Dakota, first Dr. Edward Q. Moulton and now Dr. Richard Bowen, and the Dean of the School of Medicine, Dr. George W. Knabe, Jr., for the support and sponsorship of this venture. Special thanks are due to Mrs. Dorothy E. Neuhaus for her extensive help in organization, public relations, and undaunted efforts to ascertain that all went well. We wish to thank Mrs. Myla Kampshoff for preparing the manuscripts and the South Dakota Geological Survey for the use of the IBM Magnetic Tape Selectric Composer system. Finally, we are grateful to the faculty of the Department of Biochemistry and our many friends for their unstinting help.

Otto W. Neuhaus
John E. Halver

September, 1969

CONTENTS

CONTENTS

TOPICS IN GENETICS
E. Goldberg – Chairman

TOPICS IN NUTRITION
J. E. Halver – Chairman

WELCOMING ADDRESS

Otto W. Neuhaus

Sometimes scientists in the Middle West express the feeling that nonmetropolitan areas are, scientifically speaking, hinterlands. I heard this expressed at a major institution not long ago. In my opinion, it would be better to speak of scientifically underdeveloped regions because then we are challenged to do something about it. The present symposium is an effort on our part to meet this challenge and to enhance the scientific climate in our institution and the State of South Dakota. For once we are priviledged to perform the duty of being host. We may not be expert at the job, but we sincerely hope that you will find this weekend to be pleasant, informative, and especially scientifically provocative.

Another purpose of this symposium is to challenge us, the faculty of the Department of Biochemistry, to consider seriously the potentials of using fish as experimental animals in anticipation of the possibility that research in this area may lead to opportunities for expansion of the biological sciences at this university.

I am deeply indebted to Dr. John E. Halver, director of the Western Fish Nutrition Laboratory, and his staff, especially Clarence Johnson, for their help in arranging part of this program and especially for bringing to Vermillion fascinating displays of live fish as well as their movies. We hope that you will take the time to visit these displays.

This symposium is purposely intended to be unique not only for us in the field of biochemistry, but also for you who are active in the field of fish research. We know that you have many meetings and symposia in which to report your studies on fish; therefore, this in itself is not unique. The challenge of the present symposium is not so much to review what is known concerning fish, but to determine how studies on fish can yield unique insights into biochemical and physiological phenomena.

In choosing this program we as a department are very much fish out of water. Only one of our staff is actually using fish in his research activities. In this instance, studies of collagen structure, formerly restricted to rat tail tendon, are now being expanded to include fish skin. Otherwise, our research activities center around the use of rats, frogs, *Ascaris lumbricoides*, and algae. As biochemists, we are not really interested in the welfare of rats and frogs or in learning more about algae and invertebrates. Indeed, each simply serves as a model system for the study of specific biochemical mechanisms. In other words, the primary motivation in biochemical research is to gain an understanding of chemical mechanisms of life processes. The exact model system or experimental organism used is often secondary.

I have asked myself many times why fish have not been used more extensively for studies of biochemical mechanisms and, in fact, are we really missing a good opportunity? The capacity of fish to adapt to their environment suggests numerous opportunities to study metabolic control mechanisms. For example, what is the sequence of events that allows the normally ammonotelic lung fish to become ureotelic when cut off from water? When the normal environment is restored, how do they become ammonotelic again? What mechanisms control salt balance as a fresh water adapted salmon enters its marine environment? How can trout perform their metabolic activities in a 10°C environment whereas in warm blooded animals metabolism functions best at 37°? Perhaps by using fish we could learn more about metabolic control mechanisms and the mechanism of action of hormones. Could we learn more about the mechanisms of enzyme action and the protein structure required for specific catalytic activities by using fish enzymes? Perhaps there are subtle differences in primary structure or conformation of enzyme molecules from fish adapted to low temperatures compared with the same enzymes of warm blooded animals.

The central theme of this symposium is best expressed in the deliberate choice of our title. Not "Research *on* Fish," but "Fish *in* Research." Let the central challenging question be, "What special insights can be gained in biochemical and physiological mechanisms using fish as an experimental animal?"

TOPICS IN CANCER

S. R. Wellings, Chairman

ENVIRONMENTAL ASPECTS OF NEOPLASIA IN FISHES

S. R. Wellings

INTRODUCTION

In the physical, chemical, and biological evolution of the earth, numerous naturally occurring tumorigenic agents were initially present, or subsequently appeared in soil and water. These natural agents include various heavy metals, inorganic compounds, and hydrocarbons, in addition to radioactivity, ultraviolet light, and tumorigenic parasites. The exponential acceleration of human cultural evolution, beginning perhaps with the controlled use of fire, has introduced further carcinogenic agents: new and unique chemicals, increased concentrations of radioactive substances, various hydrocarbon and other atmospheric contaminants, and insecticides and herbicides with carcinogenic properties. In addition, man has modified the spatial distribution of various carcinogens by a variety of concentration and transportation procedures. The future promises further contamination of the environment by a multiplicity of agents, especially chemical ones, of unknown potential. The general problem of environmental neoplasia has been recently reviewed (1).

The study of environmental neoplasia in animals and plants becomes of considerable importance as a means of recognizing the environmental presence of unsuspected new carcinogens, or increased concentrations of known carcinogens, before these agents damage the various populations of living organisms, including man and his food sources.

In the interest of this conference, this review will point out some environmental aspects of neoplasia in fishes. Some examples are clearly related to human modifications of the environment, whereas other examples are not. In this context it should be pointed out that naturally occurring phenomena of any kind are best studied in detail before human-generated modifications ensue. In the examples which follow, this ideal was generally not attainable; the phenomena discussed are nonetheless of considerable biological and practical interest.

HEPATOMA

The most remarkable example of neoplasia in fish resulting from human activity is the appearance in 1960 of hepatomas in epidemic proportions in hatchery-reared rainbow trout, *Salmo gairdneri*, at various localities in the United States and Europe (2). Hepatomas were also observed in cutthroat trout (*S. clarki*) and brown trout (*S. trutta*).

The appearance of the disease in epidemic proportions followed the introduction of dry, pelleted commercial feeds as a primary dietary source in private, state, and federal hatcheries. This and other circumstantial evidence led to fractionation of known hepatomagenic diets, culminating in the isolation of a group of four potent chemically related, lipid soluble carcinogenic agents, named aflatoxins B_1, B_2, G_1, and G_2. These compounds are synthesized by certain mutants of *Aspergillus flavus*, growing in vegetable dietary constituents, such as cottonseed or peanut meal products, stored under damp conditions.

Experimental studies revealed that rainbow trout fed crude aflatoxin at 0.8 parts per million of diet developed classical multinodular hepatomas. The feeding of 1 part per billion of purified aflatoxin B_1 produced hepatomas in a high percentage of fish. Careful attention to storage of dietary constituents resulted in disappearance of the problem in affected hatcheries. The lesson to be learned from this is obvious: modification of the content or methods of preparation of human and animal foodstuffs without testing of the final chemical constituents may result in the introduction into the diet of unsuspected hazardous carcinogenic substances. In man in particular, the latency period for the production of cancer may be expected to be between about 10 and 30 years. Therefore, the outcome of such dietary introductions might not be seen for many years, and then only too late for many of those affected.

The entire problem of aflatoxicosis and trout hepatomas is reviewed elsewhere in detail (2) and in annotated bibliographic form (3). It should be pointed out that other carcinogenic agents, including aminoazotoluene, DDT, p-dimethylaminoazobenzene, and dimethylnitrosamine (4), are hepatomagenic when fed to rainbow trout.

Liver tumors were induced in *Brachydanio rerio* by addition of diethylnitrosamine to the aquarium water (5). There was early necrosis of liver cells followed by the appearance of multiple regenerative foci of parenchymal and biliary cells. Macroscopically visible hepatomas and cholangiomas were noted between the 10th and 30th weeks of the experiment in 17 of 63 fish. In some instances the tumors were invasive. This study indicates that environmental neoplasia can be analyzed in the laboratory utilizing common species of freshwater aquarium fish.

Hepatic neoplasia has recently been observed in two species of bottom-feeding fish collected from Deep Creek Lake, Maryland (6). Of twelve white croakers (*Catastomus commersoni*), three had hepatic neoplasms: two cholangiomas and one invasive cholangiocarcinoma. Among one hundred brown bullheads (*Ictalurus nebulosus*), one fish had multiple liver nodules which were interpreted as minimal deviation hepatomas. Data of this sort suggest the presence of naturally occurring or human-generated carcinogenic agents.

THYROID HYPERPLASIA AND NEOPLASIA

There are a number of reports in the older literature of apparent thyroid hyperplasia, occurring in artificially raised or maintained trout of several species (7-13). In some cases the hyperplasia was possibly related to dietary iodine deficiency or to low oxygen content of the water (9-12, 14) clearly suggesting that the operation of these or other environmental factors may be of importance in the genesis of thyroid neoplasia in fishes, which doubtless occurs, even though most previously alleged examples of thyroid neoplasia were probably in fact hyperplasia.

Stolk (15) observed hyperplasia and adenoma formation in *Xiphophorus helleri* and *Lebistes reticulatus* following the addition of thiouracil to the aquarium water. Occasionally, thyroid enlargement has also been observed among naturally occurring fish populations (16), again suggesting that goiterogenic or possibly carcinogenic agents may be present in the water. The entire problem of naturally occurring and experimentally induced thyroid hyperplasia and neoplasia in fishes is in need of further experimental analysis.

Hamre and Nichols (17) observed the simultaneous appearance of thyroid hyperplasia and exophthalmos in trout fry immediately following resorption of the yolk sac at the time when the fry began to eat. Microscopically, the thyroid gland showed diffuse hyperplasia; epithelium was of the tall columnar type and the follicular lumens contained no colloid. The hyperplasia of thyroid tissue and the exophthalmos were both prevented by the addition of sodium iodide to the water.

EPIDERMAL PAPILLOMAS

Epidermal papillomas are relatively common in some species of fish. The incidence may vary in different parts of the range of a given species, suggesting variation in certain critical tumorigenic environmental agents. Early reports of this kind include Keysselitz's (18) study of epidermal papillomas on the lips of barbels, *Barbus fluviatialis*, from the Mosel River, and Breslauer's (19) report of papillary tumors of the lips, buccal mucosa, and fins of smelt, *Osmerus eperlanus*, collected in brackish water in a particular inlet of the Baltic Sea.

Transplantable epitheliomas have been observed on the lips or in the buccal cavity of large numbers of catfish, *Ameiurus nebulosus*, taken from grossly polluted waters of the Delaware and Schuylkill Rivers near Philadelphia (20). A focal area of marked hyperemia was followed by epidermal hyperplasia and finally by the development of epidermoid carcinoma. Some of the carcinomas were invasive and tumor emboli were observed in vessels. Homotransplantation to the anterior chamber of the eye was readily accomplished.

In a more recent report 10 of 353 white croakers, *Genyonemus lineatus*, were observed with epidermal papillomas located about the mouth (21). The tumors were composed of thickened epithelial folds consisting of squamous cells showing normal maturation and polarity. There was no invasion or metastasis. The white croakers were collected with trawling gear in water approximately 2 miles from the outfall of a sewage treatment plant in Santa Monica Bay, California. Over 1,000 fish of the same species were collected from unpolluted waters 50 miles away, but no tumors were observed. Instances such as these are suggestive of a relationship to pollution, but the evidence is circumstantial. Experimental analysis of these and similar problems is indicated.

Epidermal papillomas ("cauliflower disease") on eels, *Anguilla anguilla*, have become a relatively common occurrence in collections from some parts of the Baltic Sea (22-23). The disease has become a major threat to the eel fisheries in the years since 1957, although sporadic reports occurred prior to that time. The tumors tend to involve the head region and consist of lumpy cauliflower-like lesions. Microscopically, they are papillomatous, and composed of folds of epithelium supported on a fibrovascular stroma. Clear evidence of invasion is not observed. The high incidence and pattern of geographic spread over the years suggest an infectious process or progressively accumulating industrial wastes such as ship fuel oil and wastes from smelters (24), known to contain carcinogenic hydrocarbons such as benzo[a]pyrene, and heavy metals such as arsenic.

Nigrelli (25) observed epidermal papillomas on the lips of 0.5 to 1 percent of dwarf gourami, *Colisa lalia*, fish raised in outdoor pools in Florida for the tropical fish trade. This author suggested the possibility of an infectious process. Microscopically, the tumors were epidermal papillomas. Cartilaginous and bony metaplasia of the stroma frequently occurred. There were some cases which had progressed to malignant papillary epidermoid carcinoma. Although inclusion bodies may have been present, the tumors were not transmissible.

Epidermal papillomas are also relatively frequent on Japanese gobies, *Acanthogobius flavimanus* (26-29). The epidemic nature of the disease and the presence of intracellular virus-like particles by electron microscopy suggest an infectious viral etiology (30).

There are many reports of epidermal papillomata (Fig. 1) occurring among the members of the order Heterosomata. The more recent studies (Fig. 2) from parts of Puget Sound and adjacent waters (31-35) strongly suggest the operation of environmental factors in that the incidence of tumors varies unpredictably at different collection sites within the range of the same species. Moreover, at least in the instance of the flathead sole, *Hippoglossoides elassodon*, and the starry flounder, *Platichthys stellatus*, the small tumors occur on fish of the year collected in late summer and fall. Annual cycles of tumorigenesis affecting only the fish of the year suggest that chemical, physical, or biological agents are present which likewise may show a yearly cyclic variation in intensity, or which may be concentrated largely in a habitat occupied only by the young fish. Alternatively, the tumorigenic agent(s) may be present in uniform quantities throughout the year, and the yearly cycle of tumorigenesis is the result of age-dependent differences in susceptibility of the young fish.

Among the more curious aspects of the flounder papillomas is the statistically significant tendency for the tumors to occur on the eyed side in contrast to the blind side in both the flathead sole and the starry flounder. This is the more interesting in the instance of the starry flounder (Fig. 3), inasmuch as individual fish may be either dextral (right side pigmented, eyed, and uppermost) or sinistral (left side pigmented, eyed, and uppermost). The tumor incidence is still highest on the eyed side irrespective of the kind of symmetry of the individual fish.

The epidermal papillomas of flounders consist of thick branching folds of epidermal cells supported on a connective tissue stroma containing blood vessels and pigment cells (Fig. 4). The epidermal component of the larger tumors contains characteristic ovoid, hypertrophic cells with enlarged spherical nucleoli (Fig. 5). Similar but smaller cells are also observed in the stroma of many tumors. It is within these cells that objects resembling viruses have been observed by electron microscopy (32). The hypertrophic cells of the epidermis are 2 or 3 times larger than normal epidermal cells, and bear no resemblance to the enormously hypertrophic cells of lymphocystis disease of *Glugea* infestations. Similarly, the objects suspected of being intracellular parasites in the cells of the flounder papillomas are unlike lymphocystis virus particles, microsporidia, or other known agents capable of producing cell proliferation and/or hypertrophy.

At least in the instance of the starry flounder, *Platichthys stellatus*, the English sole, *Parophrys vetulus*, and the flathead sole, *Hippoglossoides elassodon*, there is evidence that the epidermal papillomas, averaging several cm in greatest dimension, develop from early 1 to 2 mm lesions which consist of proliferating dermal vascular connective tissue, and which are more suggestive of an inflammatory response than of neoplasia. These early lesions, termed angioepithelial nodules, develop a cap of hyperplastic epidermis which in time becomes folded and papillomatous. Several parasites which may serve as vectors are frequently observed on the small fish. These include at least one species of *Gyrodactylus*, a monogenetic tremotode, and *Lepeophtheirus,* a parasitic copepod. Our knowledge of the flounder papillomas to date suggests that tumorigenesis is related to as yet undefined environmental variables of a chemical, physical, and biological nature.

MESENCHYMAL TUMORS OF GOLDFISH

Mesenchymal tumors, variously described as fibromas, fibrosarcomas, leiomyomas, and neurilemmomas, have been reported on several occasions in isolated goldfish populations (36-41). The tumors were generally described as multiple, and located beneath the skin or in the abdominal cavity. In the report of Lucke, Schlumberger, and Breedis (38), 30 tumor-bearing goldfish were observed. Most of these were collected in three small ponds, and it was pointed out by the authors that once the disease was established in a particular pool, the incidence tended gradually to increase. The evidence is suggestive of the operation of hereditary factors, or environmental agents, or both.

SPINDLE CELL SARCOMAS OF PIKE–PERCH

Multiple skin sarcomas, possibly fibromas or fibrosarcomas, were observed on pike–perch, *Stizostedion vitreum*, collected in Lake Oneida (42-43). The tumors arose in the dermal connective tissue and were usually composed of cellular formations of spindle cells with variable amounts of fibrovascular stroma. Electron microscopy revealed cytoplasmic virus-like particles in abundance. The particles were unlike those of lymphocystis disease, and averaged about 1000 Å in diameter. They were not clearly polyhedral, and possessed dense nucleoids. Complete particles were observed inside the spaces of the endoplasmic reticulum and outside the cells. The virus-like particles appeared to form in relation to cellular membrane systems.

MULTIPLE OSTEOMAS OF THE RED TAI

The commonly observed vertebral osteomas of the red tai, *Pagrosomas major*, have aroused speculation as to the relative roles of genetic and environmental influences (44-47). These tumors apparently occur only on the haemal spines of the 6th to 8th caudal vertebrae in this species. Takahashi (46) reported a series of 102 fish. The tumors varied in size from about 1 to 3 cm and were sometimes multiple. Takahashi thought these related to previous fractures followed by the development of exuberant callus which subsequently ossified to form tumor masses of mature cancellous bone. Schlumberger and Lucke (47), however, found no evidence of previous fracture by means of roentgenographic and histological studies of one specimen.

LYMPHOSARCOMAS OF ESOX

In 1947 Nigrelli (48) observed lymphosarcomas of the kidney in 12 northern pike, *Esox lucius*, which had been kept together in the same aquarium for several years. Inasmuch as the bulkiest single mass of tumor involved and replaced the kidneys, this was considered to be the site of origin. There was, however, widespread involvement of the liver, spleen, and retroperitoneum in each case. The tumors were composed of cellular, lymphoblastic tissue. Ritchie (see Schlumberger [41]) observed lymphosarcoma on a number of occasions in the Muskellunge, *Esox masquinongy*. Ten percent of fish captured in Lake Scugog, Ontario, were affected.

Recently, Mulcahy (49) and Mulcahy and O'Rourke (50) have observed apparent lymphosarcoma in *Esox lucius* in Ireland, sometimes occurring in epidemic proportions. There was a tendency for the tumors to involve structures about the head, including the jaws, tongue, pharynx, and gills. In a number of instances tumors, often multiple, were externally visible involving the skin of the head, trunk, or tail. Visceral involvement might include spleen, pancreas, stomach, kidney, air bladder, and rectum. The tumors were largely composed of sheets and cords of round cells with basophilic ovoid or round nuclei and scanty cytoplasm. Some plasma cells and macrophages were also scattered through the neoplastic tissue. The tumors were generally invasive and without encapsulation and frequently caused ulceration at adjacent epithelial surfaces.

In view of the well known virus-induced leukemias and lymphomas of mice and rats, a careful search for virus-like particles and biologically active tumorigenic virus will doubtless be conducted.

EXPERIMENTALLY INDUCED CHORDOMAS IN EMBRYO TELEOSTS

A recent report (51) indicates that chordomas can be induced in *Oryzias latipes*, *Anoptichthys jordane*, *Trichogaster tricopterus*, *Aequidens portalegrensis*, and *Carassius auratus* by placing embryos in water containing beta-aminoproprionitrile. The resulting tumorous overgrowths of physalliferous cells and fibroblasts were thought to be neoplastic because they appeared after removal of the causative agent and failed to regress. Such experiments suggest that naturally occurring or human generated chemical substances may produce fish neoplasms in nature.

UNDIFFERENTIATED SARCOMAS IN LEBISTES AND PRISTELLA

Wessing (52) and Wessing and Von Bargen (53) have observed an undifferentiated sarcoma occurring in the aquarium fish *Lebistes reticulatus* and *Pristella riddlei*. Cell-free filtrates transmitted the disease to the same and other species, suggesting that a virus was the etiological agent. Tumors were most frequently observed in the kidneys, but also in the body cavities, heart, intestine, and testis. A viral etiology appears to be very likely in this instance.

MELANOMAS OF SWORDTAIL–PLATYFISH HYBRIDS

Haussler (54) first observed melanomas in swordtail–platyfish hybrids (*Xiphophorus helleri* X *Xiphophorus maculatus*). The genetics of this system has been clarified by Gordon and his associates (55-57), and the subject has been recently reviewed by Anders (58). It has been observed that platyfish in nature exhibit different distribution of cutaneous macromelanophores resulting in several different color patterns. The evidence to date suggests that each color pattern may be determined by at least one color gene, influenced in part by a number of modifier genes. Crossing of swordtails with platyfish results in hybrids which may lack a full complement of modifier genes, thus permitting in some instances the uncontrolled proliferation of macromelanophores, first leading to melanosis, and finally to invasive melanomas.

This system is important to the present discussion because a number of environmental variables may act on the genome of the susceptible hybrids in such a way as to alter the mean tumor age or influence the growth rate of the tumors (58-59). Proewig (59) observed that changes in ionic strength of the water markedly influenced the growth rate of the melanomas, and that lowering the surface tension of the water from 72 dynes per cm to 52 dynes per cm enhanced tumor growth. It is unlikely that the incidence of any tumor, even those with a seemingly firm genetic basis, is uninfluenced by environmental agents.

MULTIPLE NERVE SHEATH TUMORS

Multiple nerve sheath tumors resembling neurilemmomas of humans were observed in 0.5 to 1.0 percent of snappers (Lutianidae) of the following species collected off Tortugas: *Lutianus griseus*, *L. jocu*, and *L. apodus* (47, 60). Seventy-six tumor-bearing fish were collected. The tumors were localized along the subcutaneous nerves, especially of the head and dorsal trunk. They were white, firm, and microscopy revealed a mixture of fibrocellular and loose reticulated tissue corresponding to that observed in human neurilemmomas. Palisading of nuclei and whorled formations were also observed. A somewhat analogous situation in humans (von Recklinghausen's neurofibromatosis) has a well established genetic background, and the same could be true in the instance of the nerve sheath tumors of snappers. Experimental analysis of this system and others like it (61) would be of considerable value, and might well indicate that environmental agents are capable of modifying the tumor incidence.

TERATOMAS ASSOCIATED WITH ICHTHYOPHONUS INFESTATION

Stolk (62-63) has observed ovarian teratomas in one of 1,484 guppies, *Lebistes reticulatus*, and in one of 1,179 platyfish, *Xiphophorus maculatus*. In the instance of either species the ovary and its environs were infiltrated by the parasitic phycomycete, *Ichthyophonus hoferi*. It is possible that a chemical agent released from the parasite induced parthenogenetic development in unfertilized oocytes, leading to the formation of teratomas.

This possible relationship of parasitic infestation to hyperplasia or neoplasia has been previously mentioned in regard to epidermal papillomas of flounders (35).

DISCUSSION

A great deal of data has accumulated to implicate environmental agents as important for carcinogenesis in humans and other animals. Many of these carcinogens enter the natural waters and come in contact with fish and invertebrate inhabitants. These injurious agents of natural, industrial, and agricultural origin are very numerous, and probably include ultraviolet light, crude oil, various naturally occurring soluble metals and their salts, petroleum wastes (from power boats, oil refineries), DDT (and other insecticides), benzol, arsenic (from slag, orchard sprays), various domestic wastes (such as detergents, soaps, oils), herbicides, aromatic amines, various compounds effluent from mines and mineral refineries (cobalt, nickel, iron), and various dye stuffs. Ionizing radiations may originate from natural uranium and thorium ores, or from human activities, resulting in contamination of air, soil, ground water, springs, streams, and lakes. Many agents, such as DDT, some household detergents, and radioactive wastes, are not readily decomposed by bacterial or other means, and will continue to accumulate in the environment if they are added at present rates. Carcinogenic hydrocarbons have been found in the tissues of oysters (64) and barnacles (65) taken from harbors polluted with industrial wastes and ship oil.

The mechanisms by which these environmental agents act to generate neoplasia are unknown. That these probably do act additively and synergistically in conjunction with multiple host factors is supported by analogies with well-documented mammalian systems. The classical example is that of the mouse mammary tumor system in which there is a complex interaction of more or less interrelated multiple factors, including age, sex, genetic strain, reproductive history, hormonal status, and mammary tumor virus. It is important that there are already strong indications that this "multiple factor hypothesis" of tumorigensis is not only applicable to mammals, but also throughout the biological world. The contemporary value of such a view is that it provides an advantageous basis for both clinical analysis and experimental design.

Current industrial advance in the production of new and old products in increasing quantities to meet the demands of proliferating human populations threatens to even further increase the load of environmental contaminants. So far, humans have shown little foresight, and relatively indiscriminant destruction and contamination of the natural environment has been the rule. That it will be any different in the future seems unlikely unless human welfare is reinterpreted to include the preservation of populations of diverse kinds of living systems which may be indispensable for survival.

The presence of toxic or carcinogenic quantities of new or old injurious materials in the environment may theoretically be detected by screening of wild and domestic populations of animals and plants for hyperplastic, anaplastic, and neoplastic diseases. Such an approach might meet with some success inasmuch as the first indication of the presence of a potent carcinogen (aflatoxin) in the diet of hatchery-raised trout was the appearance of hepatomas in epidemic proportions, involving up to 100 percent of fish in some instances. As previously mentioned, this is disturbing when it is recognized that the latency period in human cancers might average around 20 years; therefore, a potent carcinogen in the human diet might be overlooked until large portions of the human population were affected.

A relatively new problem is "thermal pollution": the elevation of the temperature of natural waters when these are used as coolants for industrial complexes such as atomic piles and power plants. Such temperature changes are certain to alter the kinds and numbers of animal and plant species in affected habitats. There may be changes in reproductive cycles, food chains, and incident parasitic, bacterial, and viral diseases. It would be reasonable to assume that the activity of the chemical or other carcinogenic content of the natural waters would be altered in a most complex manner. Population structure and disease patterns should ideally be studied in detail before thermal or other pollution occurs.

As subjects for the study of experimental neoplasia, fish offer unique opportunities because of their poikilothermic nature. Moreover, some species readily and reversibly adapt to wide changes in the chemical composition and salinity of the water. The effects of experimental modification of these variables on the genesis and subsequent growth of fish neoplasms should be of great general interest.

Unlike the common laboratory animals, some adult fish (e.g., salmonids) subsist largely on the proteinaceous natural diet. In these instances protein provides the raw material for biosynthesis of most tissue components, and for energy metabolism. If cancer can be looked upon as a change in kinds and amounts of proteins, leading to an alteration of phenotype at the cell and tissue levels, then it follows that the entire genetically determined mechanism of protein synthesis is a central part of the neoplasia problem. It would seem, therefore, that studies of experimental neoplasia in fishes would be of considerable interest from the comparative point of view. Especially interesting would be comparisons of protein breakdown and biosynthesis in neoplastic cells and their normal counterparts in fish, and a comparison of these data with those from the common laboratory animals.

Fish are also unusual in that growth continues through life. In mammals and birds, of course, this is not the case: a mature size is reached following an interval characteristic for the species, after which there is relative stability of dimensions and mass. Experimental analysis of the manner in which the age variable affects the development and growth of tumors might therefore prove interesting.

Finally, the overall advantage of comparative approaches in pathology is that one can study the effect of a single inciting agent (e.g., a chemical carcinogen) on a diversity of living systems, each with its own unique metabolic, physiologic, and biochemical mechanisms of response to injury. Thus, comparative studies are analyses of nature's own experiments; as such they may enjoy greater freedom from the extremes of artificiality which are unavoidable in some laboratory investigations.

CONCLUSION

The present review emphasizes those examples of neoplasia in fishes related to proven or suspected carcinogenic agents in the environment. It seems clear that there is need for more vigorous study of the structure of natural aquatic populations and natural disease incidences, in both polluted and unpolluted waters. Changes in neoplastic disease patterns may indicate dangerous modifications of the environment, sometimes human-generated, which could threaten valuable plant and animal populations, including man. It is obvious that present practices of human environmental pollution need to be modified in order to reduce the present probability of widespread damage to animal, plant, and human populations.

ACKNOWLEDGMENTS

This study was partially supported by Public Health Service research grant CA 08158 from the National Cancer Institute.

REFERENCES

1. W. C. Hueper, Ann. N. Y. Acad. Sci. **108**: 963 (1963).

2. J. E. Halver and I. A. Mitchell, eds., *Trout Hepatoma Research Conference Papers.* Res. Report 70. Bur. Sport Fish. and Wildl. G.P.O., Wash. D. C., 1967.

3. S. R. Wellings, Neoplasia and primitive vertebrate phylogeny: Echinoderms, prevertebrates, and fishes –– A review. Nat. Cancer Inst. Monograph. (in press).

4. J. E. Halver, C. L. Johnson and L. M. Ashley, Federation Proc. **21**: 390 (1962).

5. M. F. Stanton, J. Nat. Cancer Inst. **34**: 117 (1965).

6. C. J. Dawe, M. F. Stanton, and F. J. Schwartz, Cancer Res. **24**: 1194 (1964).

7. R. Bonnet, Bayer. Fischerie Zeit. **6**: 79 (1883).

8. M. Plehn, Allg. Fisch. Ztg. **27**: 117 (1902).

9. D. Marine and C. H. Lenhart, J. Exp. Med. **12**: 311 (1910).

10. D. Marine and C. H. Lenhart, Bull. Johns Hopkins Hosp. **21**: 95 (1910).

11. D. Marine and C. H. Lenhart, J. Exp. Med. **13**: 455 (1911).

12. D. Marine, J. Exp. Med. **19**: 70 (1914).

13. H. R. Gaylord and C. Marsh, Carcinoma of the thyroid in the salmonoid fishes. G.P.O., Wash., D. C., 1914.

14. P. A. MacIntyre, Zoologica. **45**: 161 (1960).

15. A. Stolk, Proc. Koninkl. Ned. Akad. Wetenschap. **58C**: 313 (1955).

16. O. H. Robertson and A. L. Chaney, Physiol. Zool. **26**: 328 (1953).

17. C. Hamre and M. S. Nichols, Proc. Soc. Exp. Biol. Med. **26**: 63 (1926).

18. G. Keysselitz, Arch. Protistenk. **11**: 326 (1908).

19. T. Breslauer, Arch. Mikrobiol. Anat. **87**: 200 (1916).

20. B. Lucke and H. G. Schlumberger, J. Exp. Med. **74**: 397 (1941).

21. F. E. Russell and R. A. Koten, J. Nat. Cancer Inst. **18**: 857 (1957).

22. M. Christiansen and A. J. C. Jensen, Rep. Danish Biol. Sta. **50**: 29 (1947).

23. W. Schaeperclaus, Z. Fischerie 2: 105 (1953).

24. E. A. Finkelstein, Arkh. Patol. 22: 56 (1960).

25. R. F. Nigrelli, Ann. N. Y. Acad. Sci. **54**: 1076 (1952).

26. K. Oota, Gann. **43**: 264 (1952).

27. S. Nishikawa, Collecting & Breeding **16**: 236 (1954).

28. M. Iwashita, Collecting & Breeding **17**: 50 (1955).

29. I. Kumura, T. Miyake, and Y. Ito, Proc. Jap. Cancer Assoc. Oct., 154 (1967).

30. T. Imai and N. Fujiwara, Kyushu J. Med. Sci. **10**: 135 (1959).

31. S. R. Wellings, H. A. Bern, R. S. Nishioka, and J. W. Graham, Proc. Amer. Ass. Cancer Res. **4**: 71 (1963).

32. S. R. Wellings and R. G. Chuinard, Science **146**: 932 (1964).

33. S. R. Wellings, R. G. Chuinard, and M. Bens, Ann. N. Y. Acad. Sci.**126**: 479 (1965).

34. R. F. Nigrelli, K. S. Ketchen, and G. D. Ruggieri, Zoologica, **50**: 115 (1965).

35. G. E. McArn, R. G. Chuinard, R. E. Brooks, B. S. Miller, and S. R. Wellings, J. Nat. Cancer Inst. **41**: 229 (1968).

36. A. H. Roffo, Neoplasma **3**: 231 (1924).

37. J. Montpellier and R. Dieuzeide, Bull. Franc. Cancer **21**: 295 (1932).

38. B. Lucke, H. G. Schlumberger, and C. Breedis, Cancer Res. **8**: 473 (1948).

39. H. G. Schlumberger, Amer. J. Pathol. **25**: 287 (1949).

40. H. G. Schlumberger, Cancer Res. **12**: 890 (1952).

41. H. G. Schlumberger, Cancer Res. **17**: 823 (1957).

42. R. Walker, Lymphocystic warts and skin tumors of walleyed pike. Rensselaer Polytech. **14**, Troy, 1948.

43. R. Walker, Amer. Zool. **1**: 395 (1961).

44. Y. Kazama, Gann. **18**: 35 (1924).

45. E. Sagawa, Gann. **19**: 14 (1925).

46. K. Takahashi, Z. Krebsforsch. **29**: 1 (1929).

47. H. Schlumberger and B. Lucke, Cancer Res. **8**: 657 (1948).

48. R. F. Nigrelli, Zoologica, **32**: 101 (1947).

49. M. F. Mulcahy, Proc. Roy. Irish Acad. **63B**: 103 (1963).

50. M. F. Mulcahy and F. J. O'Rourke, Irish Naturalists' J. **14**: 312 (1964).

51. B. M. Levy, Cancer Res. **22**: 441 (1962).

52. A. Wessing, Naturwissenschaften **46**: 517 (1959).

53. A. Wessing and G. von Bargen, Arch. Ges. Virusforsch. **9**: 521 (1959).

54. G. Haussler, Klin. Wochschr. **7**: 1561 (1928).

55. M. Gordon, *The Biology of Melanomas*. N. Y. Acad. Sci. **4**: 216 (1948).

56. M. Gordon, Growth **10**: 153 (1951).

57. M. Gordon, Cancer Res. **11**: 676 (1951).

58. F. Anders, Experientia **23**: 1 (1967).

59. F. Proewig, Z. Krebsforsch. **60**: 470 (1954).

60. B. Lucke, Arch. Pathol. **34**: 133 (1942).

61. G. A. Young and P. Olafson, Am. J. Pathol. **20**: 413 (1944).

62. A. Stolk, Proc. Koninkl. Ned. Akad. Wetenschap. **56**: 28 (1953).

63. A. Stolk, Nature **183**: 763 (1959).

64. H. J. Cahnmann and M. Kuratsuhe, Proc. Am. Assoc. Cancer Res. **2**: 99 (1956).

65. M. B. Shimkin, B. K. Koe, and L. Zechmeister, Science **113**: 650 (1951).

COMMENTS

DR. WELLINGS: The young fish collected in late autumn months in shallow water are frequently parasitized by Gyrodactylus and by Lepeoptheirus. Later in the year the parasites are more difficult to find. It has been suggested that there may be a yearly cycle of parasitism which correlates with the tumorigenic cycle.

We have statistically analyzed populations of normal and tumor fish and it appears that tumors are more frequent on smaller fish. The tumor fish weigh less for their age group and are shorter. Tumor fish may disappear from the population at a more rapid rate than non-tumor fish. Moreover, the location of the tumors on the fish suggest that the tumors impair survival. For example, tumors are sometimes observed on the gills, eyes, or in the oral cavity.

So far we have not been able to transplant the tumors or transmit the disease in spite of numerous attempts. We have tried cell free filtrates and purified DNA preparations. Autotransplants to the anterior chamber of the eye show limited survival. Homotransplants uniformly fail.

DR. NIGRELLI: Why do you have to assume that there are multiple factors involved in production of tumors? Why don't you think that a virus is probably responsible? In terms of time from an evolutionary point of view, do you think that this is a recent disease or has it been present for hundreds of thousands of years?

DR. WELLINGS: Adherence to the multiple factor notion to my mind is the best way to explain the experimental data in those systems which have been explored in detail. An example of such a system would be the mouse mammary carcinoma system in which we find a viral agent (the milk factor), the presence or absence of which will alter the tumor incidence. There is still a base line incidence, even in the absence of the milk agent. One can also modify the incidence of mammary tumors in mice by force breeding. Also, the factor of genetic strain is a prominent one. In some strains of mice nearly 100% of the females develop carcinomas; whereas, in other strains carcinoma of the mammary gland is extremely rare. It seems to me that the multiple factor idea is useful in all of the experimental systems which have been analyzed in detail so far.

DR. SCARPELLI: In the slippery dick, which is a coral reef fish, the same type of a series of events develop. An angeoepitheleal type lesion appears first and then develops into papilloma. Is this common in the eel and other types of species?

DR. WELLINGS: I am not sure about the eel. In the catfish example, some descriptions indicate that the earliest changes include vascular proliferation, followed by epidermal proliferation leading to the development of a papilloma. Therefore, there is possibly a common pattern in the development of papillomas in fish.

DR. SCARPELLI: We found that this type of papilloma, starting out with angeoepitheleal type of growth, is present practically wherever this species is found starting at Bermuda and going as far south as the northern tip of South America. In the Bahama areas we found it in 2½% of the population, so it is a relatively high incidence type of disease. We suspect it is a viral disease and would like to see more work done on the virus type of etiology.

DR. SINCLAIR: I was wondering whether this incidence was related to changes in the environmental system of the fish. In the Salton Sea when it was first seeded, there was a relatively high incidence of diseased fish, but these died out within a few years time. The population then became stable and the ecological balance became established. Is there any suggestion in Puget Sound that changes are relating to the environment of these fish?

DR. WELLINGS: There is a correlation between tumor incidence and low salinity. This is especially true in the instance of young starry flounder. Tumor incidence in these fish is in excess of 50% in some estuaries. No relationship has yet been found between oxygen demand or the presence of various pollutants. We have not studied some specific environments which would be of interest. For example, one could study the incidence of tumors in relation to smelters, oil refineries, or other industrial complexes in the region. We expect to sample widely in Puget Sound, and to select specific sites for intensive ecological study.

DR. MALINS: Have you any ideas on the involvement of lipids in these neoplastic epitheleomas?

DR. WELLINGS: We have not done any biochemical studies of any importance, so have no idea about involvement of lipids.

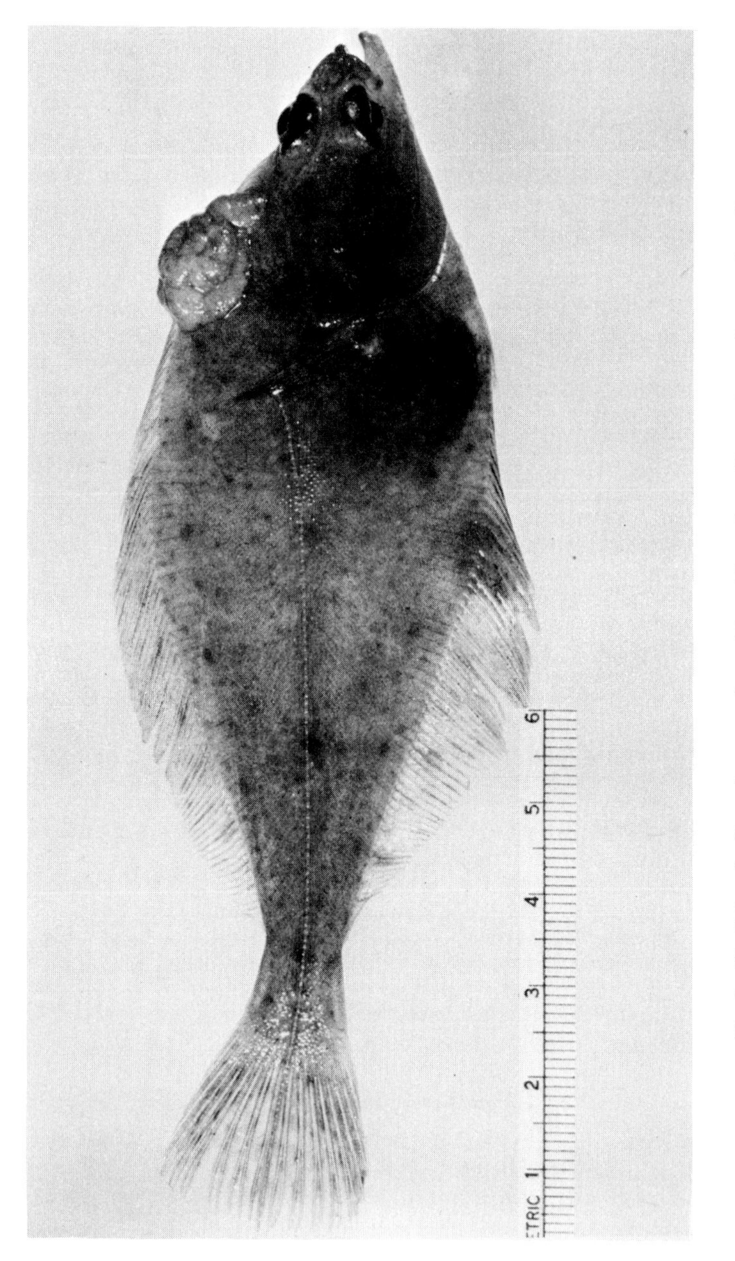

Fig. 1: English sole, Parophrys vetulus, with epidermal papilloma, located anterodorsally. Growth is soft, velvety, yellow-pink, and papillomatous.

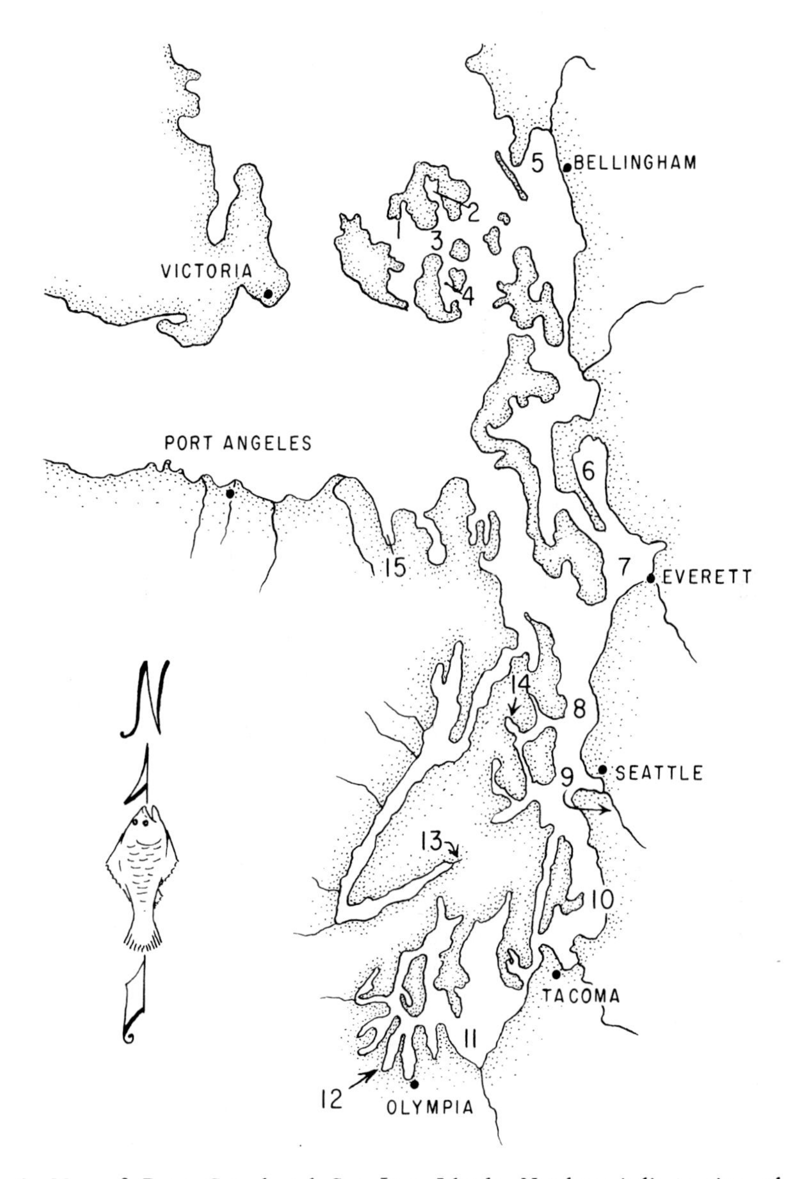

Fig. 2: Map of Puget Sound and San Juan Islands. Numbers indicate sites where pleuronectids afflicted with epidermal papillomas have been collected. Although detailed studies have been carried out only at sites 2 and 5, the distribution of the disease appears to be widespread.

Fig. 3: Starry flounder, Platichthys stellatus, with epidermal papilloma, located dorsally on pigmented side.

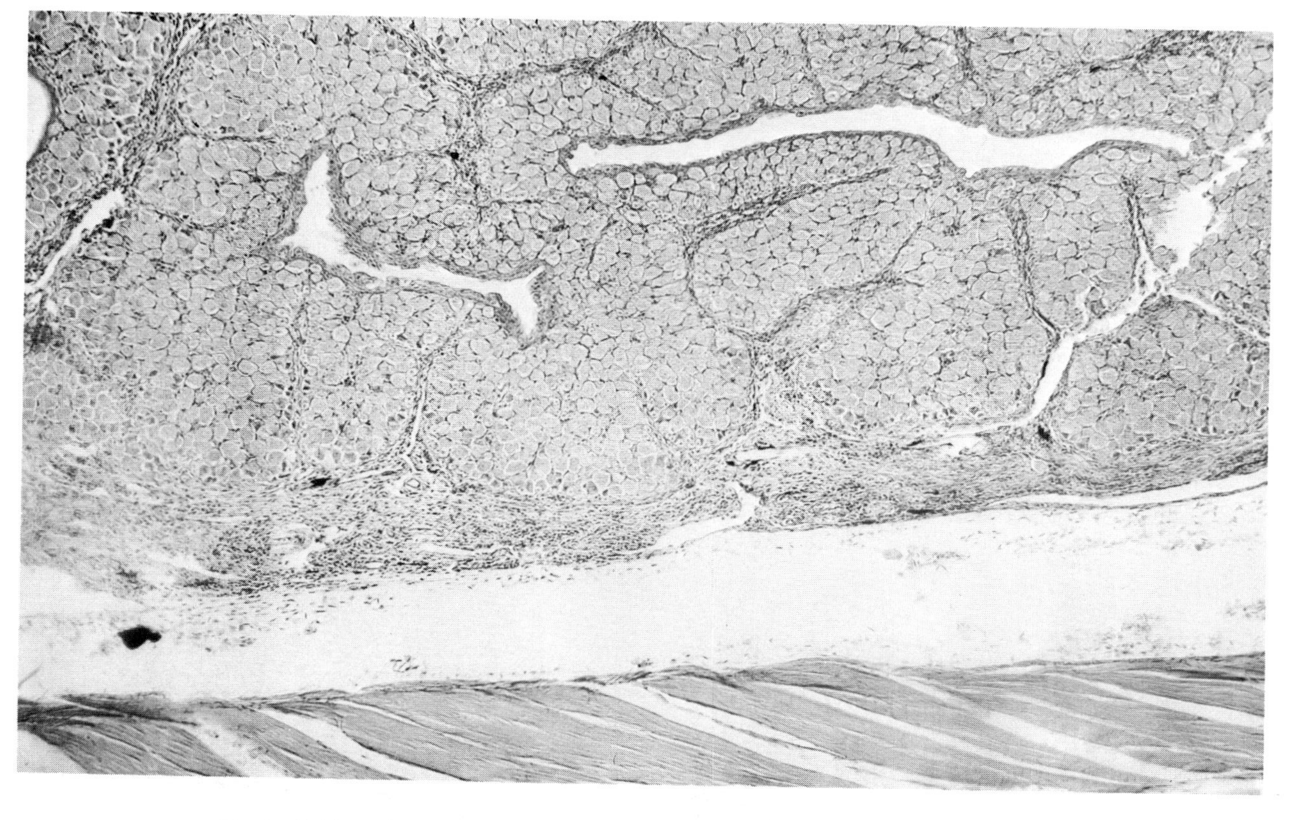

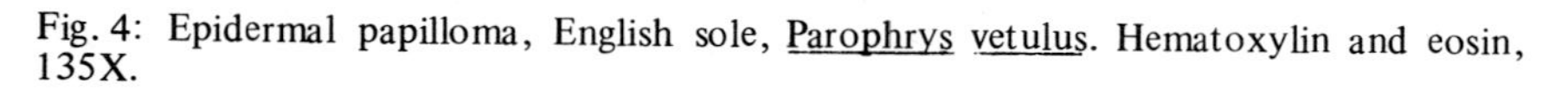

Fig. 4: Epidermal papilloma, English sole, *Parophrys vetulus*. Hematoxylin and eosin, 135X.

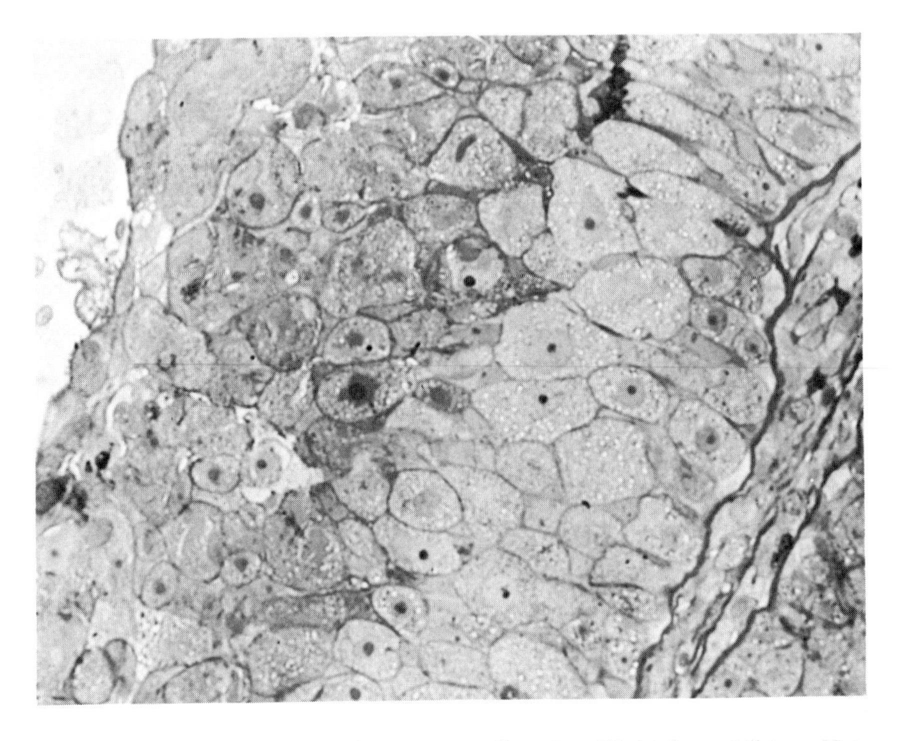

Fig. 5: Part of epidermal papilloma, starry flounder, Platichthys stellatus. Note characteristic ovoid hypertrophic cells with pale nuclei and very dark spheroid prominent nucleoli. Epon embedded, 1 μ section, stained with toluidine blue. 630X.

EXPERIMENTAL FISH NEOPLASIA

Laurence M. Ashley

INTRODUCTION

During the past 65 years, histopathology of neoplasia in fishes has undergone rather sporadic growth. This situation is now changed. In 1966 the U. S. Smithsonian Institution joined with the National Cancer Institute in establishing a REGISTRY OF TUMORS IN LOWER ANIMALS. Its purpose – to collect specimens, the study of which will aid in gaining an understanding of neoplastic processes. The University of Oregon Medical School, Department of Pathology headed by Dr. S. R. Wellings, has recently begun a similar registry for tumors of poikilothermic vertebrates in wild populations. Similar projects may now exist, or may soon be initiated, in various parts of the world. Such efforts should result in significant progress in both descriptive and experimental oncology. As additional specific causes of cancer are discovered, more effective prevention and more rational cancer therapy will become available.

Tumors and other atypical growths of fishes are generally comparable to those reported for mammals. While descriptive histopathology of fish neoplasms began in the late 19th century (1883), no published reports of experimental oncology of fishes occurred until about 1910 when Marine and

Lenhart (1, 2), based on controlled experiments, challenged the opinions of several authors who had reported thyroid carcinoma in fishes. Even later, in an extensive work, Gaylord and Marsh (3) reported many cases of simple hyperplasia (or goiter) as thyroid carcinoma. These diagnostic errors were due to the fact that fish thyroid is a diffusely arranged collection of follicles distributed along the ventral aorta and the first two or three pairs of afferent branchial arteries (Fig. 1). Hyperplasia of the scattered follicles closely resembled neoplasia, but only a few of these cases have been confirmed as thyroid carcinoma (Fig. 2). One of these included metastasis from the primary neoplasm to the rectum in a mature brook trout (*Salvelinus fontinalis*). Marine and Lenhart (1) showed that fish goiters would regress completely when traces of iodine were included in the diet, and that goiter occurred mostly in young fish while true thyroid neoplasia occurred mostly in older animals. Their notable research led to the general use of iodized salt for the control of endemic goiter in man and in other animals.

Rainbow trout (*Salmo gairdneri*), carp (*Cyprinus carpio*), goldfish (*Carassius auratus*), guppies (*Poecilia reticulata*) and others are known to be sensitive species which respond readily to toxicants, carcinogens and many other stimuli. Some species such as coho salmon (*Oncorhynchus kisutch*) and channel catfish (*Ictalurus punctatus*) are refractive to certain toxins while others, like the mosquito fish (*Gambusia affinis*) when exposed to repeated sublethal doses of certain toxicants, soon develop highly resistant strains (4). Hatchery and aquarium fish are easily raised in large numbers under controlled conditions of temperature, diet and sanitation at less cost than is required for laboratory mammals. Thus, one can use more fish than mammals for a given experiment without increasing expense. Noxious odors are also relatively absent in a wet laboratory for fish. The great diversity between various fish species permits the use of a large assortment of different biological model-systems for research. Much investigation remains to be done to determine which fish species are best suited for specific research objectives.

LYMPHOCYSTIS

The work of Sandeman (5) in 1892 on "multiple tumors in plaice and flounders" initiated experimental studies on fish affected with Lymphocystis virus. Sandeman interpreted the so-called "cells" of lymphocystis disease as eggs laid by parasites in the skin of affected fishes. A different interpretation was soon placed on these sporozoan parasites which Woodcock (6) named *Lymphocystis johnstonei*. Woodcock in England and Awerinzew (7) in Russia concurred in calling the gigantic cells protozoa. Weissenberg (8) in Berlin reported his research on Lymphocystis disease in living sea perches (*Acerina cernua*). He showed the infectious nature of the disease and was able to transmit it at will to healthy fish by a variety of methods. From his studies of living Lymphocystis tumor cells growing on the caudal fins of Acerina, Weissenberg found that these gigantic tumor cells were actually hypertrophying infected fibroblasts from the connective tissues of the fish. These slender cells, gradually enlarged until the nucleus was plump, vesicular and enclosed by an abundant cytoplasm. These later became encapsulated, uninucleate cells measuring up to 2 mm in diameter. In three later papers, Weissenberg (9-11) presented the virus theory for lymphocystis which he based on experiments in which filtered tumor material from walleye-pike (*Stizostedion vitreum*), passed through a Chamberland–Pasteur filter L-5, was used to transmit the disease to a specimen of *Fundulus heteroclitus* -- a top minnow. Besides this he had already demonstrated basophilic, osmiophilic inclusion bodies in large lymphocystis tumors from flounders (12). Walker (13) and more recently Walker and Weissenberg (14) showed that these virus inclusion bodies were filled with actual virus particles when carefully prepared material was viewed with the electron microscope. Thus, for the first time, the viral-induction of neoplasia in fish fibroblasts was proven.

MICROSPORIDIA

Similarly, by using live material, Weissenberg (15) demonstrated the induction of ganglion cell hypertrophy in the angler fish (*Lophius* sp.) by an infective microsporidian named *Glugea lophii*. This protozoan is similar to *Glugea anomala* which generally affects only mesenchyme cells in the gut of the stickleback (*Gasterosteus aculeatus*). In 1921 Weissenberg (16) reported having discovered the induction of what he later termed "xenoma" tumors in stickleback mesenchyme cells penetrated by glugea spores. He then traced the development of infected host cells through five stages during which the nucleus repeatedly divides amitotically while the parasite undergoes schizogony. The entire course of schizogony and sporogony were followed in fish larvae killed periodically from the 7-29 days of infection. Invaded host cells enlarged from 7 microns to macroscopic cysts 3 mm in diameter. These cysts or xenomas contain thousands of glugea spores about 3 X 4 microns in size and with a polar filament about 30 microns long. Rupture of mature cysts (xenomas) releases the spores which infect other fish (17). In 1968 Wellings *et al.* (18) tentatively identified *Glugea hertwigi* in a young English sole (*Parophrys vetulus*). The walls of the visceral cavity were darkened and distended by hundreds of tapioca-like xenoma cysts (Figs. 3-5). The cyst walls and the contained *Glugea* spores were histologically and cytologically similar to those described by Weissenberg for *G. anomala*.

MELANOMA

The genetic induction of pigmented tumors in swordtail (*Xiphophorus helleri*) X platyfish (*Platypoecilus maculatus*) hybrids gained wide attention among oncologists and geneticists with independent publications by Kosswig (19) in Germany and by Reed and Gordon (20) in the United States. Since then numerous papers on this subject have appeared, mostly by Gordon and associates. While melanomas received the most attention, an unusual variant of a pigmented tumor was described by Nigrelli *et al.* (21). It was a xantho-erythrophoroma with secondary invasion by a malignant melanoma. These rare hereditary neoplasms were conditioned by a series of genes for melanin, red and yellow pigments, albinism, and by certain spotting and localizing factors for the red and the black neoplasms.

Tumorous hybrids were produced by mating a male spot-sided, ruby-throated platyfish (*Platypoecilus maculatus*) to an albino swordtail (*Xiphophorus helleri*). The striking histopathology of one of these mixed neoplasms may be briefly described as follows: A primary xantho-erythrophoroma typically arises in the dermis of the head. Its epithelial surface often thins and sloughs as a result of tumor growth. Adjacent subcutaneous tissues may proliferate, sometime breaking through membranes and destroying skeletal muscles and bone. Xantho-erythrophores comprise the bulk of the tumor and are dendritic, rounded or fusiform cells that contain granular erythropterin and solubilized lutein and zeaxanthin pigments when examined as fresh tissues with light and phase microscopy. These cells are devoid of pigments in paraffin sections prepared from fixed and dehydrated tissue. A delicate vascular and connective tissue stroma supports the pigment cells. An occasional macromelanophore occurs, usually around a blood vessel. Inflammation does not occur. The red pigment cells vary in size from 15-90 microns and the cytoplasmic granules vary from coarse to fine. The nuclei stain weakly although mitotic figures are common. The condition of the nuclei and cytoplasm is often unrelated in that cells with normal nuclei and vacuolated cytoplasm, or with finely granular cytoplasm but highly degenerated nuclei, occur in equal numbers. The cell boundaries are often indistinct.

In more advanced tumors most cells are spindle-shaped, the dendritic variety being limited to the periphery of the neoplasm. Melanocytes are few and scattered. A vascular reticulum is prominent but collagen is sparse. Nerve fibers are numerous in the central regions. The epithelium is thickened and papillary except where sloughed. Mitotic figures are rather common in the epithelial elements. Few macrophages are present.

In xantho-erythrophoromas invaded by macromelanophores, the latter come from the posterior or lateral body regions. These progress along the outer dermal layer and then move in among the neoplastic cells of the "red" tumor causing them to degenerate as they are replaced by the macromelanophores. By means of tumor biopsies various stages in development of the invading melanomas can be traced in one of these fishes. The red tumor is still undisturbed anteriorly while its posterior regions are being invaded by melanoma cells. Mitotic figures and binucleate cells of the red tumors increase in number as macromelanophores continue to invade them and spread inward, filling the interstitial spaces. These highly dendritic cells surround those of the red tumor, many of whose cells soon degenerate. A few melanin-bearing macrophages and many unpigmented macrophages appear. In its postero-ventral zone the invaded tumor now appears to be a melanoma. Melanoblasts are increasingly evident and invasive with extensive destruction of surrounding muscle, bone, scales, sense organs and tissues of the original red tumor which may eventually be entirely replaced by melanoma. Other details of genetic induction and tumor modification are omitted here.

SARCOMA

Li and Baldwin (22) inadvertently induced testicular tumors in swordtail fishes that had received "retroperitoneal" injections of sesame oil as a vehicle for the injection of fat-soluble endocrine preparations. Voegtlin (23) had proposed the induction of subcutaneous sarcomas by this method and Conrad *et al.* (24) reported a case of subcutaneous tumors in the arms of a woman treated with estrogen dissolved in sesame oil. Steiner *et al.* (25) found that overheated (350°C) sesame oil is carcinogenic. With this oil he induced three sarcomas in nine mice that lived more than one year post-injection. In this connection Sinnhuber *et al.* (26) reported as cocarcinogens with aflatoxin, a number of cyclopropenoid fatty acids that were fed with aflatoxin for the induction of trout hepatoma; both tumor incidence and growth rate were increased.

Typical of the testicular tumors reported by Li and Baldwin (22) from several fish were those from one which had seven small nodules, up to 0.2 mm in diameter on the periphery of the testis or near the sperm ducts. Each nodule consisted of foamy tumor cells and was well demarcated from the rest of the organ. Cysts of necrotic cells were seen in pancreatic lobules of this fish and the liver exhibited a mild fatty degeneration.

Levine and Gordon (27) reported finding 11 Xiphophorine fish with so-called spontaneous ocular tumors and exophthalmia (popeye). However, in the light of present knowledge of virus-induced and of carcinogen contaminated feed-induced neoplasia one may suspect an unidentified carcinogen. The authors stated that cereal foods were included in the diets of these fish and recent research mentioned below, has shown that mold contaminated cereals may carry one or more carcinogens. Time does not permit listing the various hybrids and their parent crosses from these experiments for which the tumor incidences ranges from 1 in 20 to 6 in 15. All these fish were full grown (6-12 months) before tumors appeared and normal life span averaged one and one-half years.

HEPATOMA

A world wide epizootic of liver cell carcinoma (hepatoma) in hatchery reared rainbow trout, discovered in 1960 and extensively surveyed in 1960-1961 by both federal and state governments in this country, prompted many laboratories to begin experiments designed to determine the cause of trout hepatoma (28, 29). At the Western Fish Nutrition Laboratory, 13 classical rodent carcinogens were fed in a complete test diet at five levels over a 256-fold range at 1/16, 1/4, 1, 4 and 16 times the effective dose for rodents. After 12-20 months on test, hepatoma occurred in fish fed 11 of the 13 carcinogens. Dimethylnitrosamine (DMN) was most effective for it induced tumors in up to 100% of fish in from 12-20 months on the two highest levels which were 480 and 1920 mg/100 g dry diet ingredients (30, 31) (Figs. 6 & 7). Thioacetamide and aminotriazole failed to induce hepatoma but the former induced cataract in 90% of the fish within one year (32). Acetylaminofluorene induced several hepatomas in 6 of 14 fish after 20 months on test; whereas, both urethane and thiourea induced dramatic skin lesions but very few hepatomas (Figs. 8 & 9). Autologous implants and isologous explants of active hepatoma cells to 12 different organs in non-tumorous control trout cross matched with three different serum antigens were successful in eight different organ sites (Fig. 10) (33). In addition to hepatoma, DMN induced 11 Wilms'-type renal embryomas (nephroblastomas) and three additional embryomas of the same type appeared in trout thyroidectomized with one or more injections, at monthly intervals, of 100 μc of ^{131}I (34) (Figs. 11 & 12). Further experiments on DMN-induced nephroblastomas are now in progress. In 1963 an unidentified liver carcinogen was reported from the cottonseed meal fraction of a commercially pelleted trout feed (35).

Because of increasing interest in mycotoxins, particularly those of *Aspergillus flavus* Link, which were named "aflatoxins" (36), feeding experiments with aflatoxins were begun at the Western Fish Nutrition Laboratory in 1963 to determine whether or not these compounds would induce hepatomas in rainbow trout. Fish were reared at the Hagerman Field Station, Hagerman, Idaho, in spring water at 15°C and were fed a complete test diet to which either 0.5, 2 or 20 ppb of aflatoxin was added. Early experiments with aflatoxins employed a crude mixture containing fractions B_1, G_1, B_2 and G_2. All were effective in hepatoma induction; later, the B_1 fraction was shown to be most carcinogenic. After 20 months on test, fish receiving 0.5 ppb of B_1 had 10%, those receiving 2 ppb had 60-70% and those receiving 20 ppb of the toxin had 90-100% tumor incidence (37) (Figs. 13 & 14). In contrast, coho salmon fed 1, 5 or 20 ppb of aflatoxin B_1 for 20 months under identical conditions had no gross or microscopic liver tumors. To date, careful investigations by fishery virologists have failed to support a viral etiology for rainbow trout hepatoma. Hepatomas were also induced in zebra fish (*Brachydanio rerio*) exposed to diethylnitrosamine (38) in the aquarium water.

The writer and his colleagues had the good fortune to discover that aflatoxins are primary carcinogens that cause rainbow trout hepatoma (33, 37). Furthermore, this work has recently been recognized at the Rockefeller Institute of Medical Research as having guided other investigators to the recent discovery that human hepatoma, prevalent among certain native African people (Bantus and others who frequently consume moldy corn or peanuts) probably is caused by ingested aflatoxins. Such recognition of the far-reaching applications of basic research in experimental fish neoplasia should stimulate greater interest in the use of fish as model systems not only in experimental fish oncology but also in such areas as biochemistry, physiology and nutrition.

The usual hepatoma thus induced in trout may be classified as a minimal deviation tumor or liver cell adenoma, but carcinomas do occur and may metastasize. Classically, these adenomas have a trabecular pattern with much widened muralia compared with normal liver (29) (Figs. 15 & 16). Advanced tumors usually undergo central fibrotic change followed by increasing foci of necrosis and hemorrhage until, in large three to five-year fish, all that remains of the tumor is a thin peripheral layer surrounding a caseous and hemorrhagic mass of degenerated neoplasm (Fig. 17). Multinodular and multicentric hepatomas commonly occur and more than 20 micronodules have been seen in a single hepatomatous liver section. Grossly, the earliest hepatomas appear as slightly elevated or subsurface gray to yellowish nodules which must be checked histologically before ruling out such anomalies as cysts, necrotic foci, acidophilic or hyperplastic nodules. More advanced tumors are usually firm, yellowish, more or less lobulated nodes providing a rather positive diagnosis to the experienced liver oncologist (Figs. 6 & 18). In rare cases these tumors have been known to rupture, spilling bloody fluid and desquamated necrotic material into the peritoneal cavity. In one trout with a larger hepatoma, necrosis and subsequent degeneration had perforated the belly wall and the tumor drained to the exterior through a short fistula. Mature fish with large hepatomas are usually unable to spawn effectively and when spawned, eggs and sperms are often non-viable. One of several malignant hepatomas examined had metastasized to eight different organs in a 17 inch female rainbow trout (39). The metastasis to the kidney is shown in histological section (Fig. 19).

In October, 1968, Kermit Sneed, Director, Warm-Water Fish Cultural Laboratories, Stuttgart, Arkansas, reported an 11 pound female channel catfish with a large multicentric hepatoma –– the first reported for this species (40). Stained sections of this neoplasm were examined both by the writer and by other pathologists, all of whom confirmed the diagnosis of hepatoma. This fish, believed to be six or seven years of age, was known to have been fed a diet which included peanut meal during the first two to three years of life. Perhaps aflatoxins in the peanut meal were the primary cause of this neoplasm, but the feed was not assayed and therefore cannot be arbitrarily condemned without supporting chemistry. Aflatoxin feeding experiments are currently underway to determine if these toxins will induce hepatoma in channel catfish or in yellow bullheads (*Ictalurus natalis*).

ECCHORDOMA

Levy (41) described experimentally induced tumors of fish notochords (ecchordomas). Egg clusters (10-20 eggs in each) of the Japanese Medaka (*Oryzia latipes*) were immersed in 1 or 2 mg/100 ml solution of β-amino-proprionitrile. Of 422 eggs used 265 died before hatching. Development of the survivors was observed with a dissection microscope and the embryos were transferred to spring water at or near the time of development of the mandibular arch. One hundred fifty-seven fish were hatched in spring water and embryonic development was periodically observed with the microscope. Fish were sacrificed for histological study at regular intervals and were fixed in either Bouin's or Stockard's fixative. Serial sections were cut on all samples.

Eggs of *Anoptichthys jordane*, *Trichogaster tricopterus*, *Aequidens portalegrensis* and a mixed breed of goldfish (*Carassius auratus*) were treated identically and with similar results. Subspherical tumors proliferated as hyperplastic growths from the notochord or from its connective tissue sheath. These growths usually displaced surrounding muscle tissue and extended to just beneath the epidermis. In older fish the connective tissue component from the notochordal sheath was greatly thickened and had increased basophilia and the large vacuolated notochordal (physaliphorous) cells were separated by extremely thick septa. Numerous cells were also filled with an amorphous or finely granular acidophilic material. Such tumors were usually vascularized peripherally with the vessels engorged with nucleated red blood cells. Bone often enclosed the tumors. Rarely a tumor contained both hyperplastic notochordal and sheath cells which invaded inter-muscularly along the fascial connective tissue. All these chordomas appeared to be benign and it was often difficult to determine the line of demarcation between neoplasia and simple hyperplasia.

CATARACT

Lens cataracts were induced in rainbow trout fry fed thioacetamide at 30 mg/100 g of dry diet ingredients for 12 months (32). Ninety percent of these fish developed cataracts characterized by massive proliferation of lens epithelium which transformed, in its deeper layers, into a pleomorphic cell mass which in some cases replaced the major portion of the anterior lens cortex. A tumorous growth pattern was suggested by the disorganized proliferated epithelium and by bizarre forms similar to Darier cells and epithelial pearls (Fig. 20). Chronic irritation by one or more toxicants was also postulated as a cause of cataract. Hess (42) reported cataracts in rainbow trout fed suspect deficient diets. Mawdesley-Thomas and Jolly (43) felt that potassium permanganate therapy is one cause of fish cataracts, and are further investigating this problem.

SUMMARY

A joint effort of the National Cancer Institute and the Smithsonian Institution in 1966 set up a REGISTRY OF TUMORS IN LOWER ANIMALS to start a new era in the study of neoplasia in poikilotherms and invertebrates. The number of published papers dealing with experimental fish neoplasia, while limited, has been increasing notably in recent years. This paper reviews some classical examples of experimental fish neoplasia and attempts to point out the desirability of using various fish species which have been shown to be good model systems for research in oncology. Results of research with various classical rodent carcinogens, and in particular with the aflatoxins, has been extraordinarily rewarding. Soon after our laboratory demonstrated that ingested aflatoxins are a primary cause of rainbow trout hepatoma, other workers have investigated the role of mycotoxins in oncogenesis in higher vertebrates, including man.

REFERENCES

1. D. Marine and C. H. Lenhart, J. Exp. Med. **12**: 311 (1910).

2. D. Marine and C. H. Lenhart, Bull. Johns Hopkins Hosp. **21**: 95 (1910).

3. H. R. Gaylord and M. C. Marsh, Carcinoma of the thyroid in the salmonoid fishes. State Inst. for the Study of Malignant Diseases, G. P. O., Wash., D. C., Serial 99, 1914.

4. P. Rosato and D. E. Ferguson, BioScience. **18**: 783 (1968).

5. G. Sandeman, On the multiple tumors in plaice and flounders. 11th Ann. Rpt. Scott. Fish. Board, 391(1892).

6. H. M. Woodcock, Note on a remarkable parasite of plaice and flounders. 12th Rpt. Lancast. Sea-Fisher. Lab. **63**: 1 (1903).

7. S. Awerinzew, Zool. Anz. **31**: 881 (1907).

8. R. Weissenberg, Sitzungsber. Kgl. Preuss. Akad. Wiss. Sitz. Physik. Mathem, Jg. **1914**: 792 (1914).

9. R. Weissenberg, Biol. Bull. **76**: 251 (1939).

10. R. Weissenberg, Zoologica, **24**: 245 (1939).

11. R. Weissenberg, Zoologica **30**: 169 (1945).

12. R. Weissenberg, Sitzungsber. Ges. Naturf. Freunde Berlin. **1920**: 198 (1920).

13. R. Walker, Virology. **18**: 503 (1962).

14. R. Walker and R. Weissenberg, Ann. N. Y. Acad. Sci., **126**: 375 (1965).

15. R. Weissenberg, Arch. Protistenk. **22**: 179 (1911).

16. R. Weissenberg, Arch. Protistenk. **42**: 400 (1921).

17. R. Weissenberg, J. Protozool. **15**: 44 (1968).

18. S. R. Wellings, L. M. Ashley and G. E. McArn, J. Fisheries Res. Bd. Canada, In press.

19. C. Kosswig, Z. Abstammungslehre. **52**: 114 (1929).

20. H. D. Reed and M. Gordon, Am. J. Cancer. **15**: 1524 (1931).

21. R. Nigrelli, S. Jakowska and M. Gordon, Brit. J. Cancer. **5**: 54 (1951).

22. M. H. Li and F. M. Baldwin, Proc. Soc. Exp. Biol. Med. **57**: 165 (1944).

23. C. Voegtlin, J. Nat. Cancer Inst. **2**: 309 (1942).

24. A. M. Conrad, A. H. Conrad, Jr. and R. S. Weiss, J. Am. Med. Assoc. **121**: 237 (1943).

25. P. E. Steiner, R. Steele and F. C. Kock, Cancer Res. **3**: 100 (1943).

26. R. O. Sinnhuber, J. H. Wales and D. J. Lee, Federation Proc. **25**:555

(1966).

27. M. Levine and M. Gordon, Cancer Res. **6**: 197 (1946).

28. R. R. Rucker, W. T. Yasutake and H. Wolf, Prog. Fish-Cult. **23**: 3 (1961).

29. E. M. Wood and C. P. Larson, Arch. Pathol. **77** 471 (1961).

30. L. M. Ashley and J. E. Halver, Federation Proc. **20**: 290 (1961).

31. L. M. Ashley and J. E. Halver, J. Nat. Cancer Inst. **41**: 531 (1968).

32. L. von Sallmann, J. E. Halver, E. Collins and P. Grimes, Cancer Res. **26**: 1819 (1966).

33. J. E. Halver and I. A. Mitchell, eds., TROUT HEPATOMA RESEARCH AND CONFERENCE PAPERS. Bur. of Sport Fish. and Wildl. Res. Report 70, G. P. O., Wash., D. C., 1967.

34. L. M. Ashley, Bull. Wildl. Dis. Assoc. **3**: 86 (1967).

35. H. Wolf and E. W. Jackson, Science. **142**: 676 (1963).

36. B. Nesbitt, J. O'Kelly, K. Sargeant and A. Sheridan, Nature. **195**: 1062 (1962).

37. L. M. Ashley, J. E. Halver, W. K. Gardner and G. N. Wogan, Federation Proc. **24**: 627 (1965).

38. M. F. Stanton, J. Nat. Cancer Inst. **34**: 117 (1965).

39. L. M. Ashley and J. E. Halver, Trans. Am. Fish. Soc. **92**: 365 (1963).

40. K. Sneed, Personal communication.

41. B. M. Levy, Cancer Res. **22**: 441 (1962).

42. W. N. Hess, Proc. Soc. Exp. Biol. Med. **37**: 306 (1937).

43. L. E. Mawdesley-Thomas and D. W. Jolly, J. Small Anim. Pract. **9**: 167 (1968).

COMMENTS

DR. ASHLEY: Sometimes in the literature individual men imply that neoplasia from a single cell grows into a large structure. We mentioned tumors reached 2 and 3 mm in diameter arising from a single infected cell. This may be stretching the point but it depends on your personal choice.

Some individuals in the literature speak of some cells as though these were single celled neoplasias, but this could be debated.

DR. GREENBLATT: Your data on incidence of renal tumors in DMN fed trout is interesting because in mammals renal tumors were produced by acetylaminofluorene and DMN.

DR. ASHLEY: McGee and Barnes in England have published extensively on renal tumors produced by DMN in dogs but we have seen very few in fish fed this carcinogen.

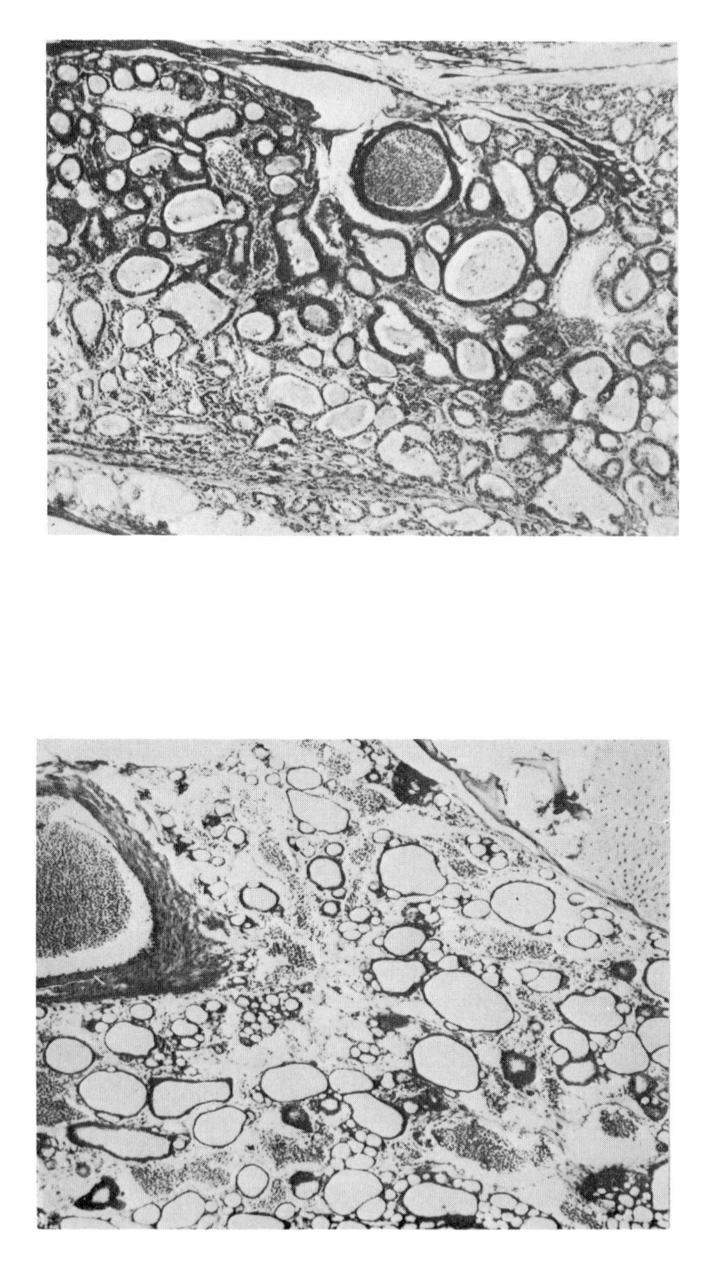

Fig. 1: Normal thyroid follicles diffusely scattered around a branchial artery in a Scotch sea trout (Salmo trutta). 60X. (From: Gaylord and Marsh, 3).

Fig. 2: Marked thyroid hyperplasia characterized by much thickened walls of the thyroid follicles in a brook trout (Salvelinus fontinalis). 60X. (From: Gaylord and Marsh, 3).

Fig. 3: Distended wall of visceral cavity seen on "eyed" side of English sole (Parophrys vetulus) heavily infested with the Microsporidian, Glugea hertwigi. 2X. (From: Wellings et al., J. Fish. Res. Bd., Canada, In press).

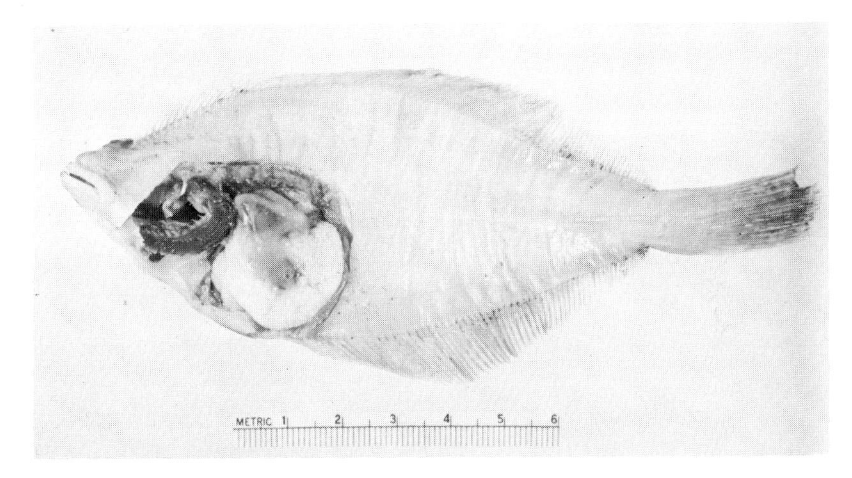

Fig. 4: Operculum and visceral wall removed from "blind" side of same fish as in Fig. 3 to show tapioca-like Glugea cysts (Xenoma tumors). 2X. (From Wellings et al., J. Fish Res. Bd., Canada, In press).

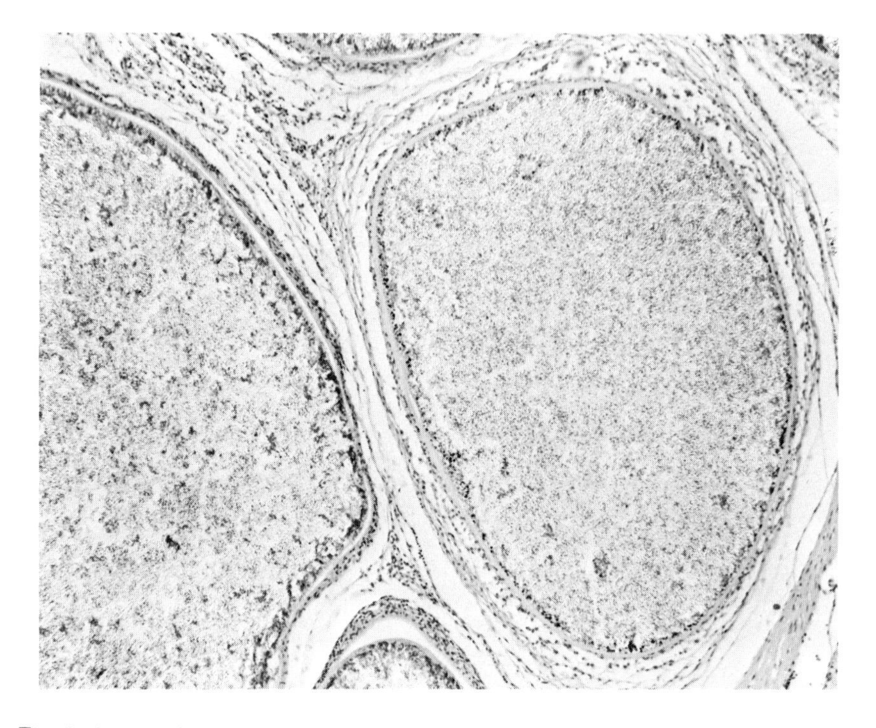

Fig. 5: Section through two Glugea cysts each containing thousands of maturing spores. Hematoxylin and eosin. 120X. (From: Wellings et al., J. Fish. Res. Bd., Canada, In press).

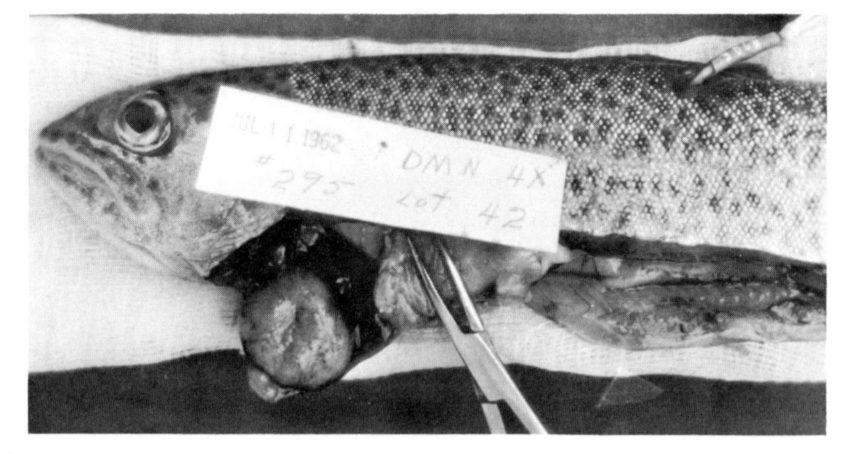

Fig. 6: Rainbow trout dissected to show liver with large, grayish colored hepatoma at left of hemostat. Fish fed dimethylnitrosamine at 480 mg/100 g of dry diet. 2X. Courtesy of Dr. J. E. Halver, Western Fish Nutrition Laboratory, Cook, Washington.

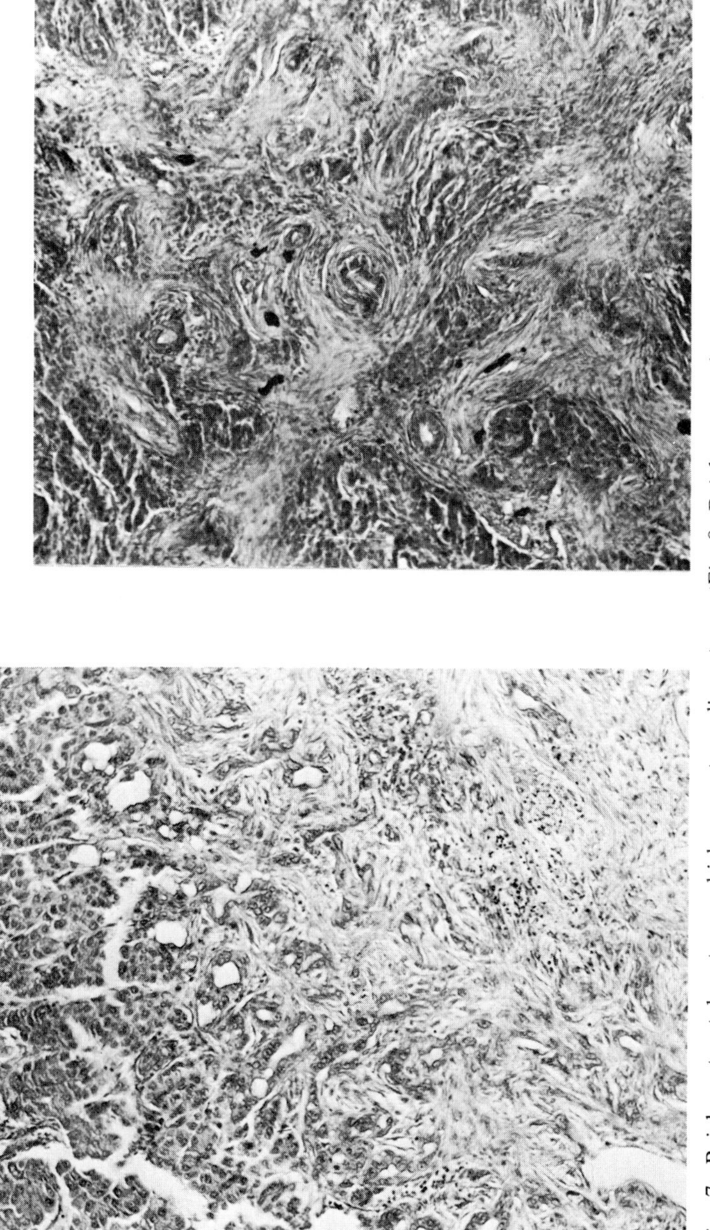

Fig. 7: Rainbow trout hepatoma which suggests malignant change (center). Dark area is classical trabecular hepatoma. Light area is fibrotic change with early necrosis (lower right). Fish fed dimethylnitrosamine at 120 mg/100 g dry diet. Hematoxylin and eosin. 400X.

Fig. 8: Rainbow trout hepatoma induced by ingested 2-acetylaminofluorene fed at 240 mg/100 g. Hematoxylin and eosin. 250X.

Fig. 9: Skin ulcers behind head in rainbow trout fed urethane at 320 mg/100 g of dry diet for six months. 1X. Courtesy of Mr. R. R. Smith, Hagerman Field Station, Western Fish Nutrition Laboratory, Hagerman, Idaho.

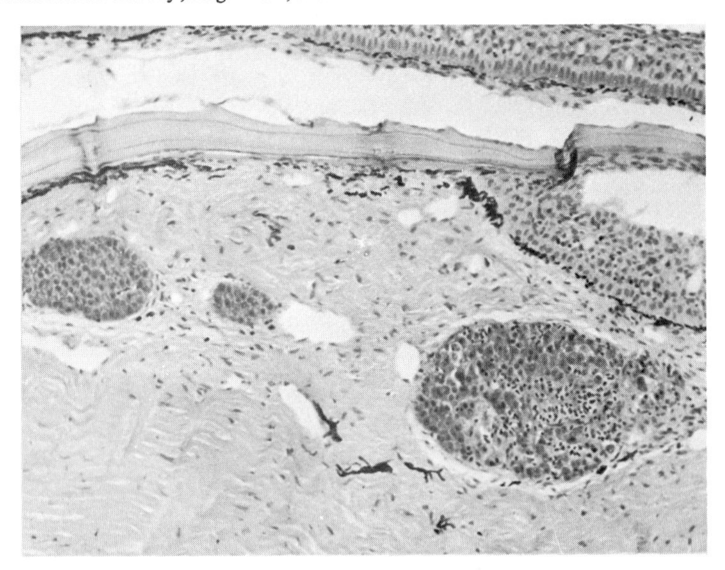

Fig. 10: Three hepatoma micronodules transplanted from primary tumor to dermis of triple cross matched rainbow trout. Hematoxylin and eosin. 400X. Courtesy of Dr. J. E. Halver, Western Fish Nutrition Laboratory, Cook, Washington.

Fig. 11: Embryoma (Wilms'-type tumor) in excretory kidney of rainbow trout which received a single injection of 100 μc ^{131}I 15 months before sacrifice. 2X. Courtesy of Mr. C. L. Johnson, Western Fish Nutrition Laboratory, Cook, Washington.

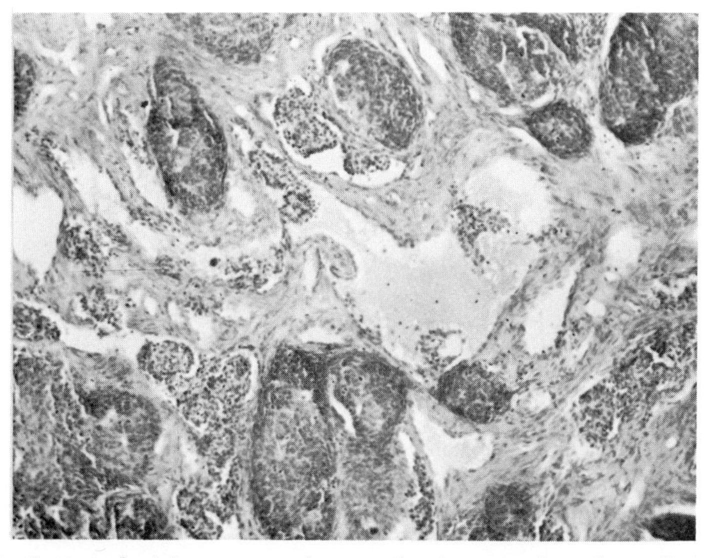

Fig. 12: Section of rainbow trout embryoma showing anomalous glomeruli and renal tubules interspersed with islands of dense neoplastic tissue. Hematoxylin and eosin. 250X.

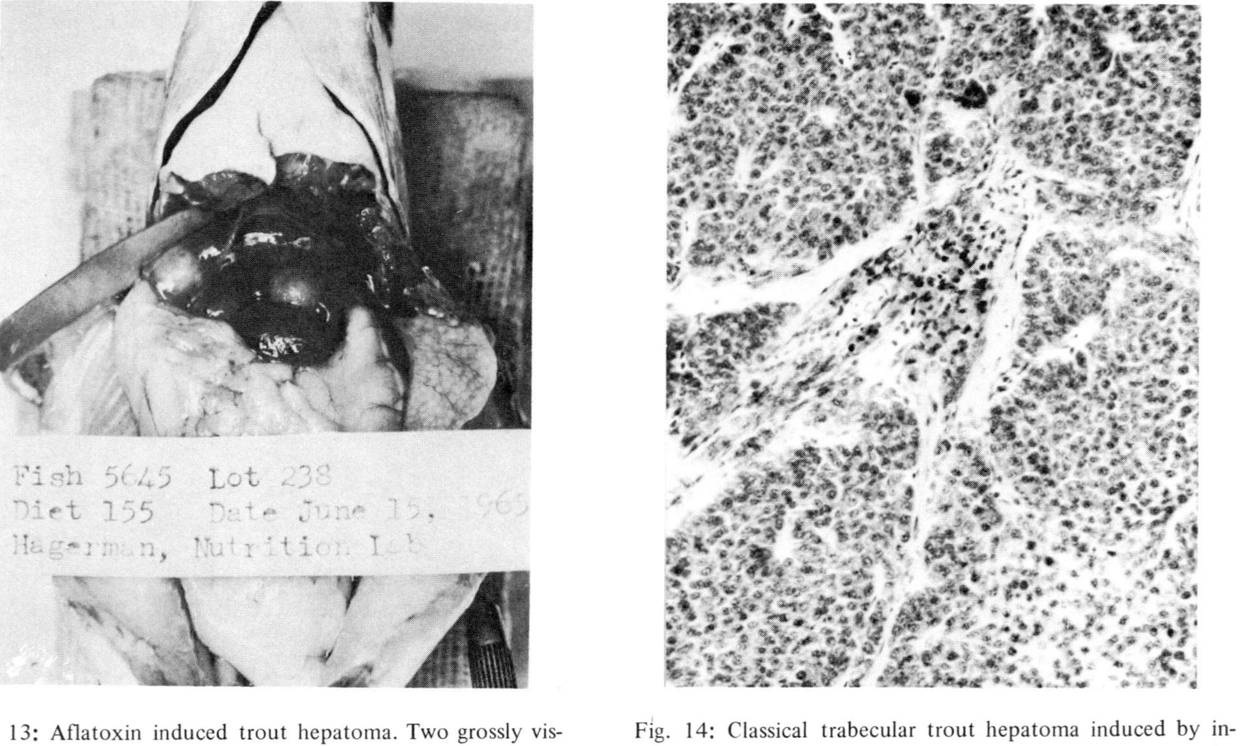

Fig. 13: Aflatoxin induced trout hepatoma. Two grossly visible tumors are evident at right of spatula in this liver. Fish fed 2 ppb aflatoxin B_1 for 20 months. 1X. (From: Dr. J. E. Halver, TROUT HEPATOMA, RES. CONF. PAPERS, G.P.O., 1967).

Fig. 14: Classical trabecular trout hepatoma induced by ingested aflatoxin B_1 fed at 20 ppb for 20 months. Hematoxylin and eosin. 450X. Courtesy of Dr. J. E. Halver, Western Fish Nutrition Laboratory, Cook, Washington.

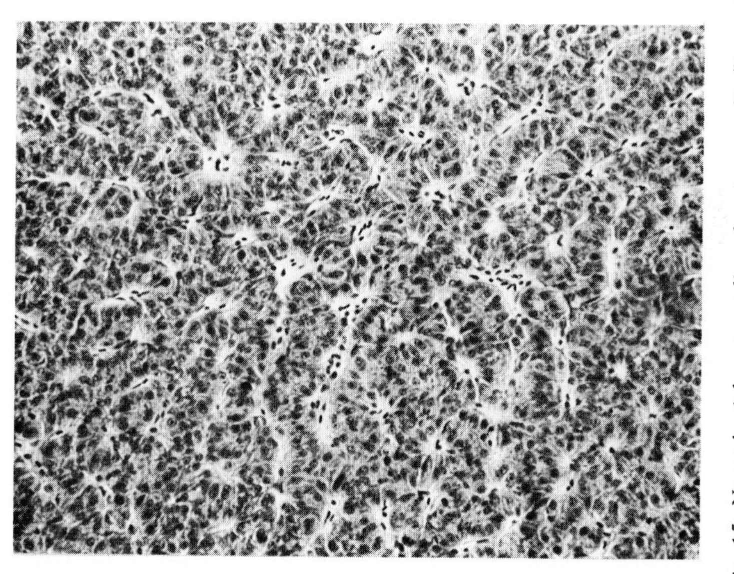

Fig. 15: Normal rainbow trout liver showing muralia (liver cell cords) alternating with sinusoids, some of which contain red blood cells. Hematoxylin and eosin. 400X.

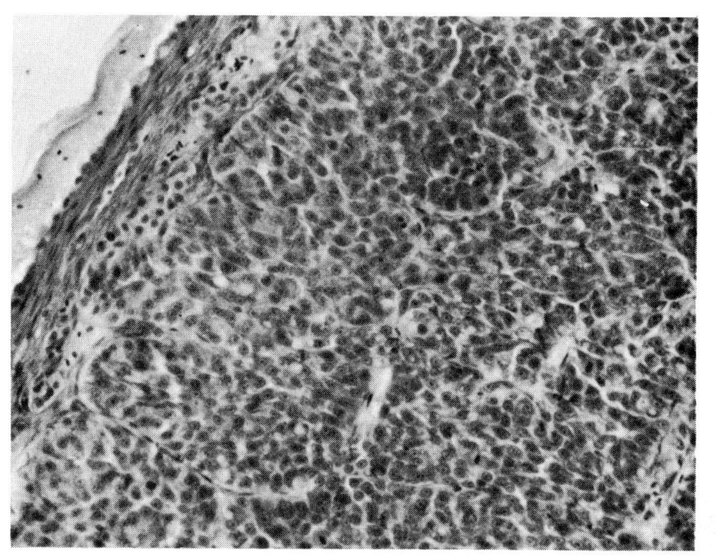

Fig. 16: Classical trabecular hepatoma in rainbow trout fed commercial trout pellets. Note basophilia, increased nuclearity, widened muralia and reduction in number of sinusoids. Mitotic figures are evident in some liver cells. Hematoxylin and eosin. 450X.

Fig. 17: Advanced hepatoma in two rainbow trout broodstock livers. Fish were approximately five years old and had been fed commercial rations including either cottonseed or peanut meals. Note caseation necrosis (white) in tumor on left and extensive hemorrhage (black patterns) in tumor on right. 1X.

Fig. 18: Multinodular trout hepatoma in fish fed dimethylnitrosamine at 1920 mg/100 g dry diet ingredients for 20 months. 2X. (From: Dr. J. E. Halver, TROUT HEPATOMA RES. CONF. PAPERS, G. P. O., 1967).

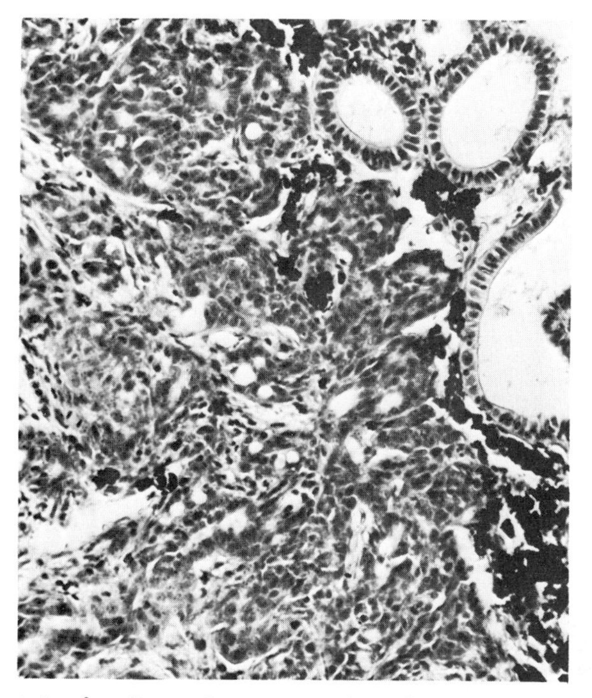

Fig. 19: Metastasis of malignant hepatoma to the kidney in a rainbow trout fed a commerical ration. Note total invasion and destruction of renal tubules and glomeruli in all but one corner of the picture where a few renal tubules remain. Hematoxylin and eosin. 450X. Courtesy of Dr. J. E. Halver, Western Fish Nutrition Laboratory, Cook, Washington.

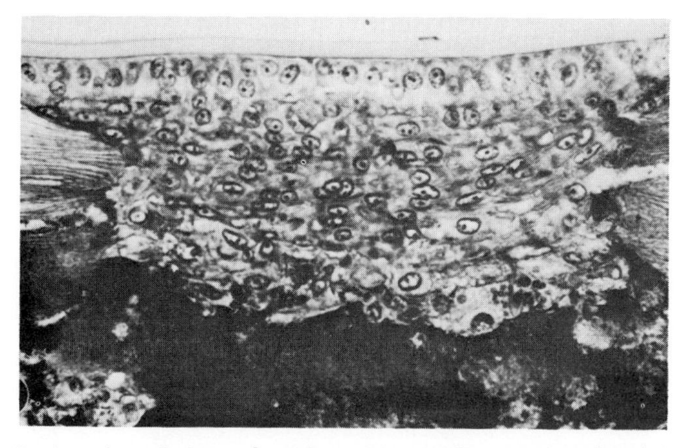

Fig. 20: Section through lens of rainbow trout with cataract induced by feeding thioacetamide at 30 mg/100 g dry diet for 12 months. Note proliferation of lens epithelium which transforms to pleomorphic cells in the deeper layers of the lens cortex. Hematoxylin and eosin. 400X. (From: von Sallman et al., Cancer Res. 26: 1819, 1966).

COMPARATIVE ASPECTS OF NEOPLASIA IN FISH AND OTHER LABORATORY ANIMALS

D. G. Scarpelli

INTRODUCTION

Since this symposium deals with the topic of fish in research, it seems appropriate to take this opportunity to point out their potential value as animal models for cancer research. Comprehensive reviews of tumors of fishes, amphibians and reptiles (1, 2) have clearly established that lower vertebrates are commonly affected by neoplastic diseases, and that some of these are essentially identical in structure and biological behavior to their counterparts in the higher vertebrates. The widespread occurrence and variety of neoplasms in fish are summarized in Table I. Some of these occur with such frequency in a given species that they must be regarded as characteristic for that species (3), others appear to represent sporadic neoplasms without any species predilection. Neoplasms of common occurrence in the bony fishes will be described and some will be compared with analogous lesions in homoiothermic animals which enjoy considerable

TABLE I.

NEOPLASMS OF FISH
(Modified after Schlumberger and Lucke)

Neoplasms	Species affected	Sites	Number reported
Papilloma	17	5	89
Adenoma	11	5	15
Dental tumors (adontoma)	3	4	5
Epidermoid carcinoma	14	6	182
Adenocarcinoma	7	6	7
Hepatocarcinoma	1	1	Thousands
Thyroid tumors	23	1	Thousands
Fibroma	33	13	39
Unclassified sarcoma	54	14	106
Myxoma	6	4	6
Chondroma	8	8	10
Osteoma	24	10	141
Osteogenic sarcoma	3	4	4
Lipoma	7	4	7
Leiomyoma	4	2	4
Neurilemmoma	3	5	154
Rhabdomyoma	8	2	8
Hemangioma	11	9	14
Lymphosarcoma	9	5	20
Melanoma	12	3	32

application in cancer research as summarized in Table II. There is little doubt that much can be learned from the detailed examination of spontaneous and induced neoplasms in fish. However, except for the past experimental work on melanoma in killifish hybrids, *Platypoecilius maculatus* and *Xiphophorous hellerii* (4-6), epitheliomas in catfish, *Ameirus nebulosus* (7), and the recent surge of studies on hepatoma in rainbow trout, *Salmo gairdneri* (8-13), fish have not been widely utilized in the experimental studies of cancer. This situation exists despite the fact that fish offer the following unique advantages which make them ideal research models: 1) The relative ease with which their total environment is subject to experimental control and manipulation, 2) the large number of viable eggs that can be derived from a single brood fish and fertilized by a single male, thus allowing for more control of genetic factors, and 3) the ease of handling, care and

TABLE II. SOME COMMON AND UNIQUE NEOPLASMS OF FISH AND CORRESPONDING NEOPLASMS OF HOMOIOTHERMS EMPLOYED IN CANCER RESEARCH

Fish		Homoiotherms		
Neoplasm	Species	Species	Strain	Tumor Designation
Epidermal Papilloma	Flathead sole, Hippoglossoides elassodon English sole, Parophrys vetulus Starry flounder, Platichythys stellatus Slippery Dick, Iridio bivattata	Mouse Rabbit	HR/De (Hairless) Cottontail and domestic Dutch strain	Papilloma Shope Papilloma
Epidermal epithelioma	Catfish, Ameiurus nebulosus Hog sucker, Hypentelium nigricans Other species	Mouse Mouse Rabbit Rabbit	"Swiss" C57L Dutch and other strains Dutch strain	Squamous cell tumor No. 1 Cowdry Squamous cell tumor No. D Cowdry Brown-Pearce Epithelioma V-2 Carcinoma
Hepatocarcinoma	Rainbow trout, Salmo gairdneri	Rat Rat Mouse Rat	 C3H Sprague-Dawley, Holtzman	Reuber Hepatoma H35 Morris Hepatoma No. 5123 Hepatoma 98/15 Hepatoma NK (Novikoff)
Melanoma (skin)	Killifish hybrids Xiphophorous hellerii x Platypoecilius maculatus	Mouse Mouse	DBA strain Transplantable to many strains	Cloudman Melanoma S91 Harding-Passey Melanoma
Lymphosarcoma	Muskellunge, Esox masquinongy Northern pike, Esox lucius Rainbow trout, Salmo gairdneri	Mouse Mouse Rat Guinea pig	Strain A mouse CFI (Carworth) Wistar Carworth Sprague-Dawley Strain 2 (Family 2)	Lymphocytic Lymphoma No. 2 Lymphosarcoma TT-8, TT-10 Murphy-Sturm A-40
Fibroma	Many species	Rabbit	Cottontail	Infectious Fibroma (Shope)

Fish		Homoiotherms
Neoplasm	Species	
Neurilemmoma	Snapper, Lutianus Goldfish, Carassius auratus	No corresponding spontaneous or induced tumors utilized as a model system in Cancer Research.
Osteoma	Salmon, Salmo salar Red tai, Pagrosomus major	No corresponding spontaneous or induced tumors utilized as a model system in Cancer Research.
Adamantinoma	Chinook salmon, Oncorhynchus tshawytscha Brook trout, Salmo fario	No corresponding spontaneous or induced tumors utilized as a model system in Cancer Research.

housing.

The incidence of spontaneous neoplasms in teleosts is largely concentrated in five families: the salmonids, cyprinoids, codfishes, flatfishes and flounders. Since these families constitute a major portion of fish consumed by man, they are caught in large numbers and inspected. Thus, the high distribution of neoplasms in these fish may be skewed and in error. The true incidence of neoplastic disease in fish populations can only be ascertained by systematic, extensive surveys in which large numbers of fish of many species are caught and subjected to careful morphologic study.

EPIDERMAL NEOPLASMS

Tumors of the epidermis are quite common and have been reported in 30 different species, 18 of these were classified as papillomas; the remainder exhibited sufficient infiltration and histologic characteristics of a malignant neoplasm to be classified as such.

Epidermal papillomas have been encountered with considerable frequency in several species of pleuronectid fishes: the flathead sole, *Hippoglossoides elassodon*, the English sole, *Parophrys vetulus*, and the starry flounder, *Platichthys stellatus* (14). Similar lesions have also been regularly observed in large numbers of the reef fish 'slippery dick,' *Iridio bivittata* (15). These lesions vary in gross appearance from small flattened nodular elevations to large cauliflower-like masses affecting a considerable portion of the body surface. Due to excessive cellular growth, the epidermis is thrown into folds which project above the integumental surface but shows no tendency to invade the underlying corium (Fig. 1). The epidermal cells are cytologically indistinguishable from normal and no mitotic figures are observed. In the earlier stages of development the underlying corium contains nodular foci of angiomatous and fibrous connective tissue overgrowth which tend to regress as the papilloma continues to proliferate.

Although a viral etiology has been suggested by the presence of intracellular inclusion bodies in some piscine papillomata (16), there have been very few systematic experimental attempts to either clearly support or negate this possibility. The studies of Wellings and his associates (14) on the presence and nature of ultramicroscopic cytoplasmic virus-like particles in epidermal papillomas of flathead sole is the first suggestive evidence of a possible viral etiology for these lesions. However, in view of the fact that epidermal papillomas are quite numerous in flounder which are bottom dwelling fish, and in catfish (7) and croakers (17) that inhabit waters heavily polluted by industrial and agricultural chemical wastes, it may be that other etiologic factors are also involved.

In warm blooded animals two epidermal papillomas occur with sufficient regularity and can be induced so readily that they are widely utilized in cancer research. These are the Shope papilloma (18), a viral-induced epidermal lesion of cottontail and domestic rabbits (Fig. 2, 3), and skin papillomas of hairless HR/De mice (19) in which the etiology remains obscure. It is of interest to note that the focal inflammatory and proliferative reaction of dermal connective tissue which characterizes the early stages of papilloma formation in pleuronectid fishes bears a striking similarity to the alterations of subcutaneous tissue which accompany the development of Shope papilloma in the rabbit. Whether this similarity is indicative of analogous etiologic factors or is merely a non-specific reaction of dermal cellular elements to epithelial injury remains to be established. The analogy of epidermal papillomas in fish to the viral-induced Shope papilloma in rabbit is obvious, yet, except for the studies cited above, our knowledge of the etiology and pathogenesis of piscene epidermal papillomas is woefully incomplete. The recent establishment of permanent cell lines in tissue culture from a number of teleost fishes (20-22) presents an additional tool for the study of viral-induced cell injury and disturbances of cell growth in these animals; further studies in this area are awaited with great interest.

Viral-induced epidermal papillomas occur in such a wide spectrum of homoiothermic and poikilothermic animals, it appears that the tropism of viruses for integumental epithelium is a phenomenon which transcends class, subclass, family and species barriers which have developed during evolution. The persistence of such pathogenetic similarities between lower vertebrates and higher ones renders the former most useful as models for comparative pathology.

Epitheliomas occur with greatest frequency on the lips, oral mucosa and skin of the head. Such lesions have been observed with greatest frequency in the *Nematognathi* (catfishes) (3). More recently, epitheliomas have been encountered in hog suckers, *Hypentelium nigricans*, from a number of lakes in the midwestern United States (23). These tumors closely resemble papillomas grossly with the exception that they tend to be larger, more fungoid and invade underlying tissue (Fig. 4). Histologically these neoplasms consist of large, compact masses of epithelial cells which show variable-sized nuclei with prominent nucleoli and occasional mitoses (Fig. 5, 6). The underlying corium is invariably invaded by the neoplasm. Occasionally intravascular tumor emboli are found; despite this, distant metastases have not been described. Fish appear to survive these neoplasms despite the fact that they often grow to a very large size. Death occurs earlier if localization of the tumor interferes with normal feeding or gill function.

Lucke and Schlumberger (7) were successful in transplanting epitheliomas of the catfish, *Ameirus nebulosus*, to the anterior chamber of the eye in this and other species of fish, and thus were able to study by direct observation the pattern of growth and the effect of intraspecies and interspecies transplantation. Since these early studies, little attention has been paid to this common piscine tumor. The generally poor results with tumor transplantation in fish is largely due to the absence of inbred strains of fish. If these were available, it is certain that experimental piscine oncology would be greatly advanced. The high incidence of epitheliomas in certain species of fish suggests that genetic factors may plan an important role in their induction. Furthermore, the predominant localization of this lesion in the lips, oral mucosa and skin of the head raises the possibility of physical, chemical or biological carcinogenic factor or factors that gain access to the host during feeding. Finally, the frequent geographic clustering of such lesions in specific rivers and lakes offers unique opportunities for the elucidation of ecological factors which may be of etiological importance. This phenomenon is an excellent example of the limited total environment in which many fish live, a unique situation which lends itself admirably to etiological studies of cancer and has been almost totally overlooked in this regard. The increasing number of pollutants, and the extent of pollution of rivers and lakes, and offshore oceans markedly increases the need for such studies (24).

Spontaneous, transplantable epitheliomas of homoiotherms that are comparable to those of fish include the squamous cell carcinomas No. 1 and No. D of the mouse (25) which, like the majority of piscine tumors, grow very slowly and show only local invasion. More aggressive tumors are the V-2 (26) and Brown–Pearce carcinomas (27, 28) transplantable to the rabbit. All of these neoplasms bear a topographical structural resemblance (Fig. 7) to epithelioma of fish, the major exception being, that depending on their degree of differentiation, the mammalian lesions may show evidence of keratinization and 'pearl' formation (Fig. 8). Since the integument of fish is not capable of synthesizing keratin, this hallmark of squamous epithelium is absent in squamous cell carcinoma arising in fish.

HEPATIC TUMORS

In the past, hepatic tumors in fish were encountered only rarely. The near epidemic incidence of hepatoma in hatchery reared rainbow trout was first recognized in 1960. Since that time numerous studies have clearly demonstrated that the major carcinogenic factors reside in the pelleted trout diets (8, 29). One of these has been identified as aflatoxin B_1, a lactone metabolite elaborated by certain strains of the fungus, *Aspergillus flavus*. This metabolite has proven to be a potent hepatic carcinogen for ducks, and rats, as well as rainbow trout (30). In fish the neoplasms appear first as minute grayish white 1-2 mm nodules in the liver parenchyma. Microscopically, these nodules consist of intensely basophilic hepatocytes which still maintain the architectural plan of normal liver. As these nodules progress in growth, the hepatocytes proliferate to the point where normal liver architecture is lost and the nodules enlarge and coalesce to form larger masses. The tumors may progress to the point where almost the entire liver is involved. Extension to adjacent organs and distant metastases occur only in advanced cases. Histologically, these tumors range from highly to poorly differentiated (Fig. 9, 10, 11) (28, 31). Cytologically, the neoplastic hepatocytes show a marked increase in rough surfaced endoplasmic reticulum, a change commensurate with the intensely basophilic nature of these cells. Additional alterations include nuclear and nucleolar enlargement, and a coarsening of the chromatin pattern. Attempts to ascertain whether the pathogenesis of this neoplasm progresses from a non-invasive adenoma to an invasive carcinoma have yielded variable results. Repeated laparotomies during the induction of the tumors suggests that the smallest nodules invariably progress to large tumors which would be classified as hepatocarcinoma. A detailed enzymologic profile study of very small nodules showed that they did not differ from the large tumors although both differed significantly from normal liver (12). On the other hand, a preliminary study of pooled samples of 1-3 mm tumor nodules induced by aflatoxin showed that these retained feedback control inhibition of cholesterol biosynthesis following prior cholesterol feeding, while larger tumors showed a loss of this control mechanism (23).

An additional feature of trout hepatocarcinoma which merits mention is the enhanced synthesis of plasma proteins by some of the tumors (12, 32). The high degree of differentiation, slow growth rate and enhanced functional capacity retained by some of these tumors is reminiscent of the so-called functional "minimal deviation" hepatomas encountered in the rat (33). Hepatomas of mammals are probably the most widely used tumors in cancer research; this, no doubt, is due to the ease with which they can be induced and the relative homogeneity of the neoplastic cell population which comprise the tumors. A spectrum of transplantable tumors of varying degrees of differentiation and function are available. A representative group listed in descending order of differentiation and function includes the Reuber hepatoma (34), Morris hepatoma (35), Hepatoma 98/15 (36), and the Novikoff hepatoma NK (37). These tumors range from those exhibiting a structure almost indistinguishable from normal liver to highly anaplastic tumors in which all semblance of liver architecture is absent. Their morphology have been so extensively described that to do so here would be superfluous. A comparative biochemical and morphological study of trout and rat hepatomas would be of great interest, since it could give valuable information on the phylogeny of carcinogenesis. However, such endeavors must await the development of sufficiently inbred strains of trout so that transplantation of tumors can be successfully accomplished.

MELANOMA

Cutaneous melanomas have been reported in 12 species of fish. The hybrid arising from the Mexican swordtail, *Xiphophorous hellerii*, and the platyfish, *Platypoecilius maculatus*, characteristically develop a diffuse melanosis which not infrequently evolve into melanotic tumors (4-6). The tumors may appear at any body site and grow to a considerable size. The tumors consist of whorled masses of large melanin-laden cells which are thought to represent macromelanophores (Fig. 12) introduced into the hybrid by the platyfish. Although these tumors show local extension, invading muscle, soft tissues and in some instances bone, distant matastases do not occur. Gordon has studied the genetics of melanosis and melanotic tumor formation in these fishes and concluded that the melanosis and tumor formation was associated with macromelanophores which are heritable as a dominant sex-linked factor. This research on the killifish hybrid stands as an important contribution to our knowledge of the biology of melanomas.

The Cloudman melanoma S91 (38) (Fig. 13) and the Harding–Passey melanoma, melanin producing tumors of mice, are widely used as models for the study of melanin producing tumors in cancer research. Their microscopic morphology in general corresponds to that of melanoma in the killifish hybrid except that the latter tumors tend to be more heavily pigmented. The biological behavior of the transplantable murine melanomas is variable; the Harding–Passey tumor does not metastasize and thus bears some resemblance to the piscine neoplasm. On the other hand, the Cloudman melanoma metastasizes almost without exception.

THYROID TUMORS

Thyroid masses are not an uncommon occurrence in fresh water fish, especially the Salmonidae; however, as pointed out by Marine and Lenhart (39), the majority of these are hyperplastic non-neoplastic overgrowths secondary to chronic iodine deficiency (40). Since the overwhelming majority of thyroid hyperplasia and goitrous masses are encountered in Salmonid fish which are known to require high oxygen levels, Duerst (41) has postulated that their propensity to goiter formation may be related to their continued need for high levels of thyroid hormone for optimal growth and development.

True neoplasms of the thyroid gland have been described in the inbred swordtail, *Xiphophorus montezumae*, by Berg and her coworkers (42). These tumors consist of masses of epithelial cells, some of which show occasional differentiation into microfollicles (Fig. 14). The tumor spreads by extension to involve the ventral musculature, ventral aorta, gill filaments and bone. Although it may be fortuitous, it is worth noting once again the striking biological similarity of this lesion to its counterpart in homoiotherms, specifically its metastases to bone. Two thyroid neoplasms in warm blooded animals have been utilized as models in cancer research. These are thyroid tumor No. 180 induced by prolonged excessive TSH levels followed chronic thiouracil administration as described by Morris and his associates (43, 44) and a spontaneous anaplastic canine thyroid carcinoma which proved to be transplantable (45).

The TSH induced tumor is slow growing, does not metastasize but kills the animal by local entension in approximately four months. The tumor consists of nests and cords of epithelial cells with hyperchromatic nuclei with occasional foci of follicle formation, some of which contain acidophilic staining colloid. This tumor is functional and it is noteworthy that three additional tumor lines have been described by these authors with varying degrees of functional activity as ascertained by uptake of ^{131}I (43). The presence of colloid filled follicles in the thyroid tumors of the swordtail fish suggests that perhaps these tumors may also be capable of thyroxin production; to the author's knowledge ^{131}I uptake studies of these tumors have not been reported.

LYMPHOSARCOMA

Malignant tumors of lymphoid tissue are not uncommon in fish, expecially in northern pike, *Esox lucius*, and muskellunge, *Esox masquinongy* (3, 46). Tumors appear to originate in most instances from the kidney, a prime hematopoietic organ; other sites of origin include the subcutaneous tissue and the branchial region. Observations which include the fact that lymphosarcoma affected all the inhabitants of the same tank at an aquarium, and that it appeared to be transmitted by tumor bearing fish to normal ones in the same tank, strongly suggests an infectious etiology. The analogy to viral induced lymphosarcoma and lymphatic leukemia in mice is immediately apparent. Thus far, attempts to identify and isolate an infectious agent from piscine lymphosarcoma have been unsuccessful. The morphology and behavior of lymphosarcoma of fish is quite similar to that described in homoiotherms. The neoplasms range from lymphoblastic lymphosarcomas consisting of large lymphoblasts with vesicular nuclei and multiple macronucleoli to lymphocytic tumors consisting of small lymphocytes with hyperchromatic dense nuclei (Fig. 15). There are numerous examples of this tumor in warm-blooded animals which are used as experimental models by oncologists. Typical examples, to name only a few, are the Murphy-Sturm (47), A-40, TT-8 and TT-10 lymphosarcomas (48) and the lymphocytic lymphoma No. 2 (49). These neoplasms, like those in fish, are usually widely disseminated with involvement of multiple organs (Fig. 16), including the integument (Fig. 17) and frequently accompanied by a lymphatic leukemia.

FIBROMA

Encapsulated benign tumors consisting of fibroblasts have been documented in 18 species of fish (2). Fibromas (Fig. 19) have been localized at numerous body sites, although the subcutaneous tissue (Fig. 18) and musculature of the trunk appear to predominate. Occasionally some of these neoplasms grow quite rapidly and are invasive and capable of local extension. They either represent aggressive fibromatoses biologically analogous to the condition described in man, or true fibrosarcomas. The fibromas are not transplantable to the anterior chamber of the eye, while their more aggressive variants are (50). The involvement of several goldfish, *Carassius auratus*, in the same pool by subcutaneous fibroma led Wago (51) to suggest that an infectious agent was responsible; attempts to demonstrate the agent or transplant the tumor were to no avail. In a recent report, Duncan and Harkin (52) describe small cytoplasmic bodies of unknown origin in superficial subcutaneous tumors of goldfish which they classified as possibly being fibromas. Although they state that the cytoplasmic inclusions did not appear to be virus particles, it would be of interest to attempt to isolate and concentrate the inclusions and to study their effect on several established cell lines derived from fish. Viral induced quasi-neoplastic proliferative lesions of the dermal connective tissue have been reported in wild cottontail rabbits (53). The subcutaneous tumor nodules consist of a florid proliferations of fibroblasts from which a virus can be recovered. The lesion is self-limited and eventually regresses. Resistance of rabbits following recovery from infectious fibromatosis to infection by myxomatosis virus suggests either that the two viruses are closely related or that the two diseases are caused by the same virus.

Several other types of tumors in fish which, though not as commonly recorded as some of those in the preceding section, occur with sufficient incidence and represent neoplasms of such infrequent occurrence in laboratory homoiotherms that they may serve as models for oncologic research.

NEURILEMMOMA AND NEUROFIBROMA

Tumors of the nerve sheath have been observed with sufficient frequency in *Lutianidae* (snappers) (54) which inhabit the waters of the Florida Keys, and in goldfish, *Carassius auratus* (55),, that they must be considered as characteristic for these species (3). The tumors appear as multiple subcutaneous sessile nodules and are most frequently located on the dorsolateral aspect of the head, trunk and caudal fin (Fig. 20). They are quite soft and not infrequently show foci of hemorrhage and cystic degeneration. Attempts to demonstrate association of these tumors with nerves have not been successful but the dorsolateral localization of the tumors correlates well with the body areas of fish which are most richly innervated by nerves of the lateral line sensory organs, the neuromasts. Histologically, approximately 30 percent of these tumors show multiple rows of palisading nuclei characteristic of Antoni type A tissue of neurilemmoma (Fig. 21). The edematous appearing loose stroma with stellate type cells and microcysts characteristic of Antoni type B tissue is less frequent (Fig. 22). In addition to foci consistent with a diagnosis of neurilemmoma there are often areas in these neoplasms that show a histological pattern consisting of dense bundles of elongate fusiform cells without evidence of palisading, findings suggestive of neurofibroma. The Bodian stain of these areas shows the presence of numerous neurites.

Recently, the classification of such neoplasms as nerve sheath tumors has been questioned on the basis of ultrastructural findings. However, these results must be accepted with caution since the neoplasms studied do not show the histological patterns of the Antoni type A and B tissue. Furthermore, since goldfish are known to develop subcutaneous fibromas, it may well be that the tumors described by Duncan and Harkin (52) are not identical to those reported by Lucke (54) and Schlumberger (55). It is of interest in this regard to point out that in addition to the tumors of the corium mentioned previously, goldfish also appear to be prone to the development of leiomyomas (56), thus, further complicating the problem of classification of neoplasms arising in this area.

ADAMANTINOMA

Odontogenic tumors have been reported in fish in only five instances. Although the number of cases are too few to enable one to draw any statistically valid conclusion, Schlumberger and Katz (57) point out that three of the five tumors reported are in Salmonid fishes and suggest that a genetic factor peculiar to that species may be involved. The lips of the fish are markedly distorted by large globular, confluent tumor masses (Fig. 23). The tumors consist of nests of columnar enamel epithelium (ameloblasts) surrounded by masses of stellate cells in a loose connective tissue stroma (Fig. 24, 25).

In laboratory homoiotherms two interesting reports have been published on tumors of dental origin. Burn and his coworkers (31) described the development of odontomas in rats on a low vitamin A diet. In 1944, Zegarelli (58) studied a group of highly inbred Slye stock mice in which there was a high incidence of spontaneous ameloblastomas. These were thought to be due to a genetic factor. However, since these initial accounts neither of these interesting models have apparently been investigated further.

OSTEOMA

Benign tumors of bone have been reported in 15 species of fish. Most of these are probably sporadic occurrences since with the exception of the Japanese food fish, the red tai, *Pagrosomus major* (59), there does not appear to be a constant association of these tumors with a single species. These lesions appear grossly as a fusiform thickening of bone. Osteomas may arise in any portion of the bony skeleton, including the mandible (Fig. 26). In the red tai there appears to be a predilection for the ventral processes of the sixth to eighth caudal vertebrae. The tumors consist of an overgrowth of cancellous bone with a very thin layer of dense cortical bone. All attempts to elucidate the etiology of this lesion have failed. Earlier work had suggested that the lesions arose as a result of trauma and fracture; however, subsequent studies failed to support this. Chabanaud (60) has suggested that perhaps some instances of localized bony overgrowth are due to metabolic changes which influence growth. It is well known that fishes are subject to periodic fluctuations of growth which may be related to seasonal changes. Though this may be so, it still does not explain the localized nature of the overgrowth.

The remarkable morphologic and biologic similarities which exist between neoplasms of fish and those of higher vertebrates renders them useful as models for the comparative study of cancer. Although neoplasms of fish are not mirrors of their counterpart lesions in homoiotherms, there is sufficient fidelity to suggest that the underlying biological processes involved are similar or perhaps even identical. From a pragmatic point of view, detailed knowledge of the epidemiology, pathology, and cell biology of piscine neoplasms will not only ensure the protection of fish populations for food and recreational purposes, but, more importantly, will serve to identify the presence, nature and mechanism of action of carcinogenic factors in the environment which often prove to be hostile for many other animals, including man.

ACKNOWLEDGMENT

The author wishes to acknowledge the kindness shown by his late mentor, Dr. H. G. Schlumberger, who made some of the specimens shown here available to him.

Research reported from the author's laboratory was supported by grant, SR01-CA-10257, from the National Cancer Institute.

REFERENCES

1. B. Lucke and H. G. Schlumberger, Physiol. Rev. **29**: 91 (1949).
2. H. G. Schlumberger and B. Lucke, Cancer Res. **8**: 657 (1948).
3. H. G. Schlumberger, Cancer Res. **17**: 823 (1957).

4. M. Gordon, Am. J. Cancer **15**: 1495 (1931).

5. M. Gordon, Growth **10**: (Suppl). 153 (1951).

6. H. D. Reed and M. Gordon, Am. J. Cancer **15**: 1524 (1931).

7. B. Lucke and H. G. Schlumberger, J. Exp. Med. **74**: 397 (1941).

8. J. E. Halver, C. L. Johnson and L. M. Ashley, Federation Proc. **21**: 390 (1961).

9. W. C. Hueper and W. W. Payne, J. Nat. Cancer Inst. **27**: 1123 (1961).

10. R. F. Nigrelli and S. Jakowska, Zoologica **46**: 49 (1961).

11. R. R. Rucker, W. T. Yasutake and H. Wolf, Prog. Fish Cult. **23**: 3 (1961).

12. D. G. Scarpelli, M. H. Greider and W. J. Frajola, Cancer Res. **23**: 848 (1963).

13. E. M. Wood and C. P. Larson, Arch. Pathol. **71**: 471 (1961).

14. S. R. Wellings, R. G. Chuinard and M. Bens, Ann. New York Acad. Sci. **126**: 479 (1965).

15. B. Lucke, Ann. Rept. Tortugas Lab., Carnegie Inst. Wash. 92 (1938).

16. G. Keysselitz, Arch. Protistenk. **11**: 326 (1908).

17. R. E. Russell and P. Kotin, J. Nat. Cancer Inst. **18**: 857 (1957).

18. R. E. Shope, Proc. Soc. Exp. Biol. & Med. **32**: 830 (1935).

19. M. K. Deringer, J. Nat. Cancer Inst. **29**: 1107 (1962).

20. W. L. Clem, L. Moewus and M. M. Sigel, Proc. Soc. Exp. Biol. & Med. **108**: 762 (1961).

21. M. Gravell and R. G. Malsberger, Ann. New York Acad. Sci. **126**: 555 (1965).

22. K. E. Wolf, M. C. Quimby, E. A. Pyle and R. P. Dexter, Science **132**: 1890 (1960).

23. D. G. Scarpelli, Unpublished observations.

24. W. C. Hueper, Ann. N. Y. Acad. Sci. **108**: 963 (1963).

25. Z. K. Cooper, H. I. Firminger and H. C. Reller, Cancer Res. **4**: 617 (1944).

26. J. G. Kidd and P. Rous, J. Exp. Med. **71**: 813 (1940).

27. W. H. Brown and L. Pearce, Proc. Soc. Exp. Biol. & Med. **18**: 201 (1921).

28. W. H. Brown and L. Pearce, J. Exp. Med. **37**: 601 (1923).

29. R. O. Sinnhuber, In: TROUT HEPATOMA RESEARCH CONFERENCE PAPERS. (J. E. Halver and I. A. Mitchell ed.) B. S. F. W. Res. Rept. **70**, 1966, p. 48.

30. J. E. Halver, In: MYCOTOXINS IN FOODSTUFFS. (G. N. Wogan ed.) M. I. T. Press, Cambridge, Mass., 1965, p. 209.

31. C. G. Burn, A. W. Orten and A. H. Smith, J. Dental. Res. **16**: 317 (1937).

32. D. G. Scarpelli, In: TROUT HEPATOMA RESEARCH CONFERENCE PAPERS (J. E. Halver and I. A. Mitchell, eds.). B. S. F. W. Res. Rept. **70**, 1966, p. 60.

33. V. R. Potter, Cancer Res. **21**: 1131 (1961).

34. M. D. Reuber, J. Nat. Cancer Inst. **26**: 891 (1961).

35. H. P. Morris, H. Sidransky, B. P. Wagner and H. M. Dyer, Cancer Res. **20**: 1252 (1960).

36. H. B. Andervont and T. B. Dunn, J. Nat. Cancer Inst. **13**: 455 (1952).

37. A. B. Novikoff, Cancer Res. **17**: 1010 (1957).

38. G. H. Algire, J. Nat. Cancer Inst. **5**: 151 (1944).

39. D. Marine and C. H. Lenhart, J. Exp. Med. **12**: 311 (1910).

40. D. Marine, J. Exp. Med. **19**: 70 (1914).

41. J. U. Duerst, In: DIE URASCHEN DER ENTSTEHUNG DES KROPFES. Bern: Hans Huber, (1941).

42. O. Berg, M. Edgar and M. Gordon, Cancer Res. **13**: 1 (1953).

43. H. P. Morris, A. J. Dalton and C. D. Green, J. Clin. Endocrinol. **11**: 1281 (1951).

44. H. P. Morris and C. D. Green, Science **114**: 44 (1951).

45. M. W. Allam, L. S. Lombard, E. L. Stubbs and J. F. Shirer, Cancer Res. **14**: 734 (1954).

46. R. F. Nigrelli, Zoologica **32**: 101 (1947).

47. J. B. Murphy and E. Sturm, Cancer Res. **1**: 379 (1941).

48. L. J. Dunham and H. L. Stewart, J. Nat. Cancer Inst. **13**: 1299 (1953).

49. E. Shelton, J. Nat. Cancer Inst. **12**: 1203 (1952).

50. B. Lucke, H. G. Schlumberger and C. Breedis, Cancer Res. **8**: 473 (1948).

51. H. Wago, Gann **16**: 11 (1922).

52. T. E. Duncan and J. C. Harkin, Am. J. Pathol. **55**: 191 (1968).

53. R. E. Shope, J. Exp. Med. **56**: 803 (1932).

54. B. Lucke, Arch. Pathol. **34**: 133 (1942).

55. H. G. Schlumberger, Cancer Res. **12**: 890 (1952).

56. H. G. Schlumberger, Am. J. Pathol. **25**: 287 (1949).

57. H. G. Schlumberger and M. Katz, Cancer Res. **16**: 369 (1956).

58. E. V. Zegarelli, Am. J. Pathol. **20**: 23 (1944).

59. K. Takahashi, Z. Krebsforsch. **29**: 1 (1929).

60. P. Chabanaud, Compt. Rend. **182**: 1647 (1926).

COMMENTS

DR. GREENBLATT: Would you amplify the term species specific? Do you mean a tumor of specific histologic structure found only in a single species?

DR. SCARPELLI: No, what I meant by this term is a very high incidence of a given type of neoplasm in a single species. Typical examples in fish are epithelioma at the lip and face in catfish and hog suckers, epidermal papilloma in pleuronectid fishes, and hepatoma in rainbow trout. Research on the environmental factors responsible for hepatoma in trout led to much of the work which firmly established that a dietary carcinogen was responsible which later was identified as aflatoxin, a potent carcinogen for both cold blooded and warm blooded animals. Thus, detailed studies of species specific tumors can lead to the elucidation of important environmental, and genetic factors involved in carcinogenesis.

DR. NEUHAUS: Is the normal plasma protein level maintained by residual liver tissue in hepatoma-bearing trout or is the tumor tissue itself capable of producing plasma proteins?

DR. SCARPELLI: Hepatomas in those fish with markedly elevated plasma protein levels contained large extracellular eosinophilic accumulations which were identified as proteinaceous and containing large amounts of arginine. Microsomes isolated from these tumors showed an increased rate of incorporation of labeled amino acid as compared to normal liver and hepatoma in fish without elevated serum protein levels. We have been trying to induce such "functional" tumors in trout with aflatoxin and have not been successful thus far.

DR. NEUHAUS: Can you demonstrate the presence of serum albumin in the normal hepatic and hepatomatous tissue by using an antiserum specific for this protein?

DR. SCARPELLI: We have succeeded in making fluorescent-labeled antibody to fish serum albumin. We have been able to demonstrate it in normal liver but have not found functional tumors yet.

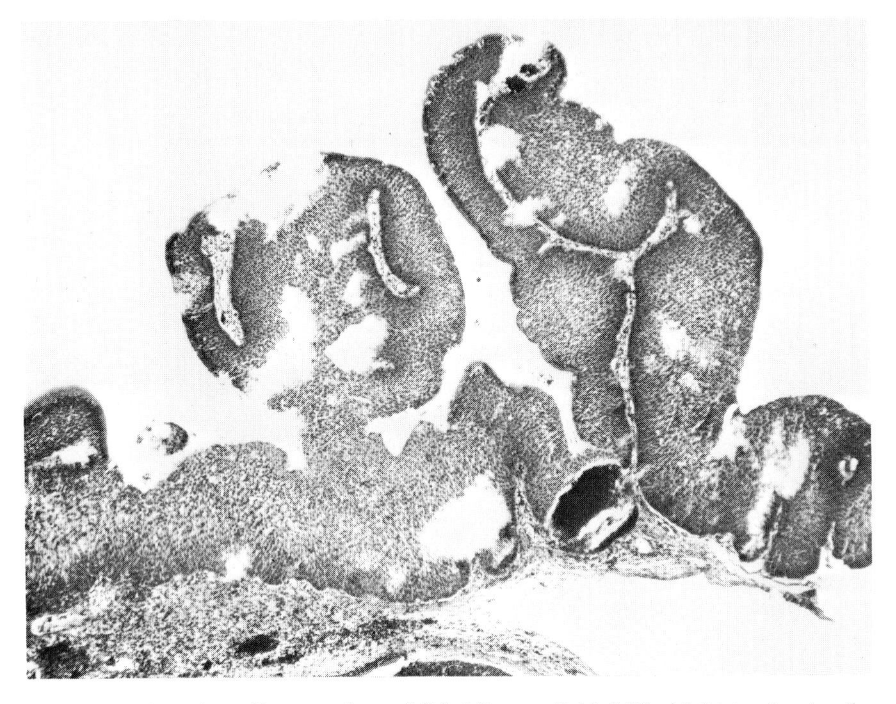

Fig. 1: Epidermal papilloma in the reef fish 'slippery dick', Iridio bivittata, showing the highly folded epithelium; note the absence of keratin. 50X.

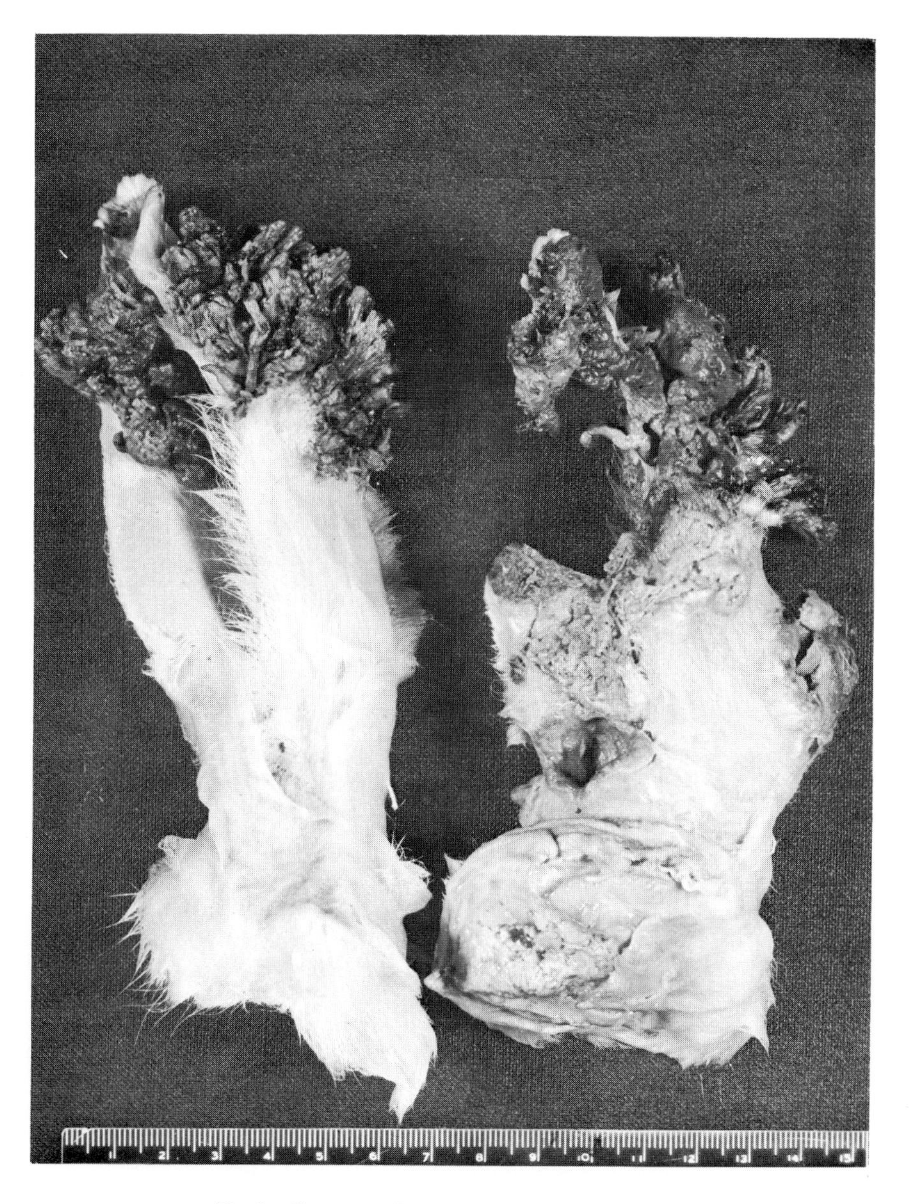

Fig. 2: Shope papilloma on ears of a cottontail rabbit.

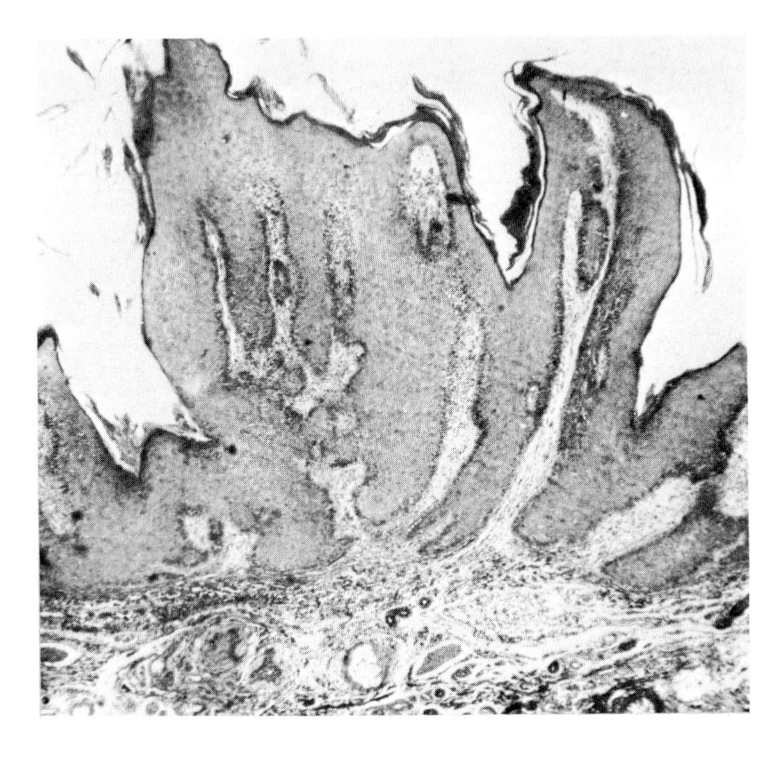

Fig. 3: Shope papilloma showing the characteristic marked hyperplasia of the epidermis with keratinization. 50X.

Fig. 4: Epithelioma in a hog sucker, *Hypentelium nigricans*, involving the head. Nodules of tumor can be seen infiltrating the skin of the face at the base of the fungoid neoplasm.

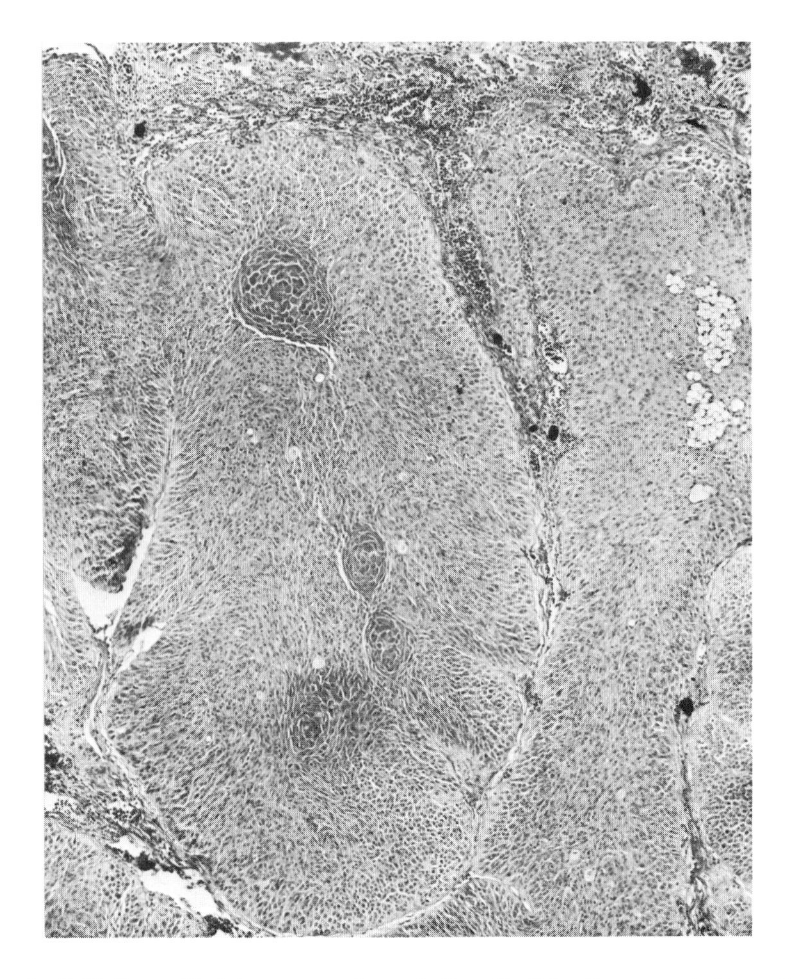

Fig. 5: Epithelioma. Nodular masses of neoplastic epithelial cells surrounded by a thin rim of vascular stroma. Note the four tightly packed nests of cells suggestive, but not diagnostic, of 'pearl formation'.

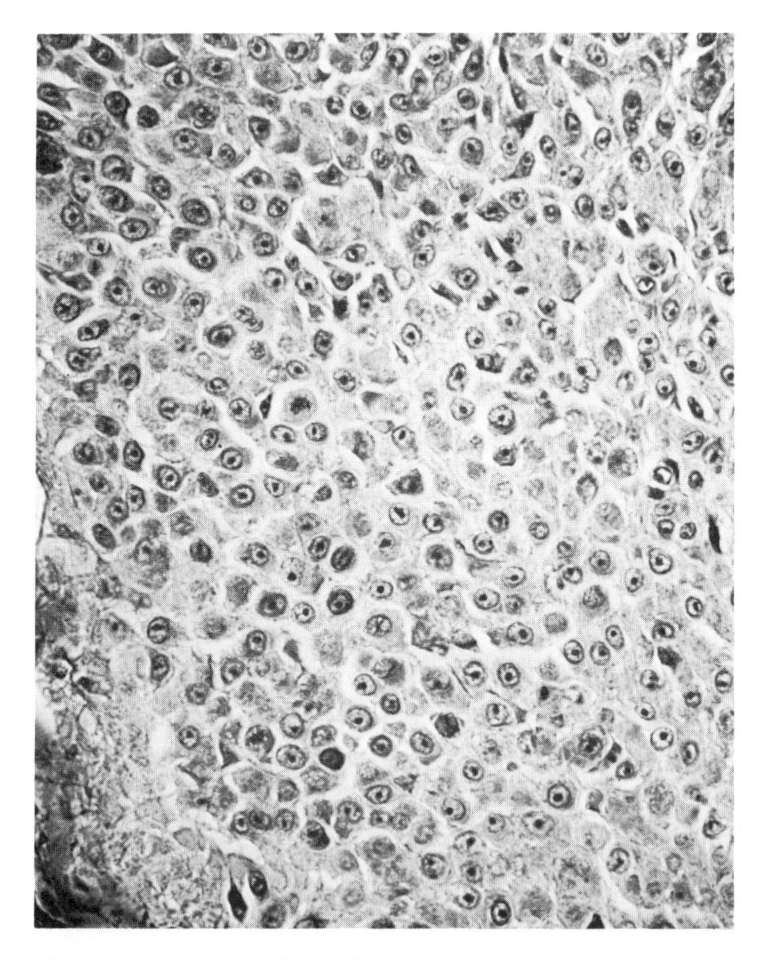

Fig. 6: Epithelioma. Cytologic detail of neoplastic cells showing large hyperchromatic nuclei, macronucleoli, a disorganized pattern of cell orientation and occasional mitoses. 330X.

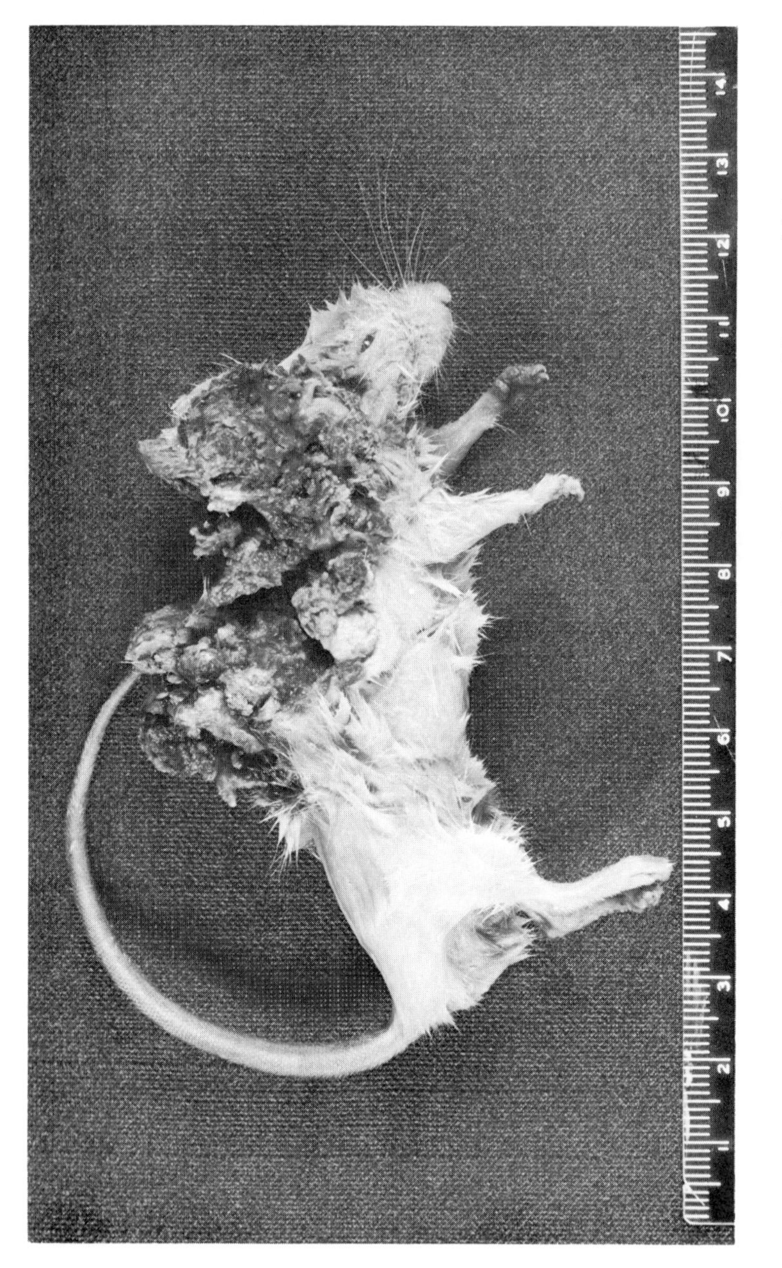

Fig. 7: Epithelioma (squamous cell carcinoma) in a C3H mouse induced by repeated 3,4 benzpyrene painting.

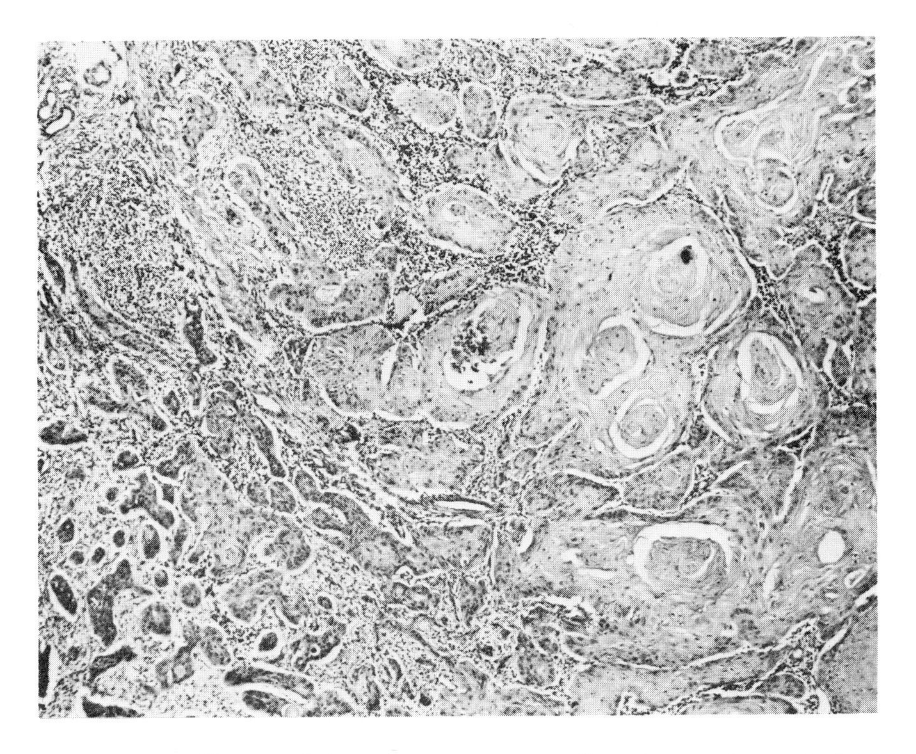

Fig. 8: Well differentiated keratinizing squamous cell carcinoma characterized by extensive epithelial pearl formation. 75X.

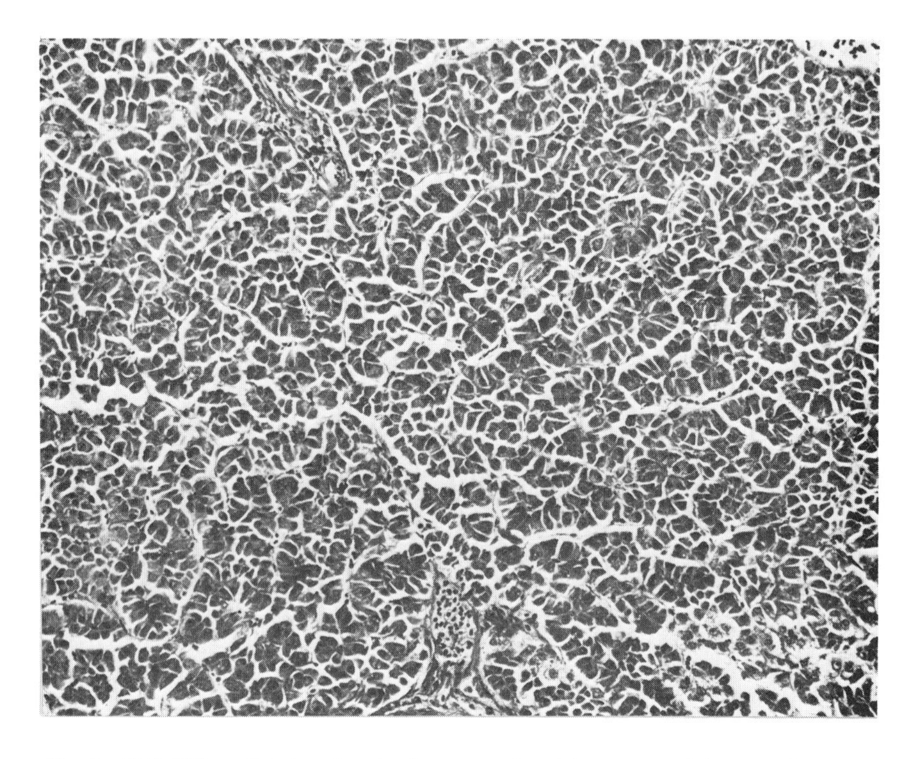

Fig. 9: Well differentiated trabecular hepatoma in rainbow trout, Salmo gairdneri. The tumor cells are oriented into nests and cords. 175X.

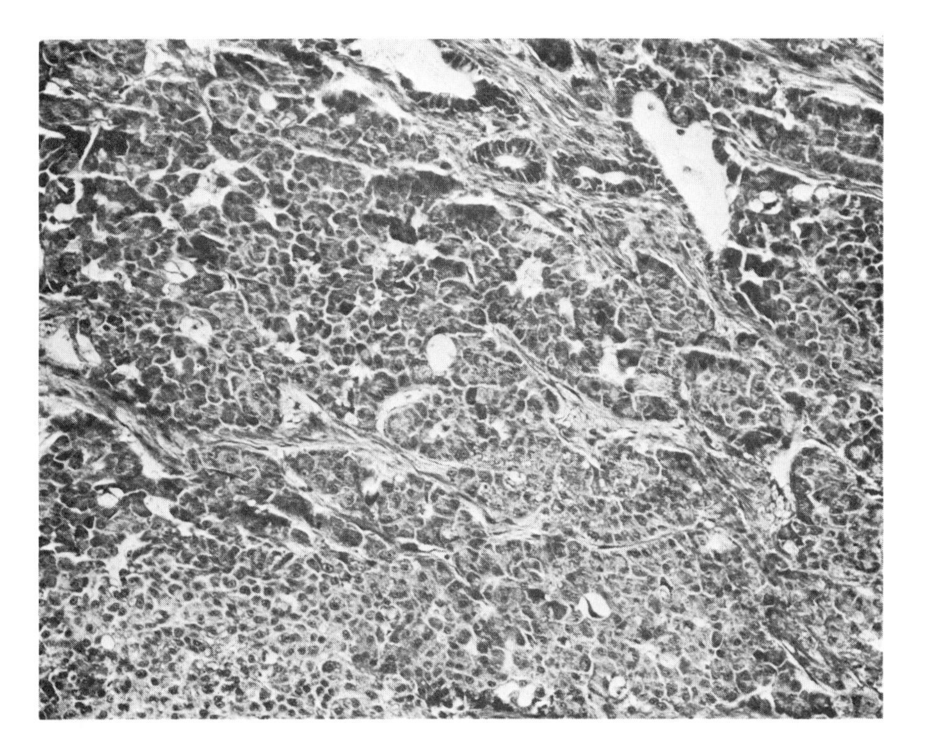

Fig. 10: Moderately well differentiated trout hepatoma in which the tumor cells are organized into highly variable nests and tubules. 175X.

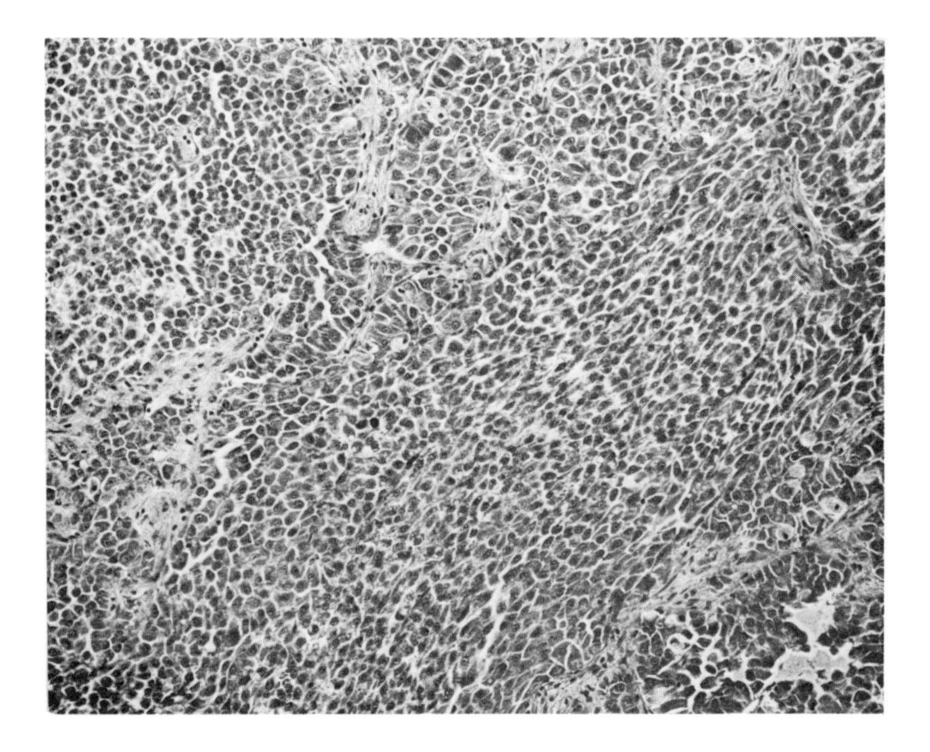

Fig. 11: Poorly differentiated trout hepatoma in which tumor cells are disposed in large masses with only rare attempts at liver cord formation. 175X.

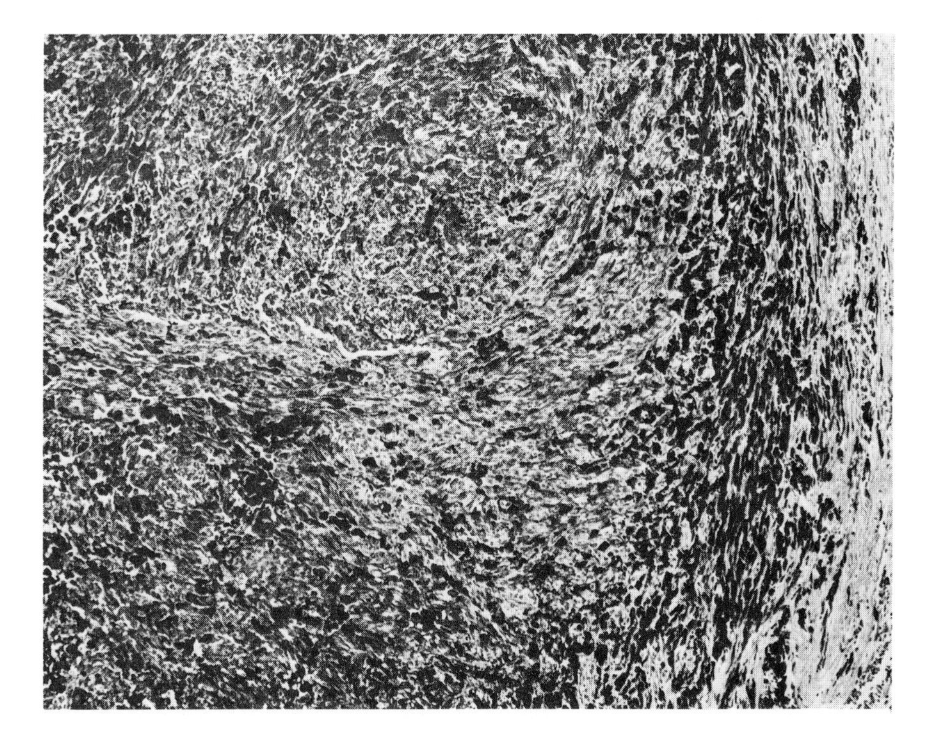

Fig. 12: Melanotic tumor in a killifish hybrid (*Xiphophorous hellerii* x *Platypoecilius maculatus*). The tumor consists of whorled masses of macromelanophores containing melanin. 110X.

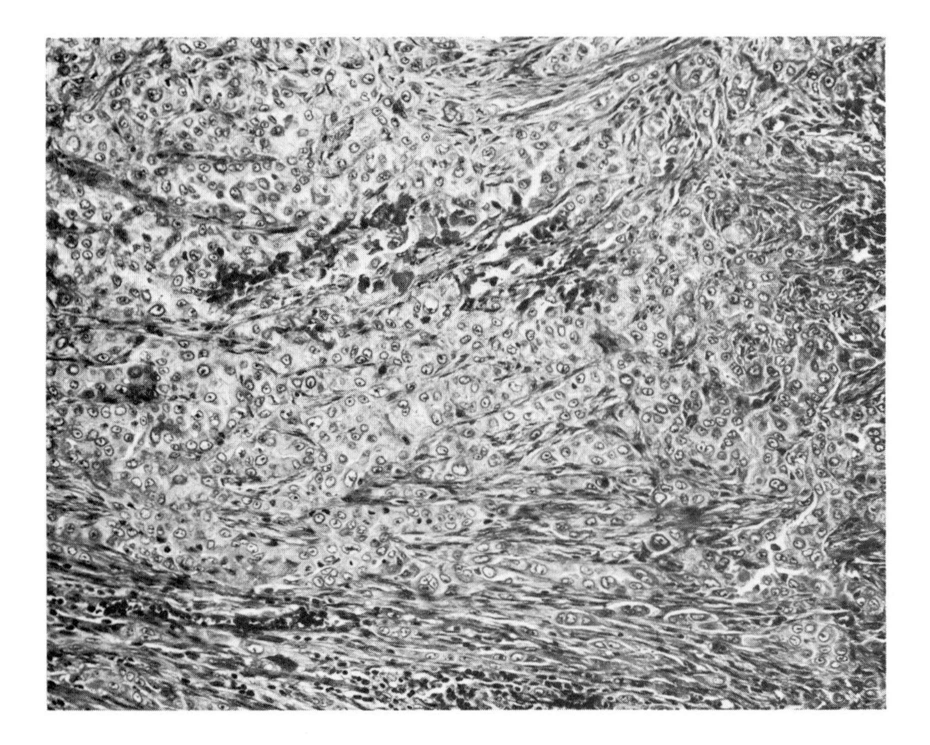

Fig. 13: Cloudman melanoma S91. Nests of polyhedral melanoblasts can be identified; occasional cells contain melanin. 150X.

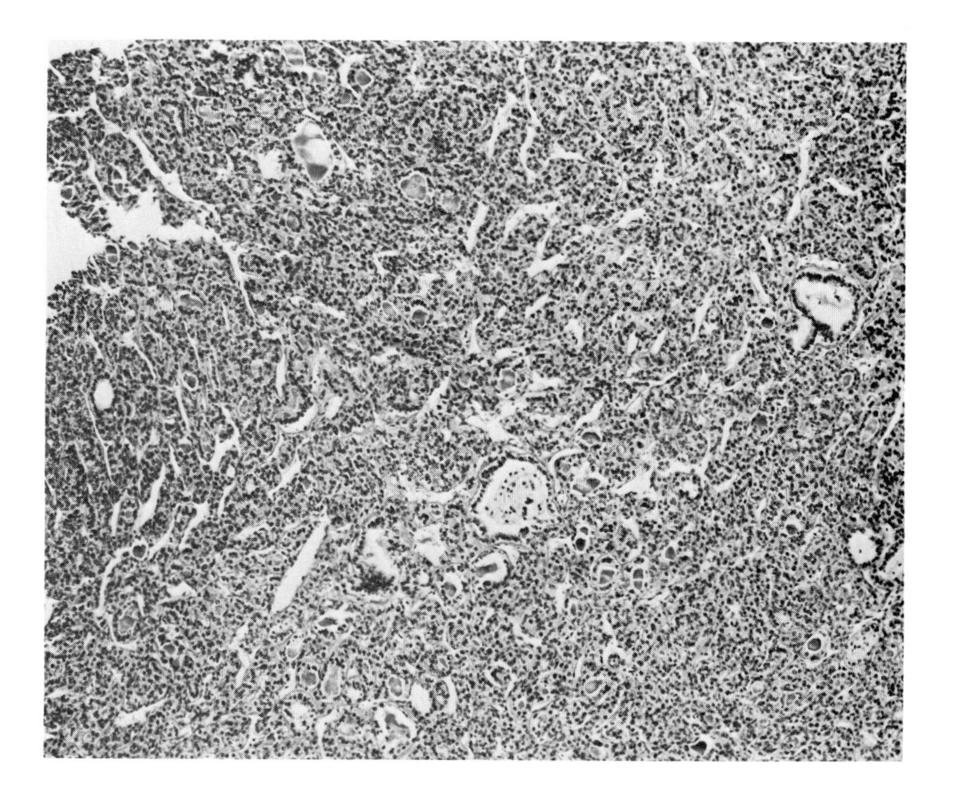

Fig. 14: Thyroid carcinoma in Xiphophorous montezumae showing masses of small cells with hyperchromatic nuclei; several well defined colloid follicles are evident. 100X.

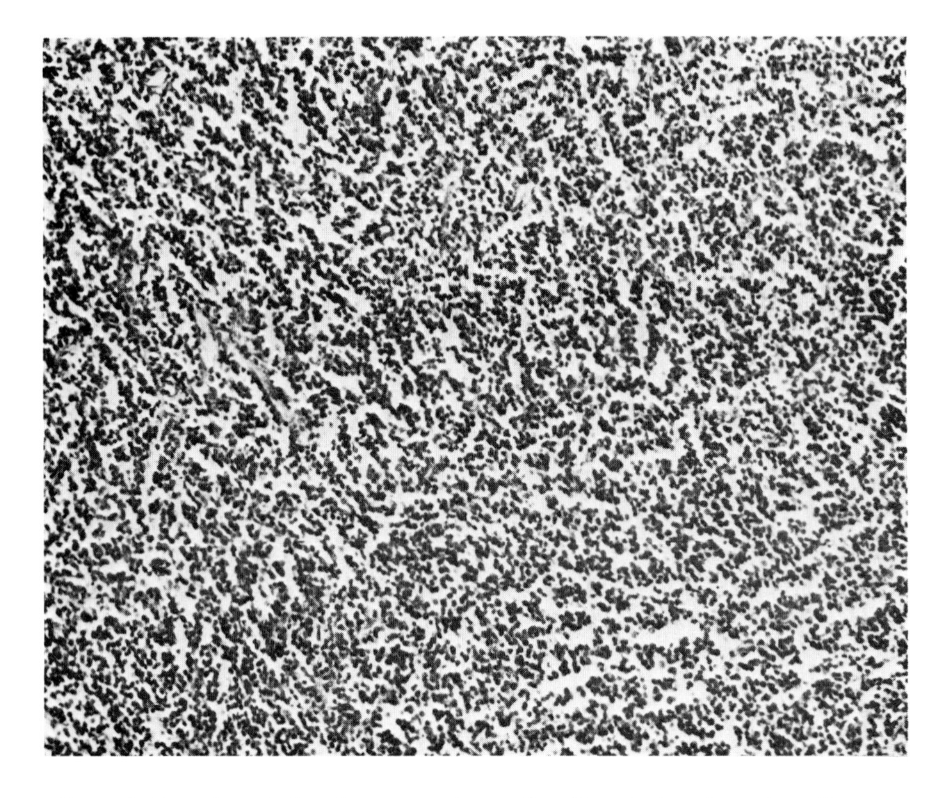

Fig. 15: Lymphocytic lymphosarcoma in a pike, Esox lucius. A section of tumor showing masses of well differentiated lymphocytes in a loose connective tissue stroma. 175X.

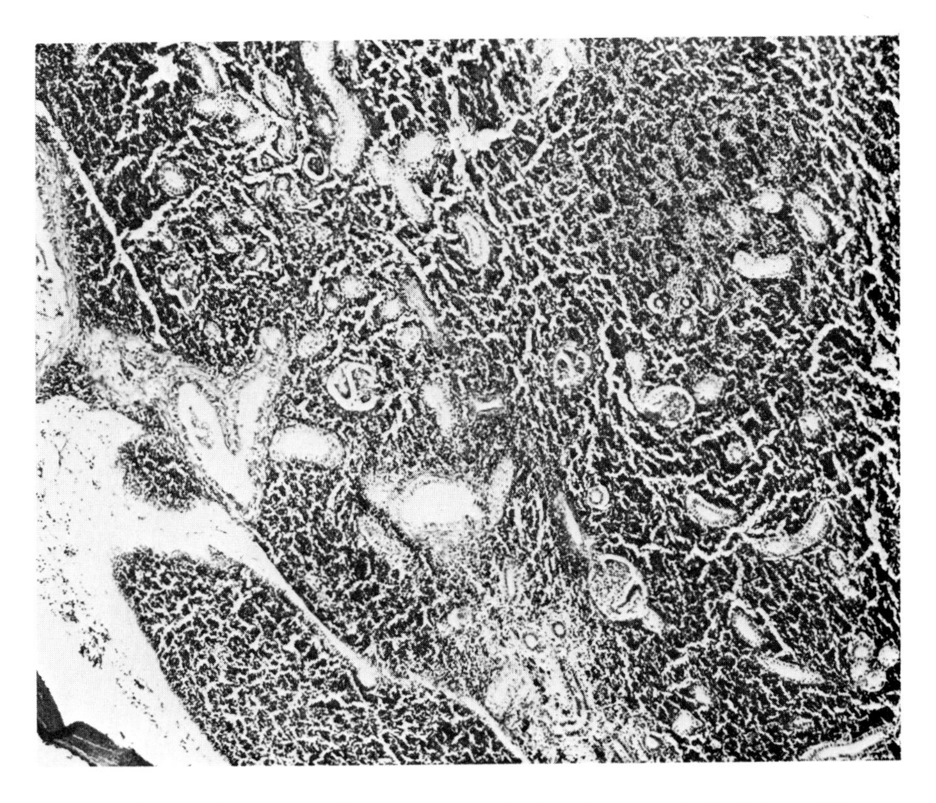

Fig. 16: Lymphocytic lymphosarcoma in a pike showing tumor in the kidney. 85X.

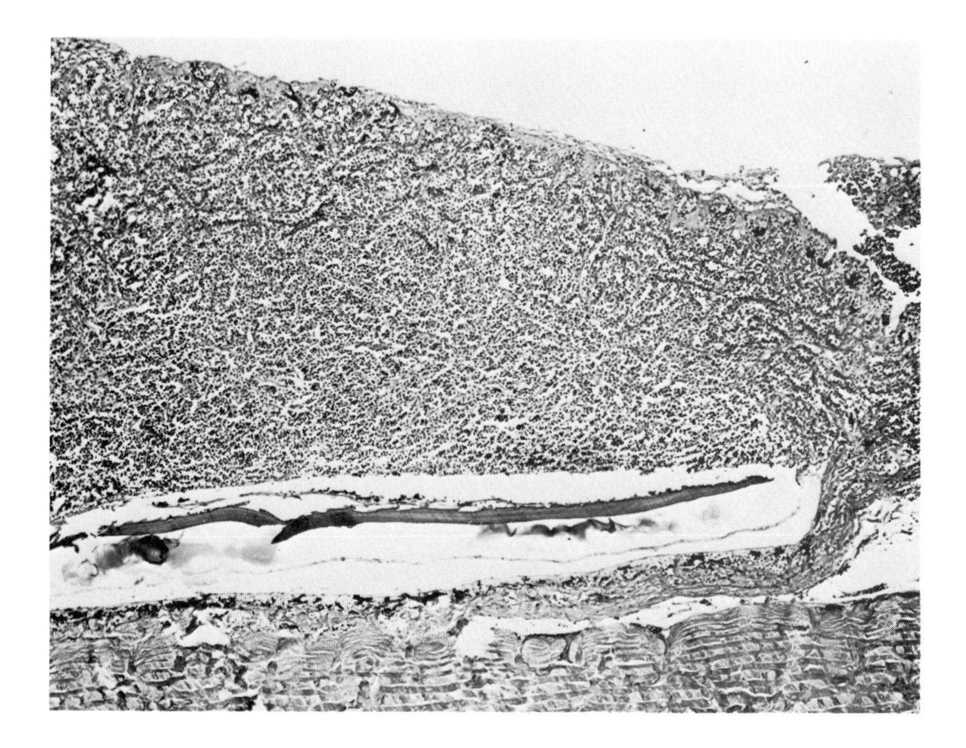

Fig. 17: Same case as shown in Figures 15 and 16. Dense tumor infiltrates in the integument and corium. 50X.

Fig. 18: Subcutaneous fibroma in Atlantic salmon, *Salmo salar*.

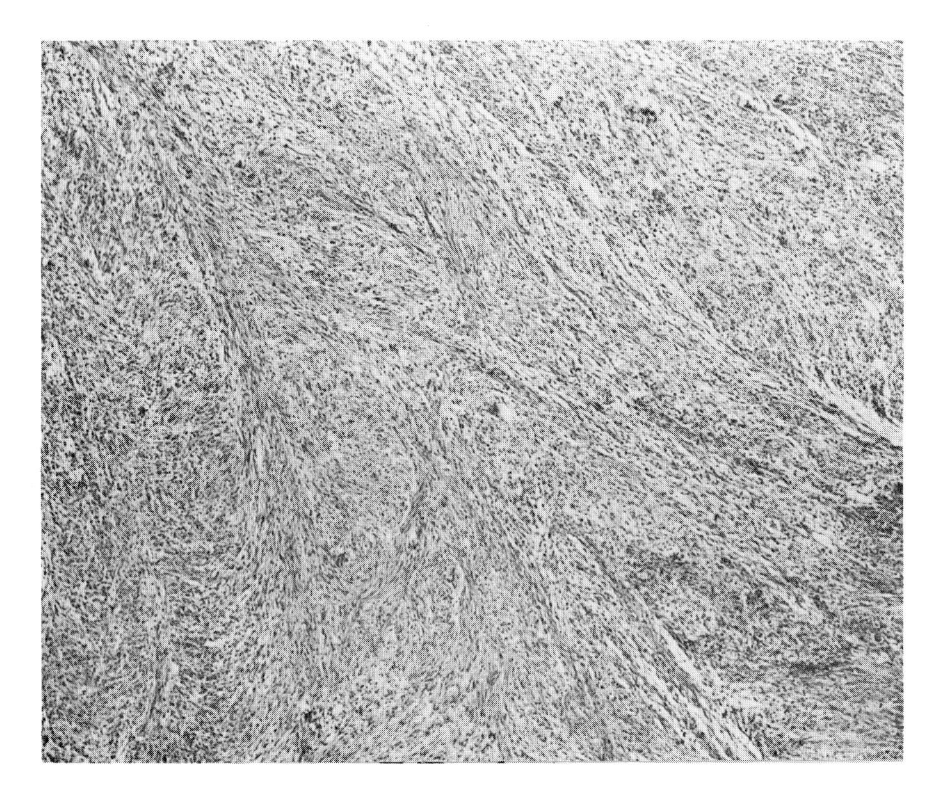

Fig. 19: Histologic appearance of the tumor in the previous figure. It consists of interlacing bands of fibroblasts and collagen. 70X.

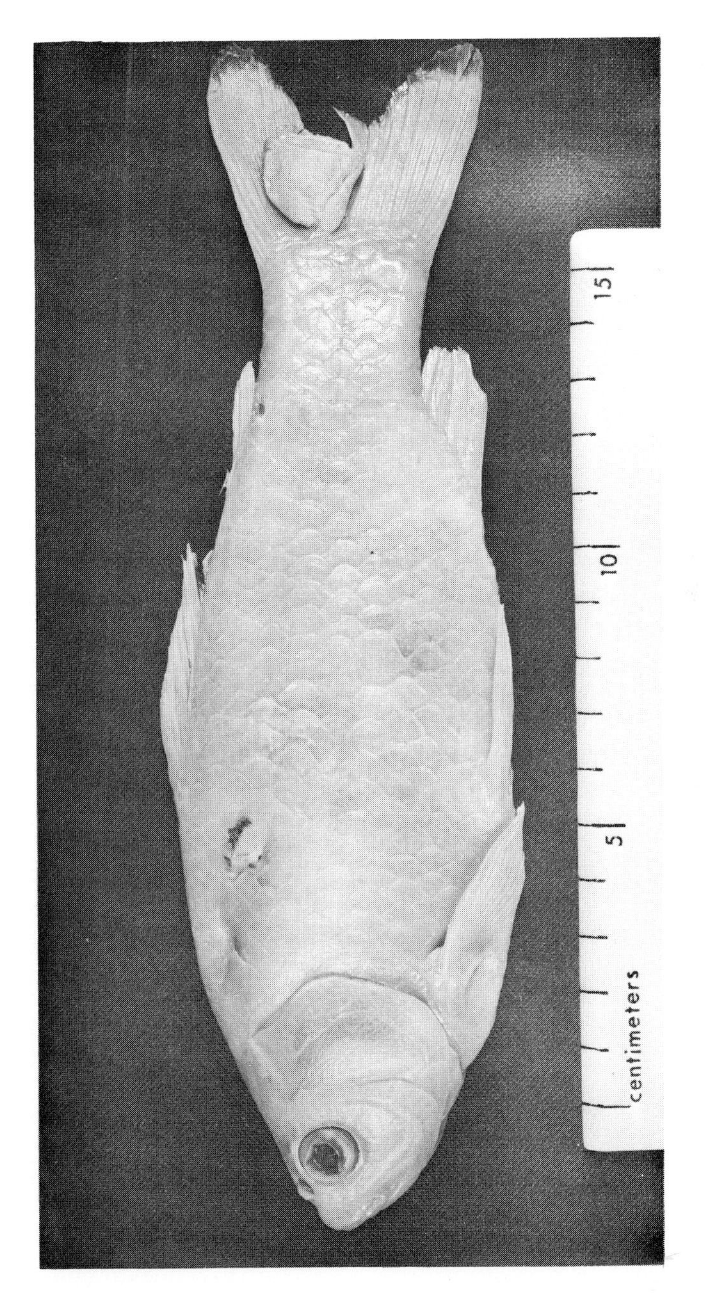

Fig. 20: Neurilemmoma on the caudal fin of a goldfish, Carassius auratus.

Fig. 21: Section of tumor shown in Figure 20. Multiple rows of palisading elongate nuclei characteristic of Antoni type A tissue. 175X.

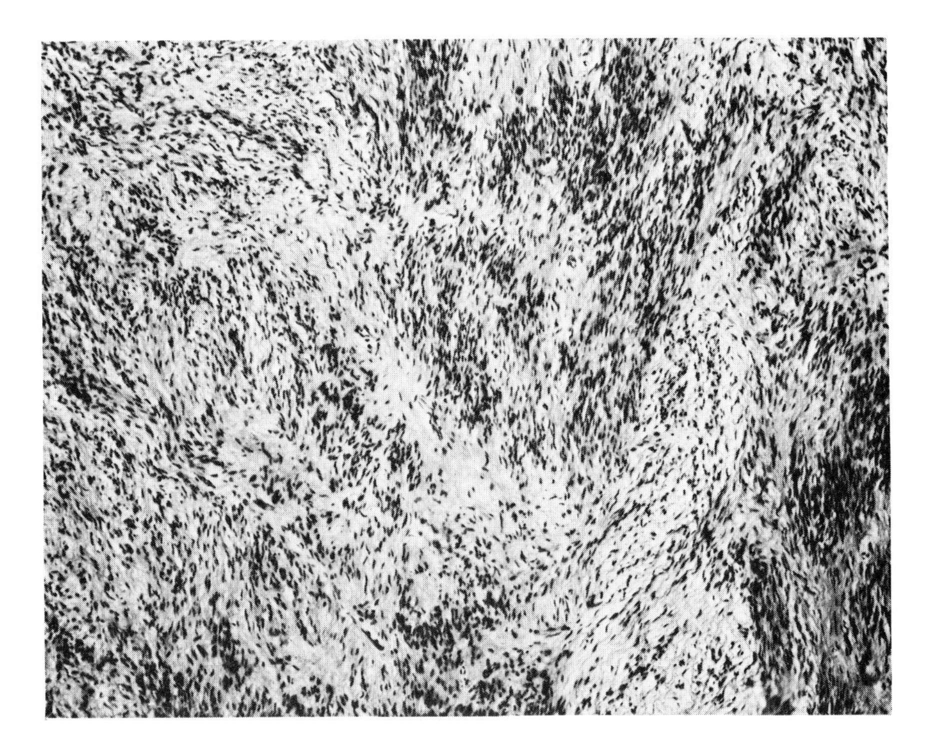

Fig. 22: An area of tumor shown in the preceding figure. Edematous, loose stroma with stellate cells resembling the Antoni type B tissue present in nerve sheath tumors of higher vertebrates. 110X.

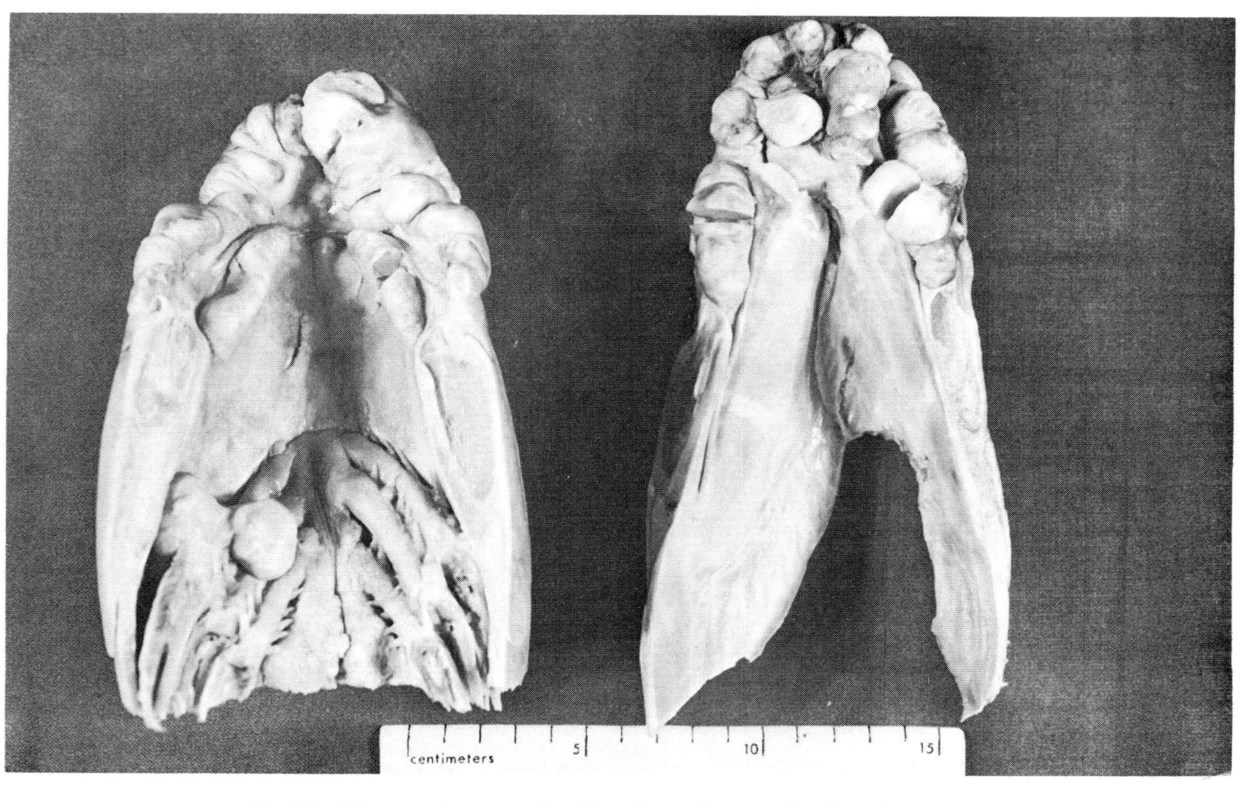

Fig. 23: Odontogenic tumors involving the oral cavity of a chinook salmon.

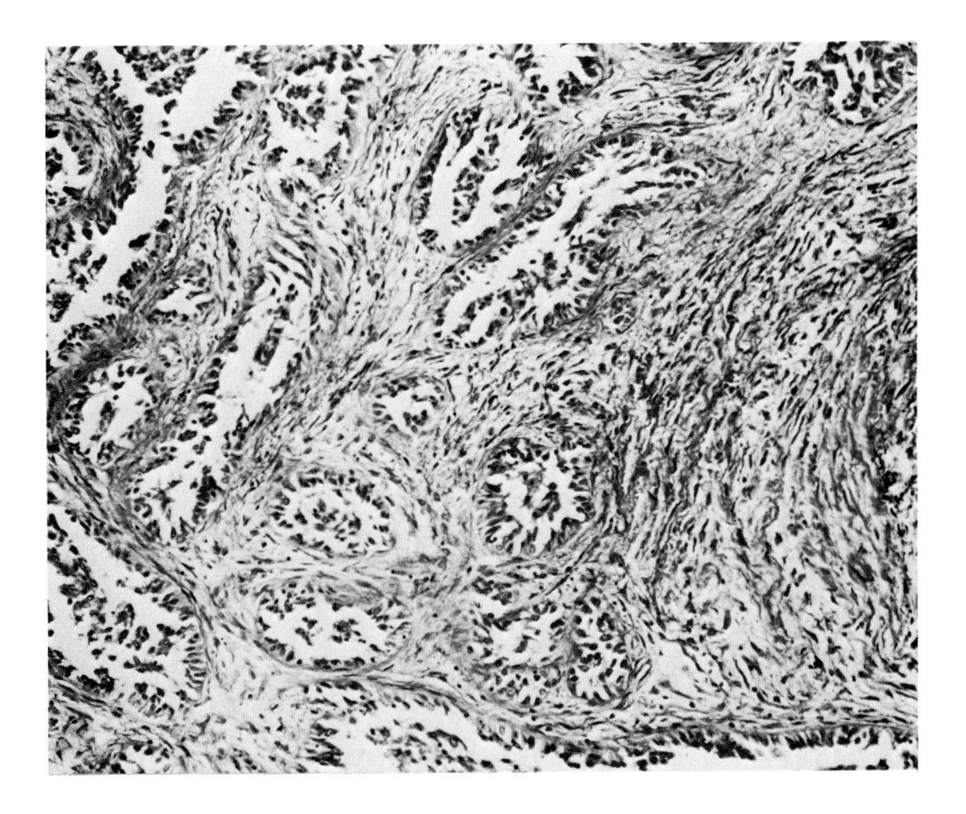

Fig. 24: A section of tumor shown in the previous figure. Columnar tall ameloblasts forming gland-like structures. The surrounding stroma consists of loose connective tissue which abuts directly on the epithelial cells. 150X.

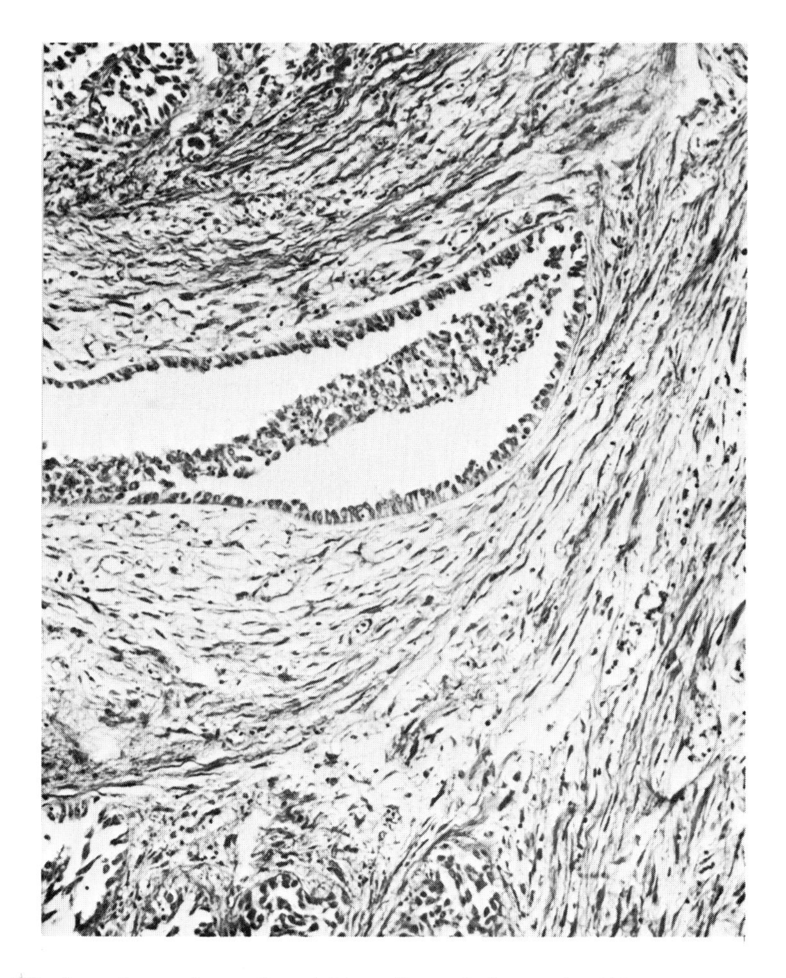

Fig. 25: A continuous layer of ameloblasts disposed along a gland-like space, the center of which contains a mass of stellate cells which resemble spindle cells of the dental lamina seen in higher vertebrates. 150X.

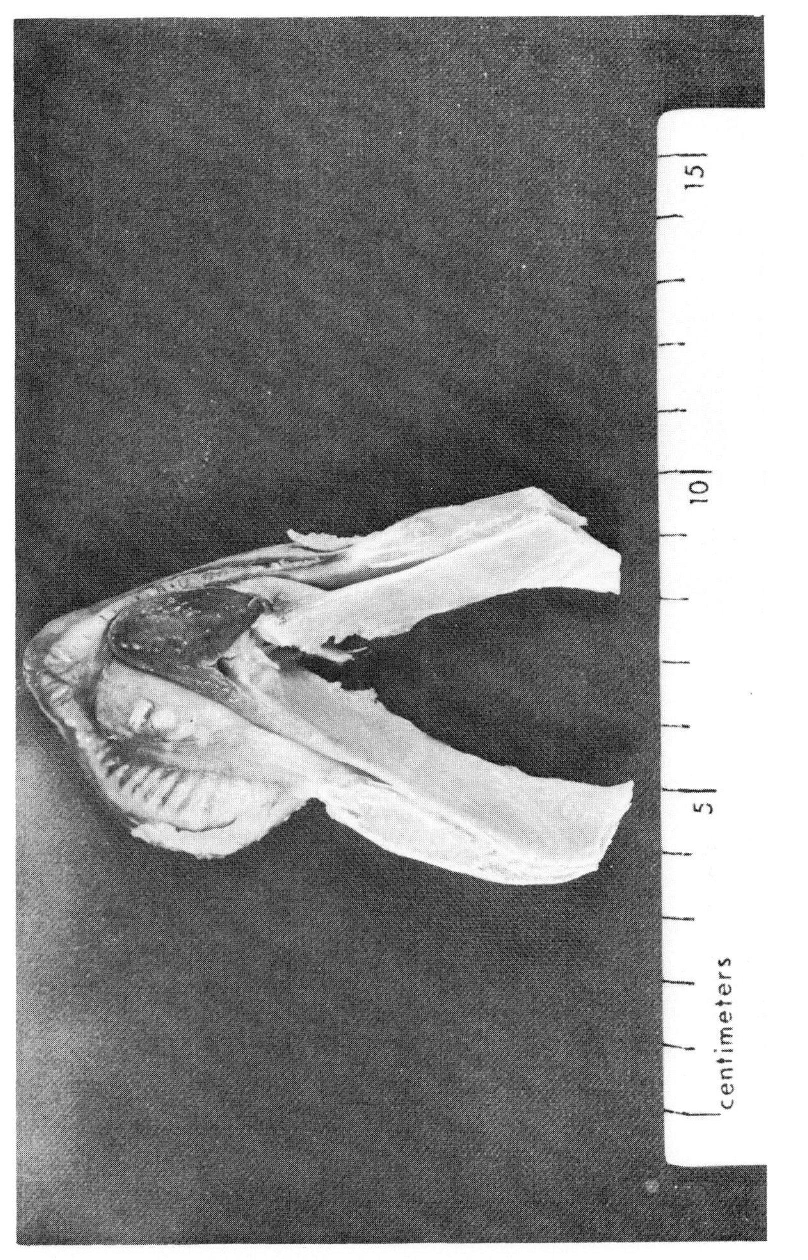

Fig. 26: Osteoma in the mandible of the salmon, *Salmo salar*.

CONTROL OF CHOLESTEROL SYNTHESIS IN NORMAL AND MALIGNANT TISSUE

M. D. Siperstein and L. J. Luby

FEEDBACK CONTROL OF CHOLESTEROL SYNTHESIS IN NORMAL LIVER AND INTESTINE

It has been known since 1950 that probably all tissues of the body are capable of synthesizing cholesterol from acetate (1). Moreover, largely through the studies of Bloch (2), Lynen (3), and Popjak and Cornforth (4), the biochemical pathway by which short chain acids are converted to cholesterol has been well worked out. By contrast, however, only recently has some insight been gained into the complex mechanisms by which the control of cholesterol synthesis is affected by several of the tissues of the body. The studies of Gould (5), Tomkins (6), Langdon and Bloch (7), and of Frantz (8), independently indicated in the early 1950's that dietary cholesterol causes a striking inhibition of *de novo* cholesterol synthesis in the liver. Subsequent studies in our laboratory (9-11), have attempted to elucidate both the biochemical and the subcellular sites of this control mechanism. As indicated in Fig. 1, data published in 1960 (9) strongly suggested that the feeding of cholesterol inhibits the overall conversion of acetate to cholesterol by blocking a site on the pathway of sterol synthesis prior to mevalonic acid but after the conversion of acetate to β-hydroxy-β-methyl glutarate.

Further studies from this laboratory (10) have amply confirmed the suggestion of this earlier study that the conversion of β-hydroxy-β-methyl glutarate to mevalonate does in fact represent the primary site at which cholesterol regulates its own synthesis. The latter experiments were based on a gas liquid chromatographic method (12), which allows a direct determination of the rate of mevalonic acid synthesis. With this technique it could be demonstrated that cholesterol feeding causes a marked inhibition of mevalonic acid synthesis while having no effect upon the synthesis of β-hydroxy-β-methyl glutarate. Such studies provided the first definitive evidence that the feedback control of cholesterol synthesis is carried out through inhibition of a single site in the overall control of sterol production, namely at the point of synthesis of mevalonic acid. This conclusion has been supported by data obtained both in cell-free enzyme studies as well as in the intact liver cell.

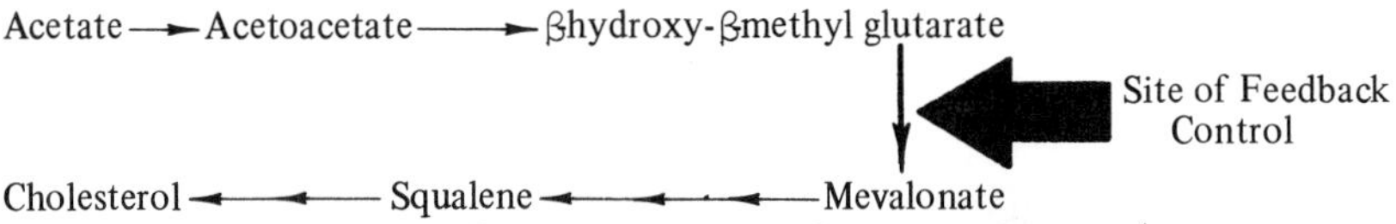

Acetate → Acetoacetate → β-hydroxy-β-methyl glutarate

↓ ← Site of Feedback Control

Cholesterol ← ← Squalene ← ← ← ← Mevalonate

Fig. 1: Site of Feedback Control of Cholesterol Synthesis

It is of interest that if cholesterol is fed for long periods of time the biochemical steps subsequent to the synthesis of mevalonate ultimately also become inhibited, a finding that was described first in our laboratory (9), and subsequently greatly extended in a study by Gould (13). There is, however, reason to believe that the inhibition of cholesterol synthesis from mevalonate that occurs only after prolonged cholesterol feeding, is secondary to the primary block in mevalonic acid synthesis, and in this sense does not represent a primary site of control of cholesterol synthesis (14).

Once it became possible to measure directly the rate of mevalonic acid production, the subcellular site at which the control of cholesterol synthesis is affected could readily be localized. While the overall conversion of acetate to cholesterol requires both the microsomal and the soluble fractions of the liver cell, it could be demonstrated that over 90% of the mevalonic synthetase of the cell is localized in the lipoproteins that make up the endoplasmic reticulum, or microsomal fraction, of the liver cell. Moreover, not too surprisingly, the feeding of cholesterol has its major inhibitory action on this microsomal enzyme (10). Hence, it may be concluded that in normal liver the subcellular site at which the entire process of cholesterol synthesis is regulated is clearly localized to the endoplasmic reticulum.

The detailed mechanism by which exogenous cholesterol can inhibit the activity of the enzyme responsible for mevalonate synthesis remains to be completely understood. It is, however, clear that this inhibition is uniquely accomplished by dietary cholesterol, since endogenous cholesterol does not inhibit the synthesis of mevalonate (15). Hence, it can be shown that when chylomicra derived from animals fed a high cholesterol diet are injected into a normal animal, cholesterol synthesis in the liver is promptly depressed; on the other hand, when cholesterol containing very low density lipoproteins derived from the liver are similarly injected, cholesterol synthesis is not affected. On this basis we believe that only specific forms of lipoprotein-bound cholesterol, probably as cholesterol esters (16), can serve as mediators of the feedback inhibition of mevalonate, and hence of cholesterol synthesis. The detailed mechanisms by which such lipoproteins specifically enter the cell and inhibit mevalonate synthesis remain to be elucidated.

While the foregoing discussion applies to the control of cholesterol synthesis in mammalian liver, it has become apparent, largely through studies being carried out by Dr. John Dietschy, that the control of sterol synthesis in the intestine differs in an important way from that which operates in the liver. It has been known for many years that cholesterol synthesis in the intestine, unlike that in the liver, is not affected by exogenous cholesterol (9). As a result, as is summarized in Table I, when animals are fed a 5% cholesterol diet, the intestine rather than the liver becomes the major site of endogenous cholesterol production. On the basis of such observations, it was

TABLE I

EFFECT OF CHOLESTEROL FEEDING ON CHOLESTEROL SYNTHESIS BY SMALL INTESTINE

	Mμmoles of Acetate-^{14}C Incorporated into Cholesterol per gram tissue	
Diet	Liver	Intestine
0	20	7.2
0.5	2	5.2
5.0	0.4	7.6

thought for many years that sterol synthesis in the intestine was not under feedback control. However, subsequent studies (17, 18) have clearly shown, as indicated in Table II, that in fact cholesterol synthesis in the intestine of the rat is under feedback regulation. This inhibition, it is now clear, is not mediated by cholesterol, but by the end product of cholesterol breakdown, taurocholic acid. It is also of interest that the biochemical site of this

TABLE II

FEEDBACK CONTROL OF CHOLESTEROL SYNTHESIS IN INTESTINE

Treatment	Acetate-2-^{14}C Converted to Cholesterol
	mμmoles/gm
1. Control	98
2. Bile Fistula	500
3. Bile Fistula + Bile Infusion	120
4. Bile Fistula + Taurocholate Infusion	130

From: J. Clin. Invest. 47: 286 (1968).

feedback control is identical to that in the rat liver, namely at the point of conversion of β-hydroxy-β-methyl glutarate to mevalonate.

In subsequent studies dealing with possible phylogenic differences in the regulation of cholesterol synthesis among different species, it became obvious that the feedback control of cholesterol synthesis differs strikingly as one moves from the mammal to the non-mammalian vertebrate. This aspect of the control of cholesterol synthesis came to light during studies on cholesterol metabolism in the goldfish. In this species, as indicated in Table III, in contrast to the rat, cholesterol feeding had a profound inhibitory effect both on hepatic and on intestinal cholesterol synthesis. Moreover, it is apparent that all portions of the goldfish gastro-intestinal tract are sensitive to such cholesterol feedback mediated regulation.

TABLE III

FEEDBACK REGULATION IN THE GOLDFISH

Tissue	Cholesterol in Diet %	Cholesterol Synthesis mμmoles added acetate-2-^{14}C/100 mg tissue
Liver	0	19.0
	5	0.4
Stomach	0	40.0
	5	1.8
Intestine	0	31.0
	5	0.4

As will be emphasized in a subsequent section of this paper, an identical intestinal feedback control mechanism will be shown to be operative in the rainbow trout. It is therefore clear that in contrast to the case in the mammal, cholesterol synthesis in the fish intestine is consistently responsive to exogenous cholesterol. Moreover, as is summarized in Table IV, such a cholesterol sensitive feedback system in the intestine is not restricted to the fish in that in all non-mammalian vertebrates studied to date, cholesterol synthesis in the intestine appears to be controlled by exogenous cholesterol.

In summary, it is apparent that the feedback control of cholesterol synthesis does not operate through a single mechanism in all tissues or in all species. In the mammal, cholesterol synthesis in the liver is regulated by the level of exogenous cholesterol, while in the intestine, bile acid plays this regulatory role. By contrast, in all species of fish so far examined, and indeed in all non-mammalian vertebrates, a "hepatic" type of feedback regulation, that is, one sensitive to exogenous cholesterol, seems to be operative in the intestine as well as in the liver.

As discussed in the next section, another major variation in sterol feedback control is observed as a result of tissues undergoing malignant change.

TABLE IV
PHYLOGENETIC DISTRIBUTION OF CHOLESTEROL FEEDBACK SYSTEM IN LIVER AND INTESTINE

Phylum or Class	Example	Hepatic Feedback Control	Intestine Feedback Control
Thallophytes	Yeast	0	---
Platyhelminthes	Planaria	---	---
Mollusca	Land Snail	---	---
Annelida	Earthworm	---	---
Arthropoda	Lobster	---	---
Echinodermata	Star Fish	---	---
Osteichythyes	Goldfish	Cholesterol	Cholesterol
Amphibia	Frog	Cholesterol	Cholesterol
Reptile	Iguana	Cholesterol	Cholesterol
Aves	Chicken	Cholesterol	Cholesterol
Mammalia	Rat	Cholesterol	Bile Acids

FEEDBACK CONTROL IN HEPATOMAS OF MAMMALS

With the discovery in the 1940's that negative feedback control mechanisms might regulate the rate of various synthetic processes in animal tissues, the possibility was raised by Potter (19, 20) that loss of feedback control might result in the development of cancer. In more specific terms, this theory of carcinogenesis suggested that deletion of a critical feedback reaction might lead to the uncontrolled production of a critical end product and hence to uncontrolled cell growth and multiplication.

In part, on the basis of this intriguing hypothesis, in 1961 (21) we began to examine the possibility that the cholesterol feedback system in liver might conceivably become defective when hepatic tissue underwent malignant change. In studies published in 1964 (11, 22), we reported that in fact cholesterol synthesis in all hepatomas of mice, rats and man without exception showed complete loss of any reponse to exogenous cholesterol.

Most significantly, a rather large number of so-called minimal deviation Morris hepatomas was examined for possible alterations in the cholesterol feedback system (22-24). The rationale for studying this tumor lies in the fact that minimal deviation hepatomas, as developed by Dr. Harold Morris at the National Cancer Institute, show most of the biochemical characteristics of normal liver, and hence biochemical alterations observed in such tumors would presumably be the result of malignant change rather than of non-specific de-differentiation.

As indicated in Table V, all minimal deviation tumors studied to date without exception have shown a complete loss of the cholesterol feedback system.

Moreover, in recent months (25) we have examined the cholesterol feedback system in two additional very highly differentiated hepatomas, the 9618A and 9633, both kindly provided by Dr. Morris. The 9618A hepatoma

TABLE V
ABSENCE OF CHOLESTEROL FEEDBACK CONTROL IN RAT HEPATOMAS

Tissue	Normal diet, mμ moles acetate-2-^{14}C	5% Cholesterol diet, mμ moles acetate-2-^{14}C
Liver	7.5	0.2
Hepatoma (minimal deviation)		
1. 7787	33.1	32.3
2. 9121	26.5	25.4
3. 5121 t.c.	9.3	7.1
4. 7795	6.1	6.7
5. 7793	3.4	4.3
6. 7794 A	7.3	17.6
7. 7316 A	4.4	8.3
8. 7800	7.4	9.1
9. H-35	6.3	9.3
Hepatoma (poorly differentiated)		
10. 7288 C	0.9	1.1
11. 3924 A	0.5	0.6
12. 3683	1.2	2.6

500 mg liver or tumor incubated in 5 ml Krebs bicarbonate buffer with 1 μc acetate-2-^{14}C at 37°C for 2 hours.

is of particular importance in that it is one of two types of minimal deviation hepatomas discovered to date that have both a normal diploid number of chromosomes, i.e. 42, and in addition, the chromosomes show no visible aberrations. It is noteworthy therefore, that as shown in Table VI, these two minimal deviation hepatomas synthesize cholesterol at extremely rapid rates, and moreover both demonstrate a complete loss of the cholesterol feedback system. Loss of the cholesterol feedback system is therefore a remarkably constant feature of all minimal deviation hepatomas regardless of the degree of differentiation.

It is of interest that the loss of cholesterol feedback control can also be demonstrated in hepatomas induced in rainbow trout by aflatoxin feeding. As shown in Table VII, such hepatomas, provided by Dr. John Halver of the Western Fish Nutrition Laboratory, failed to show any feedback response to dietary cholesterol. It should be noted in addition, that the level of cholesterol synthesis in the 4 hepatomas shown in Table VII varies markedly from 52 to 255 mμ moles. This variation, however, might well be expected since in contrast to the uniform, transplantable, minimal deviation hepatomas in the rat, each induced hepatoma in the rainbow trout represents a distinct morphologic entity, and hence cholesterol synthesis will vary

TABLE VI
CHOLESTEROL FEEDBACK DELETION IN MORRIS HEPATOMAS 9618A and 9633

Tissue	Cholesterol Synthesis	
	Normal Diet	5% Cholesterol Diet 3 days
	mμmoles/gm	mμmoles/gm
Liver	55.3	0.9
Liver	84.6	0.7
Liver	60.3	0.9
Hepatoma 9618A	151.6	125.6
Liver	37.2	0.6
Hepatoma 9633	72.4	54.1
Liver	55.5	1.0
Hepatoma 9633	23.2	29.7

TABLE VII

ABSENCE OF CHOLESTEROL FEEDBACK CONTROL IN AFLATOXIN-INDUCED TROUT HEPATOMA

Tissue	Experiment	Diet Fed	Cholesterol Synthesis mμmoles acetate-2-^{14}C/gm tissue
Hepatoma	1	Normal	52
Hepatoma	2	Normal	247
Hepatoma	3	5% Cholesterol	255
Hepatoma	4	5% Cholesterol	103
Liver	1	Normal	52
Liver	2	5% Cholesterol	5.2

widely depending upon the degree of differentiation of the tumor.

Finally, as the data in the last two lines of Table VII demonstrate, in contrast to the failure of the trout hepatomas to respond to exogenous cholesterol, the normal liver of the rainbow trout shows the expected feedback response to dietary cholesterol that is characteristic of all vertebrate species that we have examined.

In summary, then, we have concluded that hepatomas from all species, both mammalian and non-mammalian, show a consistent deletion of the cholesterol feedback system. The importance of this finding lies in the fact that while many attempts have been made over the past 10 years to demonstrate loss of various other feedback systems in a large variety of tumors, the deletion of the cholesterol feedback system remains to date the only consistent example of the loss of a feedback control system in any series of tumors. It is therefore of interest that, as demonstrated in the following section, the loss of the cholesterol feedback system is not restricted to primary tumors of the liver.

DELETION OF THE CHOLESTEROL FEEDBACK SYSTEM IN LEUKEMIC CELLS

The finding that the feedback control of cholesterol synthesis is consistently deleted in all hepatomas raised the possibility that conceivably this metabolic derangement might shed light on the basic etiology of malignant change. To support this conclusion, however, it is obviously of importance to determine whether this lesion is restricted to liver tumors or whether on the other hand cancers that are not of hepatic origin might show a similar loss of feedback control of cholesterol synthesis. We have therefore recently examined a highly malignant guinea pig leukemia, L_2C*, with the aims of: 1) determining whether such leukemic cells are capable of cholesterol synthesis, and 2) whether this tumor can respond to exogenous cholesterol with a supression of *de novo* cholesterol synthesis.

As demonstrated in the first column of Table VIII, the lymph nodes of the normal guinea pig synthesize significant amounts of cholesterol and show a response to dietary cholesterol that is quite similar to that noted in both rat and guinea pig liver. By contrast, the leukemic cell, while clearly capable of active sterol synthesis, is completely unable to respond to exogenous cholesterol with a normal suppression of sterologenesis (26). These data therefore demonstrate that loss of the cholesterol feedback system is not restricted solely to the liver cancers, and they raise the possibility that this metabolic lesion may be a more general property of malignant tumors.

FEEDBACK DELETION IN PRECANCEROUS TISSUE

The finding that the deletion of the cholesterol feedback system represents a consistent biochemical defect in each of the tumors so far examined, coupled with the fact that to date this feedback deletion remains the only example of a generalized loss of such a control mechanism in cancer

* This tumor was generously provided by Dr. Eli Nadel.

TABLE VIII
LOSS OF CHOLESTEROL FEEDBACK CONTROL IN GUINEA PIG LEUKEMIC L_2C WHITE CELLS

Tissue	Diet	
	Normal	5% Cholesterol (7 days)
	mμ moles acetate-2-^{14}C per gram (liver) or per 10^9 cells (WBC)	
Liver	5.3	0.3
Normal Lymph Nodes	9.9	0.7
Leukemic White Cells	55.0	50.0

tissues, has raised the possibility that the cholesterol feedback defect may be more than coincidentally related to the development of cancer. Clearly to support this hypothesis, it would be necessary to show first, that the loss of feedback control of mevalonic acid synthesis, and hence of cholesterol synthesis, precedes the appearance of obvious malignancy; and secondly, it should ultimately be possible to produce cancer by various mechanisms which would cause a deletion of this specific feedback response. The first of these objectives has now been accomplished and it is in these series of studies that the use of fish as an experimental model has proven to be invaluable. As is now well known, the epidemic of hepatomas that appeared in 1960 in rainbow trout was probably in large part, if not exclusively, the result of feed contamination by the heterocyclic lactone, aflatoxin, the product of the mold, *Aspergillis flavus*. The extensive studies of Halver *et al.* (27, 28), have demonstrated that the feeding of aflatoxin in very small doses can consistently induce hepatomas in rainbow trout. Rainbow trout fed low concentrations of aflatoxin should therefore provide an excellent model of a pre-cancerous tissue which with a certainty approaching 100%, will progress to overt malignancy. In close collaboration with Dr. Halver, we have therefore carried out a series of studies to determine the effect of aflatoxin upon the cholesterol feedback system in the liver and intestine of the rainbow trout. The results have been quite striking. As shown by initial experiments (Table IX) the livers of trout chronically fed the relatively small doses of aflatoxin present in a commercial diet completely lack cholesterol feedback control despite the fact that no tumor could be detected in such livers. Similarly, the livers of tumor-bearing rainbow trout also show absence of this feedback system. Finally, in confirmation of the data presented in Table VII, the hepatomas of these fish have lost all feedback responsiveness.

The finding that the feeding of aflatoxin will cause the loss of the cholesterol feedback system in precancerous liver tissues raises the question

TABLE IX
EFFECT OF CHRONIC AFLATOXIN FEEDING ON CHOLESTEROL FEEDBACK CONTROL IN RAINBOW TROUT

		Cholesterol Synthesis	
Treatment	Tissue	Normal Diet, mμ moles acetate-2-^{14}C	1% Cholesterol diet, mμ moles acetate-2-^{14}C
Aflatoxin in feed 10-100 parts per billion*	Liver	61	44
	Liver	23	53
Aflatoxin in feed 10-100 parts per billion*	Hepatoma	10	48

*After hatching, fish were fed a commercial diet containing some aflatoxin.

of how rapidly this feedback deletion occurs following aflatoxin treatment. As indicated in Table X, the earliest that we have observed this effect is approximately 5 days. In this experiment, rainbow trout were given a single dose of aflatoxin, either 100 or 300 μg/kg of body weight. In both cases, within 5 days following the aflatoxin treatment, cholesterol feeding could no longer cause an inhibition of cholesterol synthesis. It is apparent, therefore, that aflatoxin will effect a rather prompt loss of the feedback control mechanism responsible for cholesterol synthesis. The next question examined was whether this effect of aflatoxin will persist after treatment when the carcinogen is discontinued. The tentative answer to this problem is probably "no". As indicated by the data in Table XI, if aflatoxin is fed to

TABLE X
ACUTE EFFECTS OF AFLATOXIN ON CHOLESTEROL FEEDBACK CONTROL IN RAINBOW TROUT

Dose of Aflatoxin	Diet	Cholesterol Synthesis (mμ moles Acetate-2-^{14}C)	Cholesterol Concentration μg/gm
None	Normal	50.1	1,570
	Cholesterol	15.3	1,602
100 μg/Kg IP once, 5 days before study	Normal	47.1	1,648
	Cholesterol	31.2	2,004
300 μg/Kg IP once, 5 days before study	Normal	31.5	1,560
	Cholesterol	34.7	2,094

TABLE XI

RECOVERY OF CHOLESTEROL FEEDBACK CONTROL IN TROUT PREVIOUSLY FED AFLATOXIN

Trout No.	Diet	Aflatoxin Total Dose	Duration of Treatment	Cholesterol Synthesis mμ moles/gm acetate-2-^{14}C
1	Normal	0.1 μg/fish	Over a 12 week period (28 mo. prior to experiment)	74.2
2	Normal	0.4 μg/fish	Over a 4 week period (28 mo. prior to experiment)	71.9
3	Cholesterol	0.4 μg/fish	Over a 12 week period (28 mo. prior to experiment)	11.3
4	Cholesterol	0.4 μg/fish	Over a 20 week period (28 mo. prior to experiment)	44.2
5	Cholesterol	0.4 μg/fish	Over a 24 week period (28 mo. prior to experiment)	18.6

rainbow trout in doses ranging from 0.1 μg to 0.4 μg/fish given over a 3 to 8 month period, and the animals are then studied after being on a normal diet for 28 months, there remains some blunting of the cholesterol feedback system. There is suggestive evidence, however, that over this 2 year period the feedback mechanism has at least partially recovered in the trout liver cells.

The most likely explanation for this phenomenon is that only certain cells within the liver may be permanently affected by treatment with aflatoxin, and the rather crude methods currently available for detecting the deletion of the cholesterol feedback system must involve examination of the entire population of cells in a tissue slice or in a homogenate. It seems likely the "blunting" of the cholesterol feedback system observed 28 months after cessation of aflatoxin treatment may simply represent the net effect of certain hepatic cells having permanently lost the cholesterol feedback system while others have recovered this feedback control mechanism. It is apparent that a much more refined approach to the detection of the deletion of the cholesterol feedback system will be necessary before this possibility can be confirmed or refuted.

Perhaps the most intriguing correlation between loss of the cholesterol feedback system and carcinogenesis lies in the observation that in striking contrast to the effect of the aflatoxin upon the cholesterol feedback control in the liver, the intestine of the aflatoxin-treated trout retains its normal cholesterol-sensitive feedback control mechanism, Table XII. To date, no tumors have been reported in the intestine of the rainbow trout following aflatoxin treatment, and it is therefore rather striking that loss of the cholesterol feedback system occurs in the trout liver, which undergoes malignant change in response to aflatoxin, while the intestine of the fish does not show feedback deletion and does not undergo malignant change. This finding provides one further piece of evidence that loss of the cholesterol feedback system is not only a constant accompaniment of cancer, but may be a prerequisite for a tissue that is exposed to a carcinogen to undergo malignant change.

TABLE XII
CHOLESTEROL FEEDBACK IN INTESTINE OF AFLATOXIN TREATED TROUT

Exp.	Tissue	Acetate-2-^{14}C Converted to Cholesterol	
		Normal Diet	5% Cholesterol Diet
1	Pylorus	0.68	0.047
	Small Intestine	0.53	0.035
2	Pylorus	0.27	0.016
	Small Intestine	0.25	0.010

SUMMARY

In summary, the studies described in this review have demonstrated:

1. Cholesterol synthesis is controlled in normal liver and intestine through a negative feedback system, which operates through a specific inhibition at the site of conversion of β-hydroxy-β-methyl glutarate to mevalonate.

2. This control mechanism is clearly operative in the livers of all mammalian species and can moreover readily be demonstrated in all sub-mammalian vertebrates including two species of fish.

3. The mechanism of control of cholesterol synthesis in the intestine is clearly somewhat more complex than is the comparable system in liver and that in the intestine of all mammalian species so far examined, bile acids rather than cholesterol serve as the mediator of feedback control of cholesterol synthesis. By contrast, in non-mammalian vertebrates including fish, intestinal cholesterol synthesis is controlled by dietary cholesterol.

4. Loss of the cholesterol feedback system has been shown to be a constant characteristic of all hepatomas examined, including the minimal deviation hepatoma of the rat and the aflatoxin induced hepatoma of the rainbow trout. Moreover, this feedback deletion is clearly not restricted to liver tumors in that it can also be demonstrated in leukemic cells.

5. Finally, the studies carried out in the aflatoxin-treated rainbow trout would suggest that pre-malignant tissues show the absence of the cholesterol feedback system many weeks and indeed many months prior to the earliest evidence of overt malignant change.

ACKNOWLEDGMENT

This study was supported by U. S. Public Health Service Grants No. CA-08501 and CA-05200, and by Damon Runyon Memorial Fund for Cancer Research No. 747.

REFERENCES

1. P. A. Srere, I. L. Chaikoff, S. S. Treitman, and L. L. Burstein, J. Biol. Chem. **182**: 629 (1950).

2. K. Bloch, Science **150**: 19 (1969).

3. F. Lynen, In BIOSYNTHESIS OF TERPENES AND STEROLS (G. E. W. Wolstenholme and Maeve O'Connor, eds.), p. 95, J. & A. Churchill, Ltd., London, 1959.

4. J. W. Cornforth, J. Lipid Res. **1**: 3 (1960).

5. R. G. Gould, Am. J. Med. **11**: 209 (1951).

6. G. M. Tomkins, H. Sheppard, and I. L. Chaikoff, J. Biol. Chem. **201**: 137 (1953).

7. R. G. Langdon and K. Bloch, J. Biol. Chem. **202**: 77 (1953).

8. I. D. Frantz, Jr., H. S. Schneider and B. T. Hinkelman, J. Biol. Chem. **206**: 465 (1954).

9. M. D. Siperstein and M. J. Guest, J. Clin. Invest. **39**: 642 (1960).

10. M. D. Siperstein, V. M. Fagan and H. P. Morris, Cancer Res. **26**: 7 (1966).

11. M. D. Siperstein and V. M. Fagan, In ADVANCES IN ENZYME REGULATION, Vol. 2, Pergamon Press, Inc., New York, 1964.

12. M. D. Siperstein, V. M. Fagan and J. M. Dietschy, J. Biol. Chem. **241**: 597 (1966).

13. R. G. Gould and E. A. Swyryd, J. Lipid Res. **7**: 698 (1966).

14. K. Hellstrom and M. D. Siperstein, Unpublished observations.

15. H. Sakikida, C. C. Shediac and M. D. Siperstein, J. Clin. Invest. **42**: 1521 (1963).

16. M. D. Siperstein, In DEVELOPMENTAL AND METABOLIC CONTROL MECHANISMS IN NEOPLASIA, P. 427, Williams and Wilkins, Baltimore, 1965.

17. J. M. Dietschy, H. S. Saloman and M. D. Siperstein, J. Clin. Invest. **45**: 832 (1966).

18. J. M. Dietschy, J. Clin. Invest. **47**: 286 (1968).

19. V. R. Potter, Univ. Michigan Medical Bulletin **23**: 400 (1957).

20. V. R. Potter, Federation Proc. **17**: 691 (1958).

21. M. D. Siperstein and D. V. LaMarr, Proc. Fifth Intern. Congress of Biochemistry **9**: 420 (1961) (Abstract).

22. M. D. Siperstein and V. M. Fagan, Cancer Res. **24**: 1108 (1964).

23. M. D. Siperstein, V. M. Fagan and H. P. Morris, Cancer Res. **26**: 7 (1966).

24. M. D. Siperstein, Canadian Cancer Conference, Vol. 7, Pergamon Press, Toronto, 1967, p. 152.

25. M. D. Siperstein, L. Luby and H. Morris, In preparation.

26. M. D. Siperstein, L. Luby and J. Britton, In preparation.

27. J. E. Halver, In MYCOTOXINS IN FOODSTUFFS, (G. N. Wogan, editor), p. 209, M. I. T. Press, Cambridge, 1965.

28. J. E. Halver, In PRIMARY HEPATOMA, (Walter J. Burdette, editor), p. 103, Univ. of Utah Press, Salt Lake City, 1965.

COMMENTS

DR. CONTE: Would you comment on the role of the endoplasmic reticulum regarding protein synthesis? Can you also comment on the packaging of material in the cysternal spaces? Do you believe that the inhibition of cholesterol synthesis is due to cholesterol that may be transported as circulating cholesterol, or by that which is a vital part of the plasma cells, which in animal cells has a high cholesterol content?

DR. SIPERSTEIN: We strongly believe exogenous cholesterol rather than endogenous structural cholesterol bound to membrane is responsible for the inhibition of mevalonate synthesis. The studies supporting this conclusion were published two or three years ago in the JOURNAL OF CLINICAL INVESTIGATION and consisted of injecting endogenous cholesterol and exogenously derived cholesterol in their lipoprotein forms. These are very different molecules. Exogenous cholesterol, that is cholesterol absorbed through the intestine and hence packaged in a thin layer of *beta* lipoprotein, will rapidly enter the liver and quickly, within two and a half hours, switch off mevalonate synthesis. By contrast, endogenous cholesterol which is enveloped in a much thicker layer of lipoprotein does not readily enter the liver and does not turn off mevalonate and cholesterol synthesis. That does not mean that endogenous cholesterol chronically does not feed back on itself, that is, cholesterol secreted by the liver into the blood stream is excreted ultimately in the bile, is reabsorbed and in this sense becomes exogenous cholesterol. The generalization I've stated still holds; however, this is an exogenous cholesterol feedback system.

DR. CONTE: Then does mevalonate synthesis that makes the exogenous cholesterol also make the cholesterol that is attached to the membrane, or could there possibly be two types of enzymes, one for the endogenous, one for the exogenous?

DR. SIPERSTEIN: Exogenous cholesterol injected does become structural so it is likely that these two pools mix within a short time in the body.

DR. SCARPELLI: Have you tried actinomycin D in this reaction? If one injects actinomycin D weekly into rainbow trout being fed aflatoxin, one can prevent early histological changes that one sees from aflatoxin in trout liver.

DR. SIPERSTEIN: Actinomycin D, injected at varying doses into rainbow trout, killed most of them in the first experiment, so this study has gone slower than expected. The reasoning, I suspect, behind your question is that aflatoxin could possibly cause what is called by geneticists, a directed mutation. Directed mutations aren't supposed to exist; thus, one is left as an alternative with two sites of enzyme synthesis presumably directed by two genomes. Perhaps one genome produces an enzyme with a feedback sensitive site and another enzyme may be made without this site. It is mechanistically more profitable to assume that aflatoxin may turn off the synthesis of the feedback sensitive enzyme and allow a normally minor enzyme to be overproduced, in which case actinomycin D as you've anticipated should prevent this from happening. That at least is why we are doing the experiment you have suggested.

TOPICS IN METABOLISM

H. L. A. Tarr, Chairman

THE BIOCHEMICAL ASPECTS OF SALT SECRETION

Frank P. Conte

INTRODUCTION

An organism, be it a protozoan or multi-cellular animal, which in the face of diverse environmental salinities can effectively modulate the internal body fluids to within very narrow limits is described as being an "osmoregulator." Hypoosmoregulation is evidenced by those animals which can maintain the salt concentration of their extracellular fluid below that of their environment. Hyperosmoregulators, on the other hand, are animals which maintain their plasma electrolyte levels above the concentrations of the environment.

The evolution of diverse anatomical structures that specifically deal with salt and water suggests that several independent solutions to the problem of osmoregulation have arisen within animals. The development of the renal system in mammals appears to be an example of a system which couples the mechanisms for regulating both solutes and solvent within the same structure. This can be contrasted to the development of adjunct excretory organs such as salt glands of marine birds and reptiles. The latter system appears to have focused on the problem of eliminating the surplus salt which accumulates in the internal body fluids through the development of an intra- or intercellular sequestration mechanism. Comparison of the ultrastructure of several types of salt secreting cells (Figs. 1 and 2) shows a remarkable similarity in organization. The distinguishing feature in these photographs is that the "trade mark" of cells which are involved in the active ion transport appears to be an abundance of mitochondria and a highly developed labyrinth of smooth membranes. If differences are noted between the types of salt secretory cells, it appears to be restricted to the form and shape of the membranous labyrinth which serves as an extension of the cell surface.

This ultrastructure is contrasted to other types of secretory cells, such as the mucoid cell, which exports large quantities of polysaccharides (Fig. 3).

It is the purpose of this paper to consider the biochemical aspects which accompany the development of the gill epithelium that provides the mechanism for the salt secretory function in fishes.

METABOLIC ASPECTS: CATALYTIC PROCESSES ASSOCIATED WITH INTERMEDIARY METABOLISM WHICH ARE ALTERED BY EXTRACELLULAR SALT CONCENTRATIONS

A. Induction of specific enzyme activities associated with ATP synthesis: Membranes surround living cells and constitute a barrier to nutrients which are brought into the cell and to waste materials being excreted. However, this transport system is very much dependent upon the membranes having a selective permeability. In some cases, transport systems actually "pump" substrates against a higher concentration gradient which lies outside the organism. This is an energy-requiring process and has been termed "active transport," because it requires the utilization of ATP. Metabolic inhibitors, such as azide, iodoacetate, etc. which affect the production of ATP will also halt "active transport." Therefore, we conclude that synthetic processes which produce ATP are needed for the active transport of salt. Since we are interested in the development of salt secretion, we can compare the enzyme kinetics of an animal which is being actively induced to transport salt versus an animal which is repressed. Thus, the use of euryhaline fish serves the cell physiologist in the same capacity as transport-negative mutants for the microbiologists. That is, under one set of metabolic conditions, we can produce net unidirectional active transport for a specific substrate, but not under other conditions.

A series of experiments were conducted (in collaboration with one of my students, M. Tripp) (1) in an attempt to determine if certain enzymes which are primarily mitochondrial can be increased in activity during adaptation of salmon to sea water. These enzymes, succinate dehydrogenase and NADH dehydrogenase, coupled with the electron transport chain, will yield large quantities of ATP. We were led to consider these enzymes as being influenced by salt secretion because; 1) cytochemical evidence by Natochin and Bocharov (2) showed qualitative changes of succinic dehydrogenase in the gills of juvenile pink and chum salmon; 2) electron micrographs of the epithelium (3) show a cell which is rich in mitochondria; 3) Glutamic acid dehydrogenase (4) showed slight increases between steady-state freshwater (FW) and steady-state saltwater animals (SW) as shown in Table I. The

TABLE I

Specific Activity of Glutamic Dehydrogenase from Gill Tissue Isolated from Sea-Water and Fresh-Water Animals

	Specific activity of glutamic dehydrogenase	
Experimental group (number of animals)	Supernatant fraction (μmole/min/mg protein)	Mitochondrial fraction (μmole/min/mg protein)
Fresh-water-adapted animals (4)	3.71	1.48
Sea-water-adapted animals (4)	4.52	1.78

results from experiments on succinic dehydrogenase are shown in Tables II and III. It can be seen that for the steady-state FW vs. steady-state SW, there were no significant differences in enzyme activity. If one were to consider the non steady-state situation, which was produced by changing the environmental conditions, there is also no significant change in enzyme activity. However, species differences do occur as shown by the comparison of V_{max} of the SW starry flounder (5.1±0.9) and the SW staghorn sculpin (8.1±1.7) to the SW salmon (3.5±1.5) (Table IV).

In summary, it appears that the mitochondrial ATP generating system does not seem to be significantly affected by the secretion of salts.

B. Induction of specific enzyme activities associated with active ion transport and/or cytoplasmic membrane: During the period when we were investigating the enzymatic system for ATP generation, several investigators were looking at the system of ATP hydrolysis, specifically the Na-K activated ATPase associated with active ion transport. Epstein *et al.* (5) reported for *Fundulus* an increase in the Na-K ATPase activity during salt adaptation. Recently, Kamiya and Utida (6) have noted large changes of Na-K activated ATPase in the gills of Japanese eels, *Anguilla japonica*, during adaptation to seawater (Fig. 4). These workers found the distribution of the sodium-potassium ATPase system to be primarily associated with the microsome fraction and that after seven days there is a 4 - 5 fold increase in enzyme activity. Thus, salt loading appears to induce this enzyme system. (See W. Zaugg, Ch. 16)

What about other enzymes associated with cytoplasmic membranes? Cytochemical studies of Pettengill and Copeland (7) on alkaline phosphatase and the acetylcholinesterase system (8) have shown qualitative increases, but they have not been investigated in a quantitative fashion. Therefore, in enzymatic adaptations which have been induced by salt-loading, there has been only one positive finding; that is in the case of the Na-K activated ATPase. Much more work is needed in this area.

TABLE II
Succinic Dehydrogenase Activity for Gill Epithelial Cell from FW-Adapted Chinook Salmon and Those Exposed for Brief Periods in Seawater

Experimental Environment	Wt. (gm)	F. L. (cm)	Total (1) Cellular Protein (mg)	Total (2) DNA (mg)	Ratio of DNA / Cell Protein	Total (3) Mito-chondrial Protein (mg)	Ratio of Mito. Protein / Cell Protein	Enzyme Activity Vmax(4)	Specific Activity x 10^{-9} Per Cell (5)
Freshwater Steady-State									
FW-1	84.7	20.7	84	8.0	.095	4.6	.055	5.3	17.4
FW-2	148.4	24.0	135	17.8*	.131	2.9	.011	1.6	----
FW-3	109.4	21.7	83	9.8	.118	5.1	.059	8.0	24.0
FW-4	77.2	19.8	64	6.9	.107	4.4	.069	5.7	20.9
FW-5	98.5	21.5	68	6.4	.094	3.6	.053	6.5	20.3
FW-6	72.7	18.8	53	2.6	.049	2.5	.047	2.5	13.6
FW-7	77.9	19.5	56	3.0	.053	3.1	.055	2.2	12.6
FW-8	87.0	20.0	58	3.3	.057	3.6	.062	5.6	34.7
Average					.088		.051	4.8±2.1	20.5±7
Transition 4 hr. Exposure in SW									
F→SW-2	82.2	21.0	64	5.8	.090	6.2	.096	3.2	19.8
F→SW-3	81.6	20.0	48	3.7	.079	4.5	.093	4.4	30.4
F→SW-4	66.1	18.2	31	2.7	.089	8.3	.260	1.7	30.0
F→SW-5	50.6	16.5	34	2.6	.079	3.8	.110	2.4	15.1
Average					.084		.139	2.9±1.3	23.8±5
Transition 12 hr. Exposure in SW									
F→SW-1	95.1	21.8	57	3.4	.060	3.3	.057	4.0	22.0
F→SW-6	45.4	15.8	41	4.0	.099	3.3	.080	4.8	22.5
F→SW-7	103.9	21.5	74	5.6	.075	6.3	.085	7.7	49.4
F→SW-8	99.8	20.0	52	4.0	.079	4.8	.092	7.2	49.4
F→SW-9	65.9	17.8	62	4.8	.077	4.4	.079	2.8	15.8
Average					.078		.079	5.2±1.9	31.8±11.

*omitted due to DNA value
± = 1 SD
(1) = Total cellular protein calculated from assay of unit vol. of brei x total vol. of brei.
(2) = Total DNA calculated from assay of unit vol. of brei x total vol. of brei
(3) = Total mitochondrial protein calculated from unit vol. of suspended mitochondria x total vol. of suspension.
(4) = V_{max} = μM cytochrome-c-reduced/min/mg mitochondrial protein
(5) = Specific activity per cell = $\frac{(\text{Total mito. protein} \times V_{max})}{(\text{Total DNA} \times \frac{\text{no. of cells}}{\text{gm DNA}})}$ = μM product/min/cell

TABLE III

Succinic Dehydrogenase Activity for Gill Epithelial Cell from SW-Adapted Chinook Salmon and Those Exposed for Brief Periods in Freshwater

Experimental Environment	Wt. (gm)	F. L. (cm)	Total (1) Cellular Protein (mg)	Total (2) DNA (mg)	Ratio of DNA/Cell Protein	Total (3) Mito-chondrial Protein (mg)	Ratio of Mito. Protein/Cell Protein	Enzyme Activity Vmax (4)	Specific Activity x 10^{-9} Per Cell (5)
Saltwater Steady-State									
SW-1	80.3	19.5	106	9.4	.089	4.3	.040	3.3	8.6
SW-2	128.8	23.3	123	15.6*	.126	4.5	.039	1.4	----
SW-3	104.7	21.3	83	8.2	.099	4.3	.051	4.2	12.6
SW-4	62.3	18.6	58	6.2	.107	3.3	.057	5.9	18.0
SW-5	120.0	23.0	99	7.4	.074	5.7	.057	5.7	25.1
SW-6	72.5	18.5	66	4.7	.071	2.8	.042	2.0	6.8
SW-7	72.0	19.0	61	4.0	.065	3.1	.051	2.8	12.4
SW-8	57.7	17.8	63	4.9	.077	2.4	.040	2.4	6.7
Average					.089		.047	3.5±1.5	12.9±6
Transition 4 hr. Exposure in FW									
S→FW-1	97.4	21.5	58	4.2	.072	4.1	.070	3.2	17.7
S→FW-2	47.4	16.5	49	2.7	.055	2.8	.056	1.5	8.9
S→FW-3	54.1	17.8	36	2.1	.058	1.5	.041	2.9	11.7
S→FW-4	45.1	17.0	41	2.6	.063	1.9	.046	4.8	19.3
Average					.062		.053	3.1±1.4	14.4±4
Transition 12 hr. Exposure in FW									
S→FW-5	30.4	14.5	31	1.7	.054	1.7	.054	3.6	20.4
S→FW-6	45.4	18.2	48	2.0	.042	1.8	.037	4.2	21.0
S→FW-7	173.1	25.5	135	8.7	.064	4.9	.036	4.6	14.8
Average					.053		.043	4.1±0.5	18.7±2

* omitted due to DNA value

± = 1 SD

(1) = Total cellular protein calculated from assay of unit vol. of brei x total vol. of brei.

(2) = Total DNA calculated from assay of unit vol. of brei x total vol. of brei.

(3) = Total mitochondrial protein calculated from unit vol. of suspended mitochondria x total vol. of suspension.

(4) = V_{max} = μM cytochrome-c-reduced/min/mg mitochondrial protein

(5) = Specific activity per cell = $\dfrac{(\text{Total mito. protein} \times V_{max})}{\dfrac{(\text{Total DNA} \times \text{no. of cells})}{\text{gm DNA}}}$ = μM product/min/cell

TABLE IV
Succinic Dehydrogenase Activity Per Gill Epithelial Cell from Starry Flounder and Staghorn Sculpin

Experimental Environment	Wt. (gm)	F. L. (cm)	Total (1) Cellular Protein (mg)	Total (2) DNA (mg)	Ratio of DNA / Cell Protein	Total (3) Mito-chondrial Protein (mg)	Ratio of Mito. Protein / Cell Protein	Enzyme Activity V_{max} (4)	Specific Activity x 10^{-9} Per Cell (5)
Saltwater Steady-State									
Flounder 1	90.6	20.9	73	0.8*	.011	5.0	.069	6.5	----
Flounder 2	98.1	21.6	96	2.1	.021	7.1	.070	5.4	124.0
Flounder 3	91.7	21.1	60	0.8*	.013	5.1	.085	5.4	----
Flounder 4	54.3	18.4	60	3.6	.060	6.3	.105	4.4	52.2
Flounder 5	52.7	18.9	55	4.1	.074	7.4	.136	4.0	49.3
Average					.052		.093	5.1±0.9	75.1
Sculpin 1	93.7	19.3	45	3.0	.066	5.2	.114	8.0	86.6
Sculpin 2	106.3	20.2	45	2.2	.048	3.9	.086	10.5	117.1
Sculpin 3	131.2	21.8	52	2.6	.050	7.6	.146	8.0	144.7
Sculpin 4	106.2	20.0	48	2.8	.059	6.7	.139	5.0	74.4
Sculpin 5	104.1	19.5	46	2.7	.058	7.6	.165	8.4	68.9
Average					.056		.130	8.0±1.7	98.3
Freshwater Steady-State									
Flounder 1	82.2	20.5	79	2.1	.027	4.8	.061	4.0	61.9
Flounder 2	84.7	21.0	52	1.4	.026	7.9	.153	3.9	147.7
Flounder 3	54.9	18.3	46	0.9*	.018	4.5	.099	4.5	----
Average					.023		.101	4.1±0.2	104.3

* omitted due to DNA value
± = 1 SD
(1) = Total cellular protein calculated from assay of unit vol. of brei x total vol. of brei
(2) = Total DNA calculated from assay of unit vol. of brei x total vol. of brei
(3) = Total mitochondrial protein calculated from unit vol. of suspended mitochondria x total vol. of suspension
(4) = V_{max} = µM cytochrome-c-reduced/min/mg mitochondrial protein
(5) = Specific activity per cell = $\frac{(\text{Total mito. protein} \times V_{max})}{\left(\frac{\text{Total DNA} \times \text{no. of cells}}{\text{gm DNA}}\right)}$ = µM product/min/cell

STRUCTURAL ASPECTS: CHEMICAL PROCESSES ASSOCIATED WITH THE CELLULAR DISTRIBUTION OF MACROMOLECULAR COMPLEXES WHICH ARE AFFECTED BY HALOPHILIC BEHAVIOR

A. Formation of endoplasmic reticulum and plasma membrane: Prior to the Na-K activated ATPase findings, Conte and Morita (9), utilizing an immunochemical approach to the problem, studied the changes in cellular proteins which would be induced in the gill epithelium. We decided to see if cellular changes affected by halophilic behavior of fishes would manifest themselves as changes in the antigenic properties of the cytoplasm of the epithelial cells. Figure 5 shows the procedure that was followed in the experiments. The results (Table V) showed that there was a 4 - 8 fold increase in antigenicity of the cellular components of the SW-adapted gills. In addition, Ouchterlony plates resolved the antibody-antigen complexes into four distinct populations. Three antigenic fractions were found associated with the microsomes and one with the mitochondrial pellet. However, nothing is known about the chemical identity of these antigens, other than preliminary evidence which shows them to be protein.

TABLE V

Titers of Antisera vs. Cellular Antigens of Adapted and Non-Adapted Gills and Non-Cellular Antigens

	Cellular antigens		
Antisera	AG_{swh}	AG_{fwh}	Non-cellular antigen (normal serum)
$AB_{sw\text{-}No.\ 1}$	1 : 512T	1 : 512T	1 : 128T
$AB_{sw\text{-}No.\ 2}$	1 : 512T	1 : 512T	1 : 128T
$AB_{fw\text{-}No.\ 3}$	1 : 128T	1 : 512T	1 : 8T
$AB_{fw\text{-}No.\ 4}$	1 : 64T	1 : 128T	1 : 0.5T

Ring precipitin method: sw, adapted to salt water (salinity, $30^{o}/oo$);
fw, adapted to fresh water ($0^{o}/oo$); T, thousand (1 : 1T - 1 mg protein/ml);
h, whole gill homogenate.

In summary, the formation of intracytoplasmic membrane and/or plasma membranes, with associated enzymes, appears to be quantitatively increased somewhere between 2 - 8 fold during the conversion of the epithelial cell to a salt secretory cell. Much more work is needed to identify the types of protein, especially if they are membrane proteins, to determine which ones are involved in active ion transport.

B. Mitochondriogenesis: No studies have been made with regard to the gill epithelium.

REPLICATIVE ASPECTS: CHEMICAL PROCESSES INVOLVED IN PERPETUATING THE CONTINUITY AND INTEGRITY OF MACROMOLECULAR COMPLEXES WITHIN THE CELLULAR FABRIC THAT ARE INFLUENCED BY HALOPHILISM:

A. DNA synthesis and cell division: It occurred to me that for a biological system to be functional for long periods of time, consideration must be given to chemical processes that can perpetuate it. Therefore, we began an investigation of deoxyribonucleic acid (DNA) synthesis in an effort to determine the sites and length of time required for cellular renewal of the gill epithelium. Using ^{3}H-thymidine, which is a specific precursor for DNA, Conte and Lin (4) reported that the renewal sites were located in replacement cells which lie between the respiratory leaflets (Fig. 6). In addition, these same sites are the regions where the so-called "chloride" cell is located. Lastly, the rate of DNA synthesis and degradation is dependent upon environmental salinity. The half-time of DNA turnover ($t_{1/2}$) in epithelial cells in sea water is 5.5 days, whereas in fresh water it is 16 days. These findings suggested that if one were to repress the promotion of DNA-dependent syntheses in euryhaline fishes, marked alterations of osmoregulation should ensue. Indeed, this is the case, as shown by the studies of Conte (10) using x-rays and Maetz *et al.* (11) utilizing antibiotics, such as Actinomycin D. They found that these agents interfered with development of hypoosmoregulation in euryhaline fishes and that treated animals died with pronounced hypernatremia and hyperchloremia.

B. Translational processes involving ribonucleic acids: In addition, experiments with puromycin conducted by Maetz *et al.* (11) indicated that protein biosynthesis is involved in the development of salt secretion. This can be seen in Fig. 7 where the sodium turnover rate is 30% prior to injection of puromycin. Following injection, it rises to about 60% followed by a rapid decline and then recovery. This is interpreted to mean that release of nascent peptide fragments augments the ion transport system. Much more research is needed on the translational and transcriptional aspects of protein synthesis during salt secretion before we can completely understand its role in the osmoregulation in fishes.

REFERENCES

1. F. Conte and M. Tripp, Submitted to Comp. Biochem. Physiol. for publication.

2. Y. Natochin and G. Bocharov, Vopr. Ikhtiol. 2: 687 (1962).

3. L. Threadgold and A. Houston, Exp. Cell Res. **34**: 1 (1964).

4. F. Conte and D. Lin, Comp. Biochem. Physiol. **23**: 945 (1967).

5. F. Epstein, A. Katz and G. Pickford, Science **156**: 1245 (1967).

6. M. Kamiya and S. Utida, Comp. Biochem. Physiol. **26**: 675 (1968).

7. O. Pettengill and D. Copeland, J. Exp. Zool. **108**: 235 (1948).

8. W. Fleming and F. Kamemoto, Comp. Biochem. Physiol. **8**: 263 (1963).

9. F. Conte and T. Morita, Comp. Biochem. Physiol. **24**: 445 (1968).

10. F. Conte, Comp. Biochem. Physiol. **15**: 293 (1965).

11. J. Maetz, J. Nibelle, M. Bornancin and R. Motais, Submitted to Comp. Biochem. Physiol. for publication.

COMMENTS

DR. CHOUDARI: Did you say there was any difference between the antigens from gill epithelium of salt water and fresh water adapted fishes?

DR. CONTE: There is no spurring of the Ouchterlony plates so that there does not appear to be difference in the antigens when you compare them, only in titer.

DR. CHOUDARI: Did you do any absorption studies?

DR. CONTE: Yes, we performed a study using heterologous absorption, in which the antigen from fresh water fish was absorbed against salt water antibody. Then we ran the homologous reaction ($SW_{antigen}$ vs. sorbed SW_{AB}) and still we obtained a positive reaction.

DR. SINCLAIR: Did you imply then that the energy mechanism is the main adaptive step?

DR. CONTE: No, I hope I didn't imply that. The metabolic steps involved in the generation of ATP, such as the utilization of succinate or NADH, apparently were not influenced by the presence or absence of large quantities of salts in the environment. The amount of ATPase, however, which is the hydrolytic enzyme that splits ATP, seems to be drastically affected. This enzyme seems to be associated with membrane proteins, especially those in the microsomal fraction. It appears to be the one that's involved in some way with the "pump" mechanism.

DR. SIPERSTEIN: In view of the difference in antigenicity, I wonder whether you have attempted to purify the unique protein in the salt water adapted fish, perhaps by absorbing with "fresh water" antibody, throwing away the precipitate, taking the supernate and reprecipitating, with the salt water antibody. Perhaps, if you then split this precipitate with mercaptoethanol, for example, you might then have the unique proteins. Might that be ATPase?

DR. CONTE: Yes, we're in the process of doing that. Unfortunately, because of the illness of my associate, Dr. Morita, we were a little held up on this problem, but we are pursuing this approach. The other thing that we're pursuing is the formation of any protein that can be shown to be induced. We are using a number of techniques for the separation of proteins.

DR. NEUHAUS: Have you taken a microsomal or polysomal fraction and studied the uptake of radioactively labelled amino acid into a specific protein that you can precipitate with your antibody?

DR. CONTE: Yes, we have done *in vivo* studies and find a difference in the polysomal labelling. We have now also extracted the t-RNA and the aminoacyl synthetases and, we think that we have a cell-free system to test this hypothesis. I think the main thing is, that the pulse labelling of animals in salt water or fresh water showed a very definite difference in the polysomal **profile.** In the salt water adapted fish we have a disappearance of the polysomal region. The monoribosomes still have *peptidyl* residues attached which is hard to understand in view of our current concepts of this process. We think that this breakdown might be caused by enzymatic degradation.

DR. NEUHAUS: Was there a difference in ribonuclease activity?

DR. CONTE: No, we stopped ribonuclease activity by using bentonite. This possibility was also checked by taking fresh water, labelled polysomes and mixing them with cold salt water tissue, homogenizing them and determining the FW–polysomal profile. There was no change.

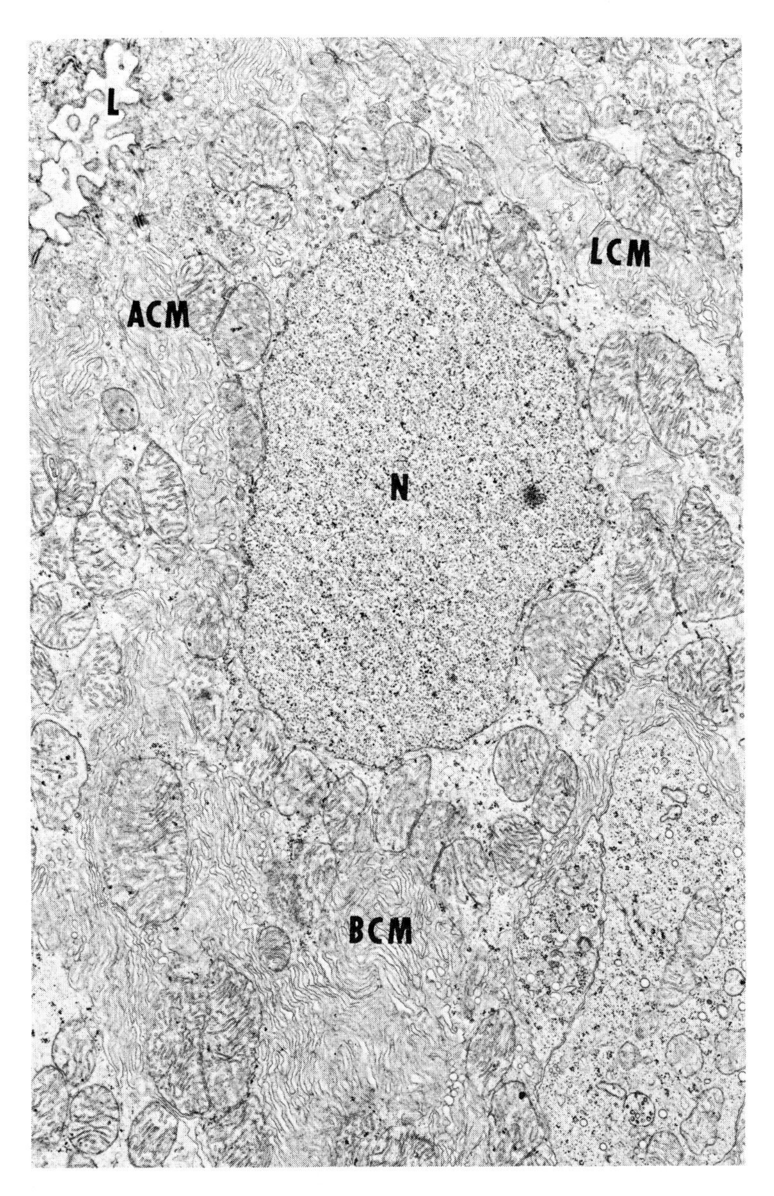

Fig. 1. An electron micrograph of a tubular cell from the rectal gland of the spiny dogfish, Squalus acanthus, which demonstrates one type of cellular form for a salt secretory cell. The lateral cell membrane (LCM) demonstrates repeated infoldings. The basal cell membrane (BCM) undergoes a process of infoldings and then secondary infolding and interdigitation. The apical cell membrane (ACM) is specialized to surround interdigitating compartment, alternate ones connecting to the cytoplasm of the cell beneath. Numerous large mitochondria and a large oval nucleus (N) are present. (L) is the lumen of the gland. Fixed with osmium tetroxide, stained with Millonig's lead stain 7,220X. Courtesy of Dr. R. Bulger, University of Washington.

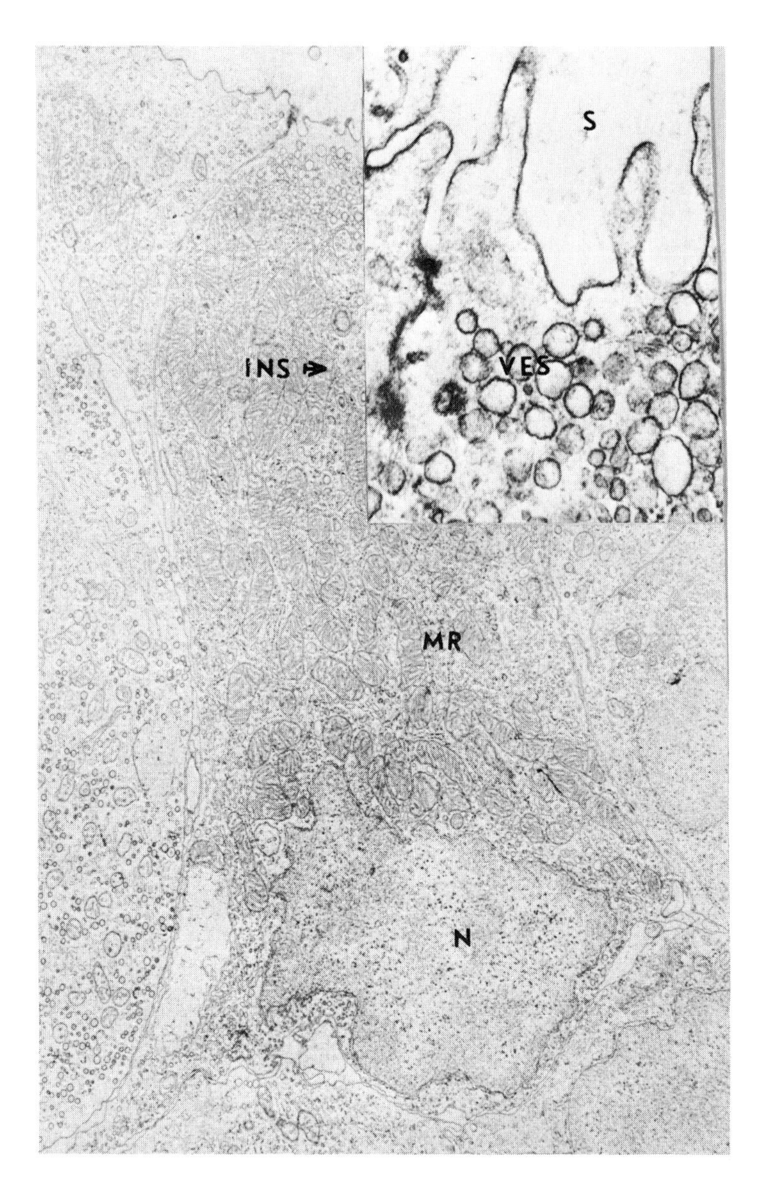

Fig. 2. An electron micrograph of the epithelial cell from the gill filament of the Coho salmon, Oncorhynchus kisutch, adapted to sea water for thirty days. (S) is the area of the external sea water adjacent to the apical region of the cell. The insert (INS) reveals that this region is packed with vesicles (VES). Note the absence of the interdigitating membranes of either the basal or lateral cell membranes. The cytoplasm is rich in large type mitochondria (MR) and the nucleus lies close to the basal lamina. Fixation with osmium tetroxide buffered with collidine 6,650X. Courtesy of Dr. J. Newstead, University of Saskatchewan.

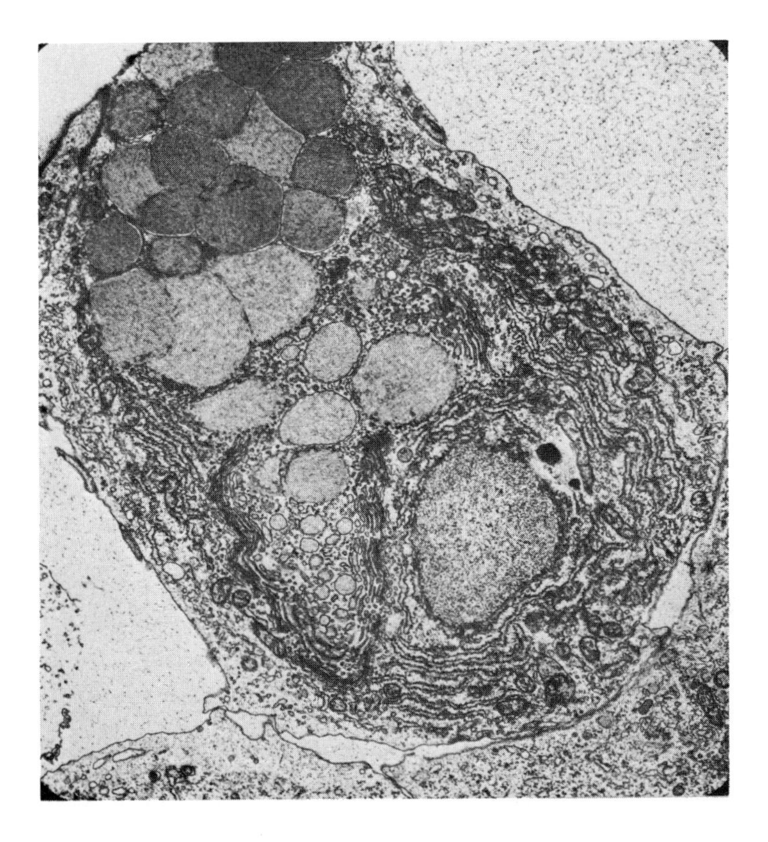

Fig. 3. An electron micrograph of an epithelial cell in the identical gill filament which is secreting mucoid polysaccharides. Note the distinctive "packaging" device which is associated with the prominent Golgi apparatus.

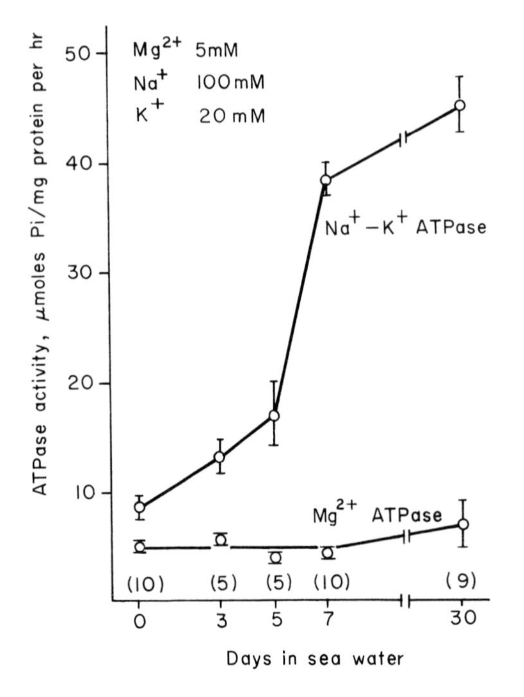

Fig. 4. Changes in Na^+–K^+ ATPase activity in gills of eels following transfer from fresh water to sea water. Vertical lines indicate standard errors at each point and the number of animals is given in parenthesis. Courtesy of Professor Utida, Ocean Research Institute, Tokyo University.

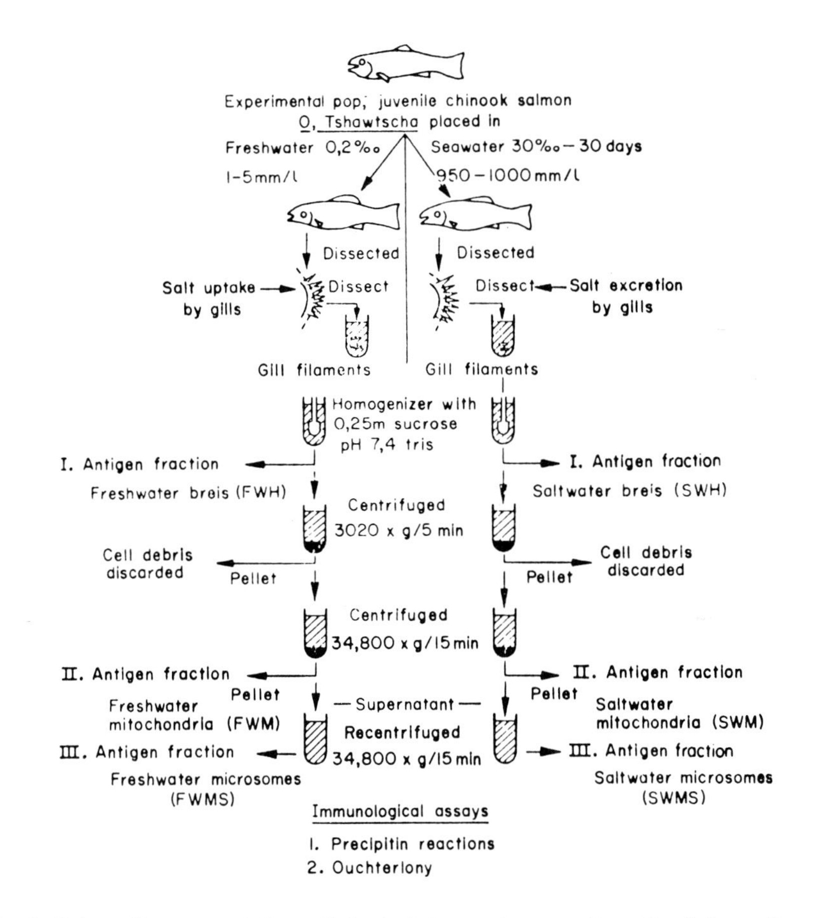

Fig. 5. Schematic representation of the isolation and preparation of cellular antigens from gills. Antigens were injected into rabbits to produce anti-gill antibodies.

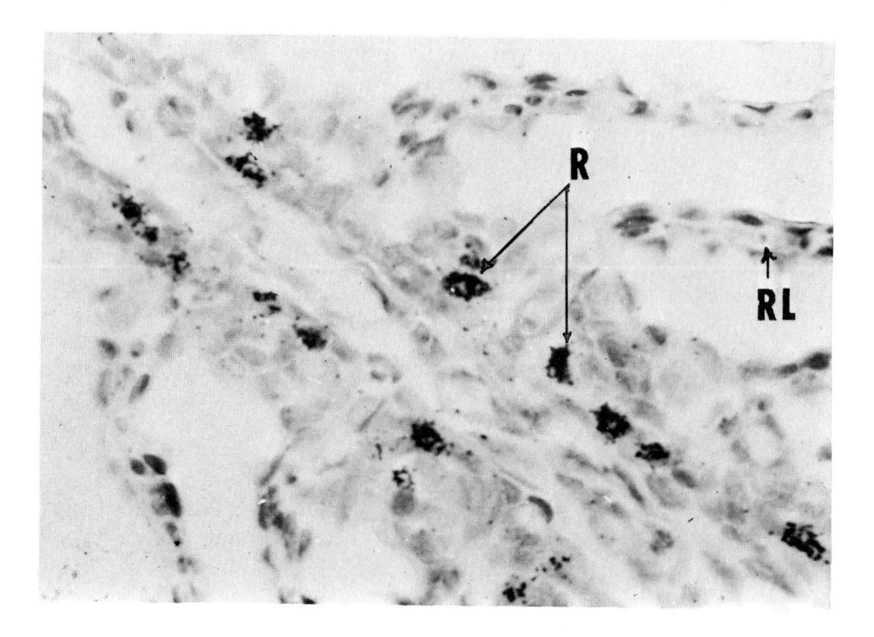

Fig. 6. Longitudinal section of gill filament one day postinjection showing the major site of the replacement cells (R) as being within the interlamellae epithelium. Note the absence of labeled cells occurring in the respiratory lamellae (RL). X576.

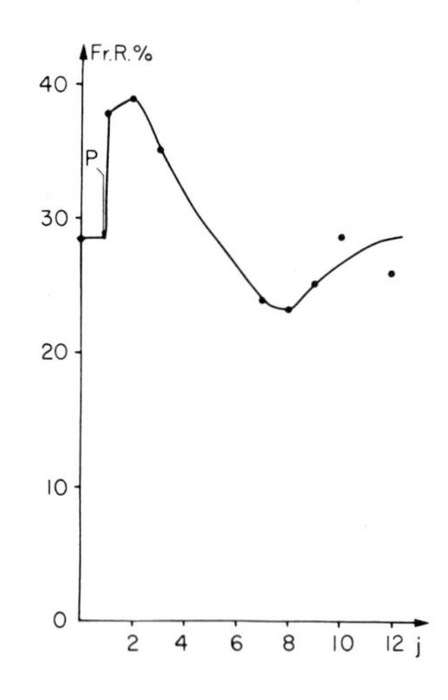

Fig. 7. Stimulation of sodium turnover rate in seawater-adapted eel (A. anguilla) following intraperitoneal injection of puromycin. Abscissa: in hours. Ordinate: in percent/hr. Courtesy of Dr. J. Maetz, Villefrance, France.

STEROIDOGENESIS IN FISH

D. R. Idler

INTRODUCTION

Comparative endocrinologists have emphasized the similarities of steroids and biosynthetic pathways in non-mammalian and mammalian vertebrates. On the contrary, when one considers the limited scope of steroid transformations in animals, hydroxylations, and hydrogenations and dehydrogenations for the most part, it is perhaps remarkable that so many differences have been found with the expenditure of so little effort. I shall refer to a few notable dissimilarities between fish and other vertebrates and mention a few reasons why fish have been both interesting and valuable experimental animals for steroid research.

WHY USE FISH AS EXPERIMENTAL ANIMALS?

Special Anatomical Characteristics of the Animal. For those interested in ion transport the salt gland of elasmobranchs is worthy of consideration. Many species are easily obtained, live well in captivity, are relatively inactive, are easily operable and survive surgery well, and the gland in question can be readily cannulated. While the salt gland has received attention, particularly in recent years, more effort has been expended on the toad bladder and the nasal gland of the duck.

It has recently been demonstrated that bovine medullary tissue is capable of carrying out many of the steroid transformations normally associated with adrenal tissue (1). In most mammals these tissues are difficult to separate quantitatively, and surgically it is quite a feat to remove one tissue without the other. By contrast, the interrenal tissue of elasmobranchs is well separated from the medullary tissue. If the species (e.g., *Raja radiata*) is carefully selected, it is a relatively simple matter to remove the interrenal quantitatively while leaving the medullary tissue undisturbed. I know of no instance where the medullary tissue has been removed from the living animal; this would be a challenge because so many bodies are involved and their location is far from ideal.

Some species of fish are particularly suitable experimental animals because of the ease with which physical characteristics associated with sexual maturation can be quantitated. Perhaps the best known examples are the five species of Pacific salmon where there is an easily quantitated increase in snout length, a decrease in flesh pigmentation, a skin thickening and rapid gonad development accompanying the spawning migration (2).

Because of the large number of eggs from a single parent, it is also possible to control genetic variables in a large experiment without breeding.

Special Phenomena Influenced by Steroids. There are several interesting phenomena associated with salmon which may be influenced by steroids. It has been speculated and some evidence has been presented to suggest that changes in sex hormones in the blood may act as a trigger for commencement of the freshwater phase of the spawning migration (3). An impaired hormone metabolism has been observed in moribund Pacific salmon, all five species of which die after spawning (4). Spawned Atlantic salmon, on the other hand, which frequently survive spawning exhibit no such impairment. Similar impairment has also been observed in moribund cod. During the freshwater phase of the spawning migration of Pacific salmon there is a switch from a predominantly fat metabolism to one where protein is favored; it will be of interest to see whether any of the various changes which have been observed in the steroid levels during spawning migration can be related to this phenomenon. The same observations would apply to changes which have been observed in specific metabolites such as liver glycogen (5).

The control of sexual maturation of salmonid fishes may well prove to be one of the most important problems to be solved to improve the economics of artificial rearing. When such fish mature, energy is diverted to gonad development; characteristics such as flesh color and texture can be adversely affected. It has been established that sex hormones play a role in the maturation of fish and a more detailed understanding of these actions will permit more effective manipulation of these animals in captivity.

Special Steroid Metabolism. There are several aspects of steroid metabolism in fish which present unique opportunities for investigating phenomena which either do not occur in other forms investigated to date or which occur infrequently. To mention a few examples, we have the occurrence of 11-ketotestosterone as a normal testicular metabolite in testis of salmonids while 11-hydroxylation does not normally occur in mammalian testis (6). 1α-Hydroxycorticosterone has been shown to be widely distributed among the elasmobranchs but has yet to be reported from another natural source. The recent isolation of 1-oxygenated steroids from hypertensive infants has increased the interest in this and related steroids. In view of the relatively large amounts of corticoids in the circulating blood of spawned (moribund) sockeye salmon, it is interesting to speculate that knowledge of the pathways of steroidogenesis might have evolved more quickly had some of the effort which went into research on common laboratory animals gone into a study of these interesting fish. It is generally conceded that 17-hydroxylation does not occur in mammals once 21-hydroxylation has taken place. By contrast, evidence is recently accumulating that 17-hydroxylation of corticosterone may be an important transformation in some species of fish, notably teleosts. The conversion of cortisol to cortisone appears to occur readily, both in appropriate mammals and in teleost fish. However, the reverse reaction was not favored in the one species of fish investigated to date while it occurs readily in certain mammals (4, 7). This brings up the interesting possibility of regulating the biologically active compound (cortisol) by conversion to the biologically inactive form (cortisone) but such speculation requires that only the 11-hydroxylated steroid is biologically active in fish as in mammals. As one final example of the important areas where fish offer interesting possibilities for research, protein-binding is worthy of mention.

11-HYDROXYLATION OF C-19 STEROIDS: A TRANSFORMATION BY NORMAL FISH GONADS

Let us now turn to a somewhat more detailed examination of a few phenomena.

Figure 1 shows the structure of 11-ketotestosterone, first isolated from plasma of sockeye salmon (8). The steroid has been shown to influence male secondary sex characteristics and spermatogenesis in male sockeye salmon (*Oncorhynchus nerka*); although its effects were less pronounced in the female it influenced both skin thickness and coloration (2). A chick comb growth bioassay showed that 11-ketotestosterone had only 58% of the biological activity of testosterone propionate for this species. It is therefore of particular interest that it has been reported to be at least ten times as effective as testosterone in producing male secondary sex characteristics in the female medaka (9). Recent investigations have established the metabolic pathways to 11-ketotestosterone in fish and 11β-hydroxylation is a normal metabolic pathway in testis of some fish (6), although in mammals it occurs in testicular tumors but not in normal tissue. The sockeye salmon run to Adams River, British Columbia, is of considerable economic importance. These fish are also of special interest because they do not move from sea water into fresh water until the gonads are in a relatively developed state in contrast with most of the large Fraser River runs; sexual maturation proceeds during the several weeks spent in brackish water before the fish begin their spawning migration. During this period the gonads reach a stage of development where any delay en route can easily take the fish past their normal spawning time. The timing of this run is of particular interest because these fish, which can amount to many millions in a peak year, begin the freshwater phase of the migration on an almost exactly predictable date. Table I illustrates the levels of steroid hormones in the blood of fish captured at the mouth of the Fraser River immediately prior to and immediately after the commencement of the spawning migration. The data suggest that it might well be worthwhile to carry out a more detailed investigation of this phenomenon in view of the suggested correlation between the quantitative distribution of steroids and the beginning of the spawning migration. The possibility that sex steroids trigger the migration is suggested.

TABLE I

"TRIGGER" FOR MIGRATION OF ADAMS RIVER SOCKEYE

		Plasma Steroid (μg/100 ml)			
Sex	Time	11 Keto.*	Test. **	Cortisol	Cortisone
F	Delay	2.0	17.5	13.2	14.3
F	Migrating	7.1	3.2	13.9	21.2
M	Delay	5.4	4.3	5.4	7.7
M	Migrating	8.9	10.8	8.9	24.5

* 11-Ketotestosterone **Testosterone

1α-HYDROXYCORTICOSTERONE: A STEROID THUS FAR UNIQUE TO ELASMOBRANCH FISH

Until recently there has been very little experimental work on the circulating steroids in elasmobranchs (10). However, the little available information has been reviewed so extensively that one might be forgiven for concluding that either cortisol or corticosterone, or both, are the principal corticoids in various species. By contrast, we have found 1α-hydroxycorticosterone (Fig. 2) to be the principal corticoid of all 13 species of elasmobranchs investigated, and while deoxycorticosterone and 11-dehydrocorticosterone are probably widely distributed in elasmobranchs they have not been the predominant corticoid (11-12). At the present time it would seem that there is no well documented evidence for a significant 17-hydroxylase system in an elasmobranch and therefore the reports of cortisol in several species should be confirmed (13). Perhaps one lesson that might be learned from this is that interesting phenomena may be overlooked because of an oft-quoted notion that significant differences in steroids among classes of animals are not likely.

The metabolic clearance rate of 1α-hydroxycorticosterone by *Raja radiata* averaged 2.2 l/kg/24 hr at 6° C and when blood levels of individual animals were considered the production rate averaged 100 μg/kg/24 hr. When the glands were excised and incubated *in vitro* after determination of production rates by continuous infusion, the average production rate of the isolated gland was 53 μg/kg/24 hr. Thus, it may be concluded that under the experimental conditions the glands were good producers of 1α-hydroxycorticosterone when appropriate comparisons are made with production rates of other steroids in mammals (13).

1α-Hydroxylase is a rather interesting enzyme in that it is unlikely to be found in warm-blooded animals because it does not function at temperatures as high as 37° C (14). Since corticosterone is rapidly converted to 1α-hydroxycorticosterone by this enzyme, it is unlikely that corticosterone will predominate unless elasmobranchs are found which do not contain the enzyme.

We were rather surprised to find that this steroid occurred in the blood as a glucuronoside in concentrations equal to, or higher than the free steroid (15). The significance of this observation from the viewpoint of biological activity, or, for that matter any other criteria, has not yet been established.

Attempts to determine the effects of interrenalectomy on male *Raja radiata* brought to light a particularly interesting phenomenon. In order to establish that interrenalectomy had resulted in a decrease of 1α-hydroxycorticosterone in the plasma to negligible levels we were surprised to find a small but measurable amount of this steroid remaining in the blood for long periods of time. The animal contains some quite large fluid compartments, and in two of these, the perivisceral fluid and the pericardial fluid, high levels of 1α-hydroxycorticosterone were found in both the normal and adrenalectomized animal (Table II)(15). It now seems that the perivisceral fluid is able to "bleed" this residue of steroid, which is frequently many times the concentration in the blood plasma, into the blood in order to maintain a small but possibly biologically very significant level (unpublished). If we might reason by analogy with steroids such as aldosterone, it seems not unreasonable to suggest that the animal could maintain levels sufficient to satisfy his needs. It has been established that this steroid is a potent mineralocorticoid for maintaining ion balance in rodents (16) *in vivo* and ion transport in the isolated toad bladder (17). To date no effect could be demonstrated on liver glycogen deposition of 1α-hydroxycorticosterone in adrenalectomized rodents (16) or of interrenalectomy in skate (18).

TABLE II

1α-HYDROXYCORTICOSTERONE
IN BODY FLUIDS OF R. RADIATA

No. of Fish	Fluid	1αOH B (μg/100 ml)
9	Plasma	2 ± 0.5
7	Pericardial	17 ± 6
12	Perivisceral	23 ± 4
4	Cranial	2 ± 0.6

It became of interest to determine whether 1α-hydroxycorticosterone was characteristic of the class *Chondrichthyes* or was confined to the sub-class, *Elasmobranchii*. There are very few living members of the other chondrichthyean sub-class, *Holocephali*, but one species, the ratfish (*Chimaerae*) is readily available on the West Coast of North America. Incubates of the interrenal tissue of the ratfish with ^{14}C-progesterone produced cortisol and there was no detectable 1α-hydroxycorticosterone (6). Thus, available evidence suggests that 1α-hydroxycorticosterone is confined to the elasmobranchs where it occurs in the order *Selachii* and the order *Batoidea*. It will be of interest to see whether the primitive sharks, sawfish and torpedos produce 1α-hydroxycorticosterone. In view of the demonstrated mineralocorticoid activity of this steroid, it will also be of interest to determine whether it occurs in freshwater elasmobranchs.

PROTEIN-BINDING OF STEROIDS IN FISH BLOOD

Protein-binding appears to be an important mechanism for the regulation of blood hormone levels. The data shown in Fig. 3 and other studies in our laboratory suggest that very strong binding of the transcortin type is either not found in fish or it is quantitatively insignificant; however, the porcupine fish warrants further study in this respect. On the other hand, the data in Fig. 4 clearly show that cortisol is bound to a significant extent in salmonid fishes under conditions of dialysis where weaker binding is detected (19). Further experiments (unpublished) suggest that cortisol in Atlantic salmon is bound less strongly than if by transcortin and more strongly than if by albumin. Sex hormone-binding-protein, on the other hand, appears to bind appropriate steroids quite strongly, but again less than for transcortin-binding of cortisol. The specificity of the binding proteins in the limited number of fish investigated to date also appears to differ from that of the mammalian counterparts. It has recently been demonstrated that a low metabolic clearance rate, presumably associated with the great extent of protein-binding, rather than a high production rate accounts for the substantial levels of testosterone encountered in a certain elasmobranch fish (20). A similar situation may prevail in certain salmonid fishes and the point is being investigated in our laboratory. To date there is so little information on the function of steroid hormones in fish that there would be little to be gained from speculating on the role that protein binding might play in determining biological activity.

THE STATUS OF ALDOSTERONE IN FISH BLOOD

The presence or absence of aldosterone in fish is a challenging riddle. This steroid is the potent mineral-regulating hormone of mammals and therefore might be expected to play a key role in anadromous species because of their changing environment with respect to mineral content. If this hormone exists in fish, and if its function were the same as that in mammals, one might reason that it would most likely reach elevated levels during the freshwater phase of the fish's life. Phillips and his co-workers isolated a substance from spawning sockeye salmon which had some characteristics expected of aldosterone (21). However, working with blood from the same fish at the same stage of maturation we were able to separate from authentic aldosterone what we believe to be the same substance studied by Phillips (quoted in 22). Significantly the fraction which should contain aldosterone from another large plasma sample showed no biological activity for the rat (23). Studies in other laboratories have similarly failed to conclusively establish the presence of aldosterone in teleost fishes. It is possible that much of this work was made more difficult by preconceptions about the amount of aldosterone that might be present. In this regard, it is of significance that aldosterone levels in mammals have, until recently, continued to decrease as the methodology has improved. It is now generally accepted that ca. 7 nanograms per 100 ml is about the average quantity in peripheral plasma from normal humans. When one considers that this substance stimulates sodium transport in the toad bladder at concentrations as low as 10^{-9} to 10^{-10} M, these concentrations are quite sufficient to meet the needs of the animal. If the levels in fish are of a similar order of magnitude to those now being found in mammals, it is not surprising that many of the studies have failed to identify this steroid; this is particularly true since there seems to be a substance in the blood of some fish which has properties similar to aldosterone but which appears to occur at higher concentrations than would be expected for this steroid in the light of recent evidence.

Two recent developments suggest that the search for aldosterone in fish is well worth continuing. Arai and Tamaoki (24) have demonstrated that a salmonid can transform deoxycorticosterone to 18-hydroxydeoxycorticosterone, and Truscott and Idler have shown that corticosterone can serve both as a precursor of aldosterone and 18-hydroxycorticosterone (25). It has thus been established that at least one species of fish can synthesize aldosterone. The question remains, does it synthesize it *in situ* and in sufficient quantity to serve as a potential mineralocorticoid? The answer to this question would appear to depend, at least in part, on the availability of corticosterone in the fish in question and in the relative rates of the reactions that might be competing for this substrate. This brings us to what may prove to be a rather unique aspect of steroid metabolism in some fish; namely, the conversion of corticosterone to cortisol. In mammals, 17-hydroxylation does not normally occur once 21-hydroxylation has taken place. By contrast, corticosterone is efficiently converted to cortisol by the interrenal tissue of the ratfish (Table III)(6). The tritium:carbon ratios observed when ^{3}H-pregnenolone and ^{14}C-progesterone were incubated for 1 hr. with herring interrenal tissue are shown in Fig. 5. These unpublished results suggest that there is not only strong competition directing synthesis away from corticosterone under the experimental conditions but, in spite of the fact that several microcuries of each substrate were employed, no radioactive aldosterone could be isolated. These findings are consistent with the fact that corticosterone does not generally occur to any significant extent in the blood of several teleost fishes. However, the results of Leloup-Hatey suggest that corticosterone can accumulate under certain undetermined circumstances (26). It thus seems entirely possible that herring, and perhaps other fish, can determine the amount of aldosterone synthesis by regulating the

TABLE III

INCUBATION OF 4-^{14}C-CORTICOSTERONE WITH RATFISH INTERRENAL FOR 11 HRS. AT 25° C (6)

Compound	% Total dpm
1α-Hydroxycorticosterone	0.0
Polar unknowns	16.3
Cortisol	37.2
Corticosterone	32.4
11-Dehydrocorticosterone	1.0
No Δ^4-3-Keto	13.1
	100

reactions leading to, and possibly from, corticosterone. It now becomes of considerable importance to attempt to rigorously identify aldosterone in the blood of a teleost and then to analyze blood samples, taken under different physiological conditions, in order to obtain a lead as to whether aldosterone synthesis takes place in quantities sufficient to produce measurable blood levels.[1]

REFERENCES

1. A. Carballeira and E. H. Venning, Steroids **4**: 329 (1964).
2. D. R. Idler, I. I. Bitners and P. J. Schmidt, Can. J. Biochem. Physiol. **39**: 1737 (1961).
3. P. J. Schmidt and D. R. Idler, Gen. Comp. Endocrinol. 2: 204 (1962).
4. D. R. Idler and B. Truscott, Can. J. Biochem. Physiol. **41**: 875 (1963).
5. V. M. Chang and D. R. Idler, Can. J. Biochem. Physiol. **38**: 553 (1960).

[1] Since this presentation was given, a steroid isolated from the blood plasma of Atlantic herring (*Clupea harengus*) has been identified as aldosterone by chromatography and crystallization of isotope derivatives to a constant specific activity equal to the supernatant. Quantitative assays for aldosterone in the blood of 4 species of teleosts gave values which were usually even lower than those reported for humans (27). The findings are consistent with the above discussion.

6. D. R. Idler, B. Truscott, and H. C. Stewart, Proc. 3rd Intern. Congr. Endocrinol. Mexico City, 1968, Excerpta Med. Found. Intern. Congr. Ser. In Press.

7. E. M. Donaldson and U. H. M. Fagerlund, Gen. Comp. Endocrinol. **11**: 552 (1968).

8. D. R. Idler, P. J. Schmidt, and A. P. Ronald, Can. J. Biochem. Physiol. **38**: 1053 (1960).

9. R. Arai and B. Tamaoki, Gen. Comp. Endocrinol. **8**: 305 (1967).

10. J. G. Phillips, J. Endocrinol. **18**: xxxvii (1959).

11. D. R. Idler and B. Truscott, Steroids **9**: 457 (1967).

12. B. Truscott and D. R. Idler, J. Endocrinol. **40**: 515 (1968).

13. D. R. Idler and B. Truscott, Gen. Comp. Endocrinol. Supplement **2**: 325 (1969).

14. D. R. Idler and B. Truscott, Proc. 2nd Intern. Congr. Hormonal Steroids. Milan, 1966, Excerpta Med. Found. Intern. Congr. Ser. 132, 1041.

15. D. R. Idler and B. Truscott, J. Endocrinol. **42**: 165 (1968).

16. D. R. Idler, H. C. Freeman and B. Truscott, Gen. Comp. Endocrinol. **9**: 207 (1967).

17. A. S. Grimm, M. J. O'Halloran, and D. R. Idler, J. Fisheries Res. Board Can. **26**: 1823 (1969).

18. D. R. Idler and B. J. Szeplaki, J. Fisheries Res. Board Can. **25**: 2549 (1968).

19. D. R. Idler and H. C. Freeman, Gen. Comp. Endocrinol. **11**: 366 (1968).

20. G. L. Fletcher, D. C. Hardy, and D. R. Idler, Endocrinol. In press.

21. J. G. Phillips, W. N. Holmes, and P. K. Bondy, Endocrinol. **65**: 811 (1959).

22. H. A. Bern, C. C. de Roos, and E. G. Biglieri, Gen. Comp. Endocrinol. **2**: 490 (1962).

23. D. R. Idler, A. P. Ronald, and P. J. Schmidt, Can. J. Biochem. Physiol. **37**: 1227 (1959).

24. R. Arie and B. Tamaoki, J. Endocrinol. **39**: 453 (1967).

25. B. Truscott and D. R. Idler, J. Fisheries Res. Board Can. **25**: 431 (1968).

26. J. Leloup-Hatey, Annales de l'institute Oceanographique **42**: 298 (1964).

27. B. Truscott and D. R. Ilder, Conference of European Comparative Endocrinologists. Utrecht, 1969, Zoological Laboratory, State University of Utrecht, The Netherlands, Abstract No. 165.

COMMENTS

DR. CONTE: Dr. Idler, have you measured aldosterone levels in the fresh water fishes, that is, in terms of the ones that you have shown here? These fish, are they not from sea water? What are the levels, if you see anything in fresh water, in regard to both the enzymatic synthesis "potential" or the circulating levels of hormone?

DR. IDLER: Yes, this is certainly a very important question, Dr. Conte, but I'm afraid that we've only just gotten this worked out and have no data on differences that may exist in fresh vs. salt water. For the moment we are going to concentrate on the attempt to isolate aldosterone from the plasma (see addenda to paper). Recently, considerable improvements have been made in the double isotope derivative assays and we think now that perhaps we can get down to the very low levels that apparently exist. Some of the quite high levels that were reported when the methodology was not well developed are now considered unreliable.

DR. NEUHAUS: How do you account for the difference in the extent of binding of cortisol by plasma between your Sephadex and dialysis experiments?

DR. IDLER: We believe that strong binding of the transcortin type may not exist in the fish which we have studied, hence the absence of complex stable to Sephadex chromatography. I think that what we are measuring in the dialysis experiments is a much weaker complex, perhaps of the albumin type. Seal and Doe have published evidence to suggest that plasma of the porcupine fish may contain a protein which binds cortisol more strongly than does the plasma of the species we work with.

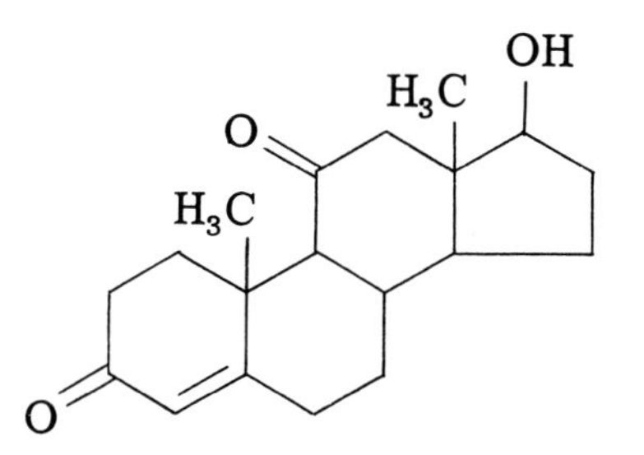

Fig. 1: 11-Ketotestosterone, a principal androgen in some fish.

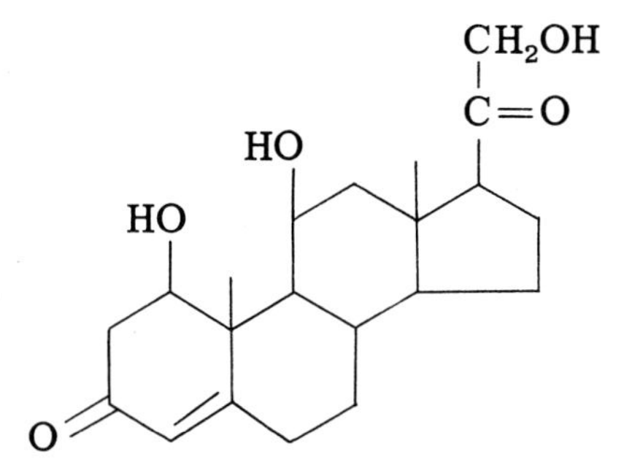

Fig. 2: 1α-Hydroxycorticosterone, the principal steroid of several elasmobranchs.

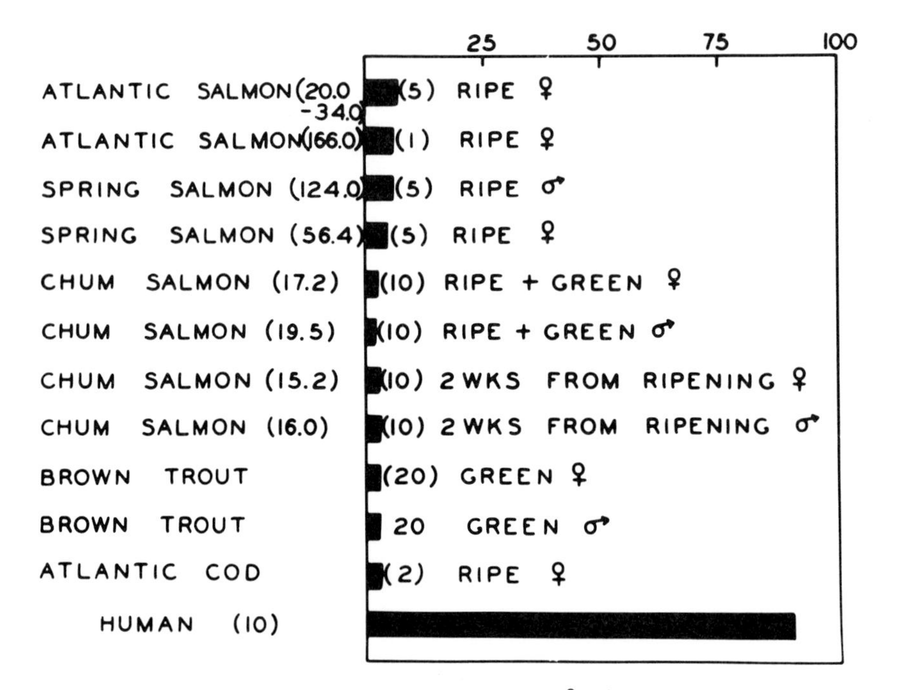

Fig. 3: Percent cortisol bound by plasma at 5° C (gel filtration).

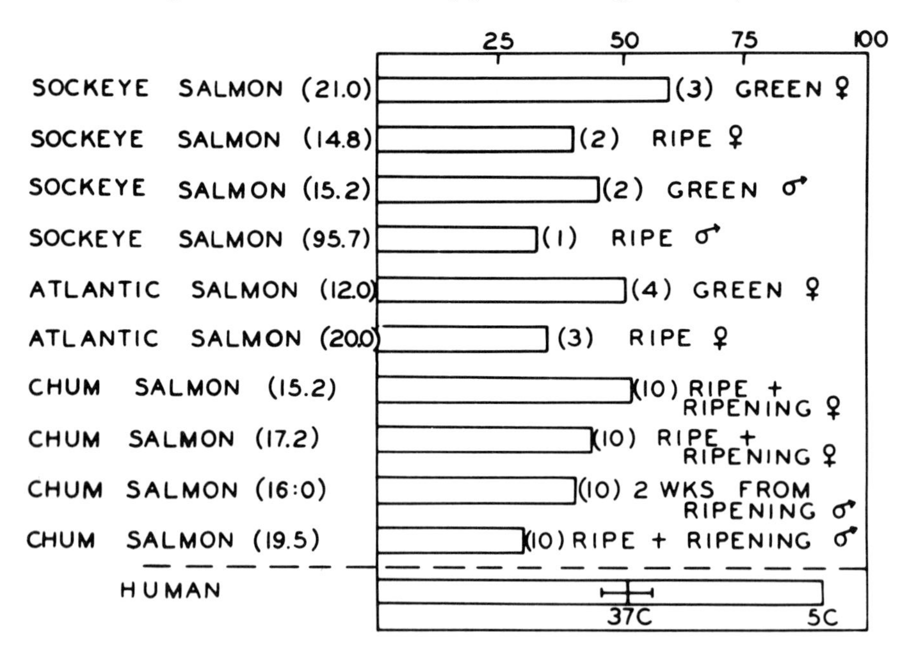

Fig. 4: Percent cortisol bound by plasma as a function of maturation (dialysis 5° C).

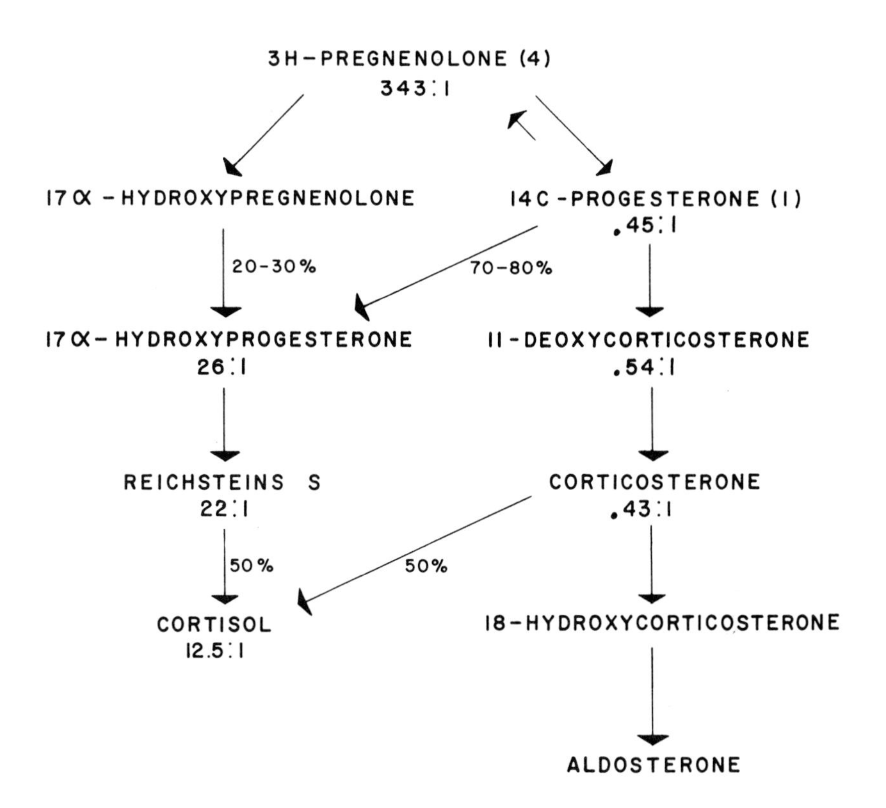

Fig. 5: Incubation of equimolar ^{3}H-pregnenolone (47 μc) and ^{14}C-progesterone (11 μc) with herring head kidney tissue for 1 hr. at 25° C. The $^{3}H/^{14}C$ ratios and estimated conversions assuming 2 major pathways are indicated. There was not enough aldosterone after 1 hr. to permit crystallization to constant $^{3}H/^{14}C$ (D. R. Idler and C. Sangalang, unpublished).

LIPID CATABOLISM IN FISH MUSCLE

E. Bilinski

INTRODUCTION

The metabolic mechanism involved in utilization by fish of lipid reserves from muscle is the main scope of this review. The investigations carried out by us in this area are summarized and discussed in relation to the current knowledge in this field of lipid metabolism. The present understanding of the function of the red muscle in energy metabolism in fish is also discussed. The catabolism of phospholipids in fish muscle is only briefly covered.

In contrast to mammals, wherein the adipose tissue serves as the main site of lipid storage, fish appear to use the liver and skeletal muscle for a similar storage function (1). The relative importance of liver and skeletal muscle as the site of lipid storage varies in different species of fish, Table I (2). Generally, the liver serves as the storage site for lipid reserves in sluggish, bottom-dwelling fish, whereas skeletal muscle plays this role in more active species (3). With respect to the deposition of fat in the muscle, it is known that the red muscle contains more lipid than the ordinary one (2). In fish, the red and white muscle fibers are more clearly distinct from one another than in mammalian muscle, where the two types of fiber are normally mixed. The red muscle is usually situated in fish under the skin, along the lateral line; some fish have developed deep-seated red muscle (4, 5).

Our studies on the mechanism of utilization of lipid reserves in fish muscle were conducted with rainbow trout (*Salmo gairdneri*) and sockeye salmon (*Oncorhynchus nerka*). In these species, the skeletal muscle is an important site of lipid storage and the red muscle is located along the lateral line. The utilization by salmon of the energy released by breakdown of lipids

TABLE I

Fat Content of Liver and Skeletal Muscle of Fish[1]

Species	Liver	Red Muscle	Ordinary Muscle
	Fat - Per Cent		
Coalfish (Gadus virens)	42.2	2.5	0.5
Herring (Clupea harengus)	1.8	28.2	13.0
Mackerel (Scomber scombrus)	7.7	29.7	13.1
Salmon (Salmo salar)	7.4	15.1	2.0

1. From Braekkan (2).

is most evident in fish that migrate without feeding. The quantitative aspects of the depletion of lipids and other body reserves during migration of salmon were the object of several studies (6-8).

OXIDATION OF FATTY ACIDS BY FISH MUSCLE

The main part of the energy, supplied to the animal by the breakdown of triglycerides from depot fat, comes from the oxidation of fatty acids. In mammals, liver has long been considered as the main site of fatty acid oxidation; but in recent years, it has been clearly established that a large portion of fatty acid is oxidized in extrahepatic tissues, including the skeletal muscle (9, 10). In fish containing a large fat reserve within the muscle, catabolic oxidation of fatty acids would be expected to provide energy for muscular activity. The first evidence indicating that fish muscle is capable of oxidizing fatty acid was obtained in a study conducted with rainbow trout (11). In this investigation, tissue slices from dorsal white muscle and from red lateral line muscle were used, and the ability of the tissue to oxidize fatty acids was determined by measuring the formation of $^{14}CO_2$ from carboxyl-labelled hexanoic, octanoic, and myristic acid, Table II. It was shown in this study that trout muscle contains an enzyme system that would oxidize fatty acids and that the red lateral line muscle is considerably more active in this respect than the white muscle. This observation was confirmed by George and Bokdawala (12) in an investigation conducted on oxidation of butyric acid by red and white muscle preparations from several species of the *Cypridinidae* family. In another study, the fatty acid oxidizing capacity

TABLE II.

Oxidative Activity of Rainbow Trout Skeletal Muscle[1]

Substrate	Dorsal Muscle	Lateral Line Muscle
	mμ moles Substrate Converted to $^{14}CO_2$ per 1 mg Tissue N[2]	
Na Hexanoate-1-^{14}C, 1.0 mM	2.29	19.70
K Octanoate-1-^{14}C, 1.0 mM	0.18	10.39
K Octanoate-1-^{14}C, 0.1 mM	0.06	3.38
K Myristate-1-^{14}C, 0.1 mM	0.09	2.94

[1]From Bilinski (11).
[2]Incubation Conditions: Tissue slices, Krebs–Ringer phosphate, pH 7.3, 60 min incubation at 25°.

TABLE III

Oxidative Activity of Various Tissues from Sockeye Salmon[1]

Substrate	Dorsal Muscle	Lateral Line Muscle	Heart	Liver
	mμ moles Substrate Converted to $^{14}CO_2$ per 1 mg Tissue N[2]			
Na Acetate-1-^{14}C, 1.0 mM	2.40	23.40	36.60	31.40
K Octanoate-1-^{14}C, 1.0 mM	0.40	14.60	22.20	24.40
K Myristate-1-^{14}C, 0.1 mM	0.14	1.40	6.24	7.36

[1]From Jonas and Bilinski (13).
[2]Incubation Conditions: Tissue slices, Krebs–Ringer phosphate, pH 7.3, 60 min incubation at 25°.

of skeletal muscle of sockeye salmon was compared to that of other tissues (13), Table III. Data on the oxidation of acetate by various tissues were also obtained. The lateral line muscle of sockeye salmon was found to have a very pronounced ability to oxidize fatty acids and acetate, showing an activity only moderately inferior to that of liver and heart, whereas the ordinary muscle had a much lower activity. These studies strongly suggest that the fatty acids may be considered as a likely source of energy in the red muscle, and they indicate that there is a basic difference between the red and white muscle of fish in their capacity for oxidative metabolism.

The differences between red and white muscle of fish, in their capacity for utilization of fatty acids, may be related to some other biochemical properties of these two types of muscular tissue. As far as enzymic activity is concerned, it was shown in several studies that the enzymes of the tricarboxylic acid cycle, which are involved in the final stages of fatty acid oxidation, are more active in the red than in the white muscle of fish (14-18). Evidence was also presented that the oxygen consumption of red muscle of fish is higher than that of the white one (19, 20). The greater oxidative capacity of red muscle of fish is also indicated by relatively higher contents of myoglobin and cytochrome C (21-23), both of which are known to be involved in oxidative catabolic processes. The presence, in red muscle, of much larger amounts of pantothenic acid is of special interest, as this compound may reflect the concentration of its derived nucleotide, coenzyme A, which is an indispensable factor for fatty acid oxidation (2). If we consider, now, the cellular structure of the red and white muscle of fish, the most important feature is the abundance of mitochondria in the red muscle (17, 22, 24), as the mitochondria are, within the cell, the site of oxidative metabolism.

The properties of the fatty acid oxidizing system present in mitochondria from lateral line muscle were studied in rainbow trout and sockeye salmon, Table IV, (25). When this work was initiated, it was already known from the study of Brown and Tappel (26) with carp liver mitochondria, that the fatty acid oxidizing system in fish liver is essentially the same as in mammals. The fatty acid oxidizing system studied in carp liver mitochondria and that isolated from lateral line muscle of salmonid fish are similar in that each required ATP, Mg^{++}, and tricarboxylic acid cycle intermediates. Such requirements for fatty acid oxidation have been demonstrated in mammalian mitochondrial systems, in which the presence of an enzymic mechanism for β-oxidation of fatty acids has been established. The addition of coenzyme A to mitochondrial preparations from lateral line muscle resulted also in a considerable increase in oxidative activity. It is noteworthy that the stimulatory effect of coenzyme A on fatty acid oxidation is generally not observed with mammalian liver mitochondrial preparations, and coenzyme A was found to have no effect on fatty acid oxidation by carp liver mitochondria (26). It thus appears that, in fish lateral line muscle mitochondria, the fatty acid oxidation system involves the same mechanism as in liver mitochondria; but a difference may exist in metabolic processes related to the generation of coenzyme A for fatty acid oxidation.

TABLE IV

Effect of Various Additives on Oxidation of K-Myristate-1-^{14}C, 0.1 mM by Particulate Fraction from Sockeye Salmon Lateral Line Muscle[1]

Compound			Omitted (-) or Added (+) to the Std. Incubation Medium[2]	Per Cent Change in $^{14}CO_2$ Formation
$MgCl_2$,	7.5	mM	-	-72
ATP,	1.5	mM	-	-85
ATP,	1.5	mM	-	-15
ADP,	1.5	mM	+	
ATP,	1.5	mM	-	-30
AMP,	1.5	mM	+	
α-Ketoglutarate,	0.5	mM	-	-40
α-Ketoglutarate,	0.5	mM	-	-6
Fumarate,	0.5	mM	+	
α-Ketoglutarate,	0.5	mM	-	-4
Succinate,	0.5	mM	+	
Malonate,	10	mM	+	-95
DNP,	0.1	mM	+	-76
Cytochrome C,	0.015	mM	+	+2
CoA,	0.01	mM	+	+400

[1] From Bilinski and Jonas (25).
[2] Standard Incubation Medium: $MgCl_2$, 7.5 mM; ATP, 1.5 mM; α-Ketoglutarate, 0.5 mM; KCl, 50 mM; NaK phosphate, 20 mM, pH 7.4; 60 min incubation at 25°.

LIPOLYTIC ACTIVITY IN FISH MUSCLE

During the various biochemical changes involved in the utilization of lipid reserves, the energy yielding process of fatty acid oxidation is preceded by the release of fatty acids from triglycerides and their transport to the site of utilization. The physiological function of lipase in mobilization of lipids consists essentially in promoting the hydrolysis of depot fat. In many species of fish, the depot fat consists largely of triglycerides containing long chain fatty acids having carbon chain lengths that generally range from 12 to 24 carbons. Fish muscle is known to contain a lipase able to catalyze the hydrolysis of short chain triglycerides (17, 27-29), but the presence of lipolytic activity toward long chain triglycerides was not well established in this tissue. It was found by George (17) that, in lateral line muscle of mackerel, lipase activity was higher than in the white muscle. Histological studies have also shown that lipase activity was mainly localized in the red muscle, both extra and intracellularly; whereas in the white, only extracellular lipase could be demonstrated (12, 30). The higher lipolytic activity found in the red muscle, rather than in the white muscle of fish, considered in conjunction with other biochemical differences, led George to observe that the red muscle is better adapted to utilize fat as a source of energy (12, 17).

In rainbow trout, the lipolytic activity in lateral line muscle was studied recently in relation to its physiological function in promoting the mobilization of lipid reserves of the muscle (31). Two aspects of this question were considered; namely, the demonstration of the ability of fish muscle lipase to hydrolyze triglycerides containing long chain fatty acids and the characterization of this activity in comparison to that responsible for mobilization of depot fat in mammals. The results indicated the presence in fish muscle of a lipolytic activity toward long chain triglycerides, which showed decreasing intensity toward tripalmitin, triolein, and tristearin in the order mentioned, Fig. 1. The proper emulsification of substrate is of critical importance to demonstrate this activity *in vitro* and it was found that the whole phospholipid fraction isolated from fish muscle acts as the proper dispersing agent. The available data suggest a rather specific phospholipid requirement for the inter-reaction between fish muscle lipase and long chain triglycerides, since various means of emulsification of substrate suitable for the assay of other more active lipases were found unsatisfactory. The demonstration of lipolytic activity against long chain triglycerides presents difficulties also in the case of mammalian skeletal muscle and it was found that triglycerides containing fatty acids with 12 carbons or more were not hydrolyzed by preparations from rat skeletal muscle (32). The demonstration of the presence, in trout muscle, of lipolytic activity toward long chain triglycerides poses a question as to whether this hydrolytic process is different from that promoting cleavage of short chain triglycerides. By having optimum activity at pH 7.3, the lipolytic activity against long chain triglyceride in trout muscle differs markedly from the short chain triglyceride lipase from lingcod muscle, which was found by Wood (27) to have an optimum activity at pH 8.2. These differences show agreement, with recent reports on esterolytic and lipolytic activities in mammalian tissues. In mammals, several dissimilarities were observed between the two types of lipolysis and it was found that, in contrast to the hydrolysis of short chain triglycerides, which showed optimum activity above pH 8, the cleavage of long chain triglycerides had optimum activity near neutrality (33-35). These observations are consistent with the existence of distinctive lipolytic activities, depending on water solubility of triglycerides (36).

As far as the metabolic process of depletion and deposition of tissue triglycerides is concerned, two types of lipolytic activities are known to occur in mammals: lipoprotein lipase, which plays a role in facilitating the transport of fatty acids from blood triglycerides into the extrahepatic tissues; and epinephrine sensitive lipase, which is responsible for hydrolysis of adipose tissue triglycerides during the mobilization of lipid reserves. The two enzymes have not been obtained in pure form and the evidence for their separateness comes from differential effects of various inhibitors and activators, and from a different pH optimum (37). In relation to the physiological function played by lipase in utilization of depot fat in fish muscle, some properties of trout muscle lipase were compared to those of epinephrine sensitive lipase from mammalian adipose tissue, Table V. Since it

TABLE V

Effect of Various Additives on Lipolytic Activity in Rainbow Trout Lateral Line Muscle[1]

Compound Added	Concentration	Relative Activity[2]
Control		100
Bovine serum albumin	1 mg/ml	111, 112
Bovine serum albumin	10 mg/ml	109, 112
Bovine serum albumin	30 mg/ml	94, 106
NaF	25 mM	66, 68
NaF	100 mM	48, 57
NaF	200 mM	38, 41
Protamine sulfate	0.3 mg/ml	19, 20
p-chloromercuriphenyl sulfonic acid	0.25 mM	51, 65

[1]From Bilinski and Lau (31).
[2]Incubation Conditions: 40 mμ moles ^{14}C-tripalmitin, 60 min incubation at 35°; NaK phosphate, pH 7.3, 100 mM - with bovine serum albumin; tris, pH 7.3, 100 mM - with other additives.

was found, with mammals, that the response of lipase to epinephrine is more apparent with tissue slices than tissue extracts (38), this type of preparation was used in our study. The lipolytic activity in trout muscle shows similarities with epinephrine sensitive lipase by having a pH optimum near neutrality and by being inhibited by NaF. It differs, however, from this enzyme by being strongly inhibited by protamine. In mammalian tissue only, the lipolytic activity determined at higher pH, which is attributed to lipoprotein lipase, is inhibited effectively by protamine. In studies conducted *in vitro* with mammalian adipose tissue, it was generally found necessary to add serum albumin in order to demonstrate maximal rates of lipolysis. The lack of pronounced stimulation of lipolysis by albumin, in the case of trout lipase, could be related to the fact that the incubation system containing muscle preparation is rich in protein material, which may serve as fatty acid acceptor. As far as the action of epinephrine on lipolysis is concerned, it was found that, though epinephrine did not stimulate lipolysis in fresh muscle, it showed some effect by partially preventing loss of activity during preincubation, Table VI. Epinephrine was also found to protect the lipolytic activity from deterioration in preparations from rat adipose tissue (38, 39). In mammals, the response to epinephrine and other lipomobilizing hormones has been found to be rather specific for the depot type of adipose tissue (40), and to vary with different species (41). In fish, in a study conducted with freshwater species, Farkas (42) found that the incubation of adipose tissue with norepinephrine did not stimulate the hydrolysis of triglycerides and that the intramuscular injection of norepinephrine did not increase the free fatty acid level in blood. Leibson *et al.* (43) found that adrenaline injected intraperitoneally into the lamprey and scorpion fish (*Scorpanea porcus*) raised the concentration of free fatty acid in blood plasma.

TABLE VI

Effect of Epinephrine on Lipolytic Activity in Rainbow Trout Lateral Line Muscle[1]

	Relative Activity
Control	100.0 ± 1.42 (20)[3]
Control plus epinephrine 10μg/ml[2]	118.9 ± 2.64 (20)
Significance level[3]	$p < 0.001$

[1]From Bilinski and Lau (31).
[2]Preincubation without substrate in Krebs–Ringer phosphate, pH 7.3, for 60 min at 35°, followed by incubation in the same medium with 40 mμ moles ^{14}C-tripalmitin.
[3]Mean value, standard error, and number of determinations.
[4]Student's t-test.

RELEASE OF FREE FATTY ACIDS IN FISH IN VIVO

In mammals, the non-esterified fatty acids in blood plasma are known to be the form in which the lipid reserve from fat depot is transported to the various tissues for utilization (44). An increase in free fatty acids in blood plasma and in blood irrigating areas rich in adipose tissues is found in mammals under starvation (45). In fish, the presence of free fatty acids in blood plasma was demonstrated in several studies and their metabolic function was investigated. Tashima and Cahill (3) isolated free fatty acids from blood plasma of toadfish but could not obtain indication that their concentration increases during fasting. Farkas (42) and Leibson *et al.* (43) determined the free fatty acid concentration in blood plasma of various species of fish and they investigated the effect of epinephrine (see under "Lipolytic Activity in Fish Muscle") and other treatments. These studies have shown that free fatty acids are present in fish blood in concentration similar to mammals and they suggest that in fish, as in mammals, they constitute a form of transport of lipid reserved.

As far as the utilization of lipids in fish muscle is concerned, a study was conducted on the effect of starvation on free fatty acid level in the lateral line muscle, white muscle, and blood plasma of rainbow trout (46). The determinations were made in trout maintained without food for 1 to 70 days. The effect of starvation on the level of plasma free fatty acids was limited, showing a more significant increase during the first two weeks without food than during prolonged periods of starvation. A trend generally similar to that found for plasma free fatty acids was observed for lateral line muscle. Starvation had no effect on the amounts of free fatty acids in the white dorsal muscle. When compared with mammals, the effect of starvation on free fatty acid levels in trout was found to be very moderate. This might be partially due to the low metabolic rate of fish under prolonged starvation. Trout (47, 48) and other species of fish can resist long periods of starvation, and it has been shown that during prolonged starvation, the resting metabolism of fish decreases markedly (49, 50). Another explanation is provided by a direct utilization of fat reserve of muscle without the participation of the blood stream. This is an interesting possibility, especially in lateral line muscle which is known to contain a rich lipid reserve in conjunction with an active enzyme system for utilization of fat (3). There is strong evidence suggesting that, in mammals, muscle is partially provided with free fatty acids as fuel without participation of the blood stream as carrier (51, 52). The mechanism involved in this process is not clear, since it has been shown that intracellular muscle lipids are not utilized to any significant extent as energy for contractile activity (53), and it does not appear yet established that adipose tissue cells in close anatomical relation to the muscle may be able to supply free fatty acids for the muscle *in vivo* by simple diffusion.

THE FUNCTION OF THE RED MUSCLE IN ENERGY METABOLISM IN FISH

As shown earlier in this review, it is apparent from the biochemical properties of the red and white muscle of fish that the former is better adapted to utilize fat. Studies on structure and contractile properties of fish muscle fibers (4, 5, 17, 54, 55) and recordings taken within the muscle of live fish (55, 56) suggest that the red and white muscle of fish have distinctive functions in locomotion of fish; the red fibers, which give a slow, long-lasting contraction, may be more active during prolonged swimming of fish; whereas the white, which give a more rapid contraction, are used, rather, for quick movements. Consequently, fat may have a prominent role as a source of energy during long periods of swimming in fish (17). Such an interpretation is consistent with the observation that many fish have little glycogen reserve and have to draw energy from other sources for prolonged muscular work to which they are often subjected (57). The concept – that for sustained swimming, fish utilize fat by aerobic processes in the red muscle and for short periods of vigorous swimming, they utilize glycogen by anaerobic processes – received support from the study conducted by Bone (55) on depletion of glycogen and fat during swimming of dogfish. However, the fact that more glycogen is present in the red than in the white muscle (12, 24, 55) suggests a rather complex picture regarding the metabolism of this compound in fish muscle.

In view of the important biochemical differences existing in fish between the red and the white muscle, some workers have postulated that the red muscle may perform other functions rather than muscular work (2, 58). Braekkan considered that the main role of the red muscle of fish is not contractile activity, but a metabolic function similar to that of mammalian liver. According to his interpretation, the red muscle carries synthetic processes and stores nutrients which are utilized for muscular activity by the white muscle. Braekkan's conclusions are based on the similarity between the red muscle and liver in their chemical composition and enzymic activity and also on consideration of cell structure. In view of the strong evidence for the capability of the red muscle to perform muscular work and, due to the under-estimation of this function by Braekkan, his hypothesis has not been generally accepted (55). We think that, despite some limitations, the approach taken by Braekkan is of interest and the possibility that the red muscle may have in fish a specific metabolic function should be further explored. The concept that a tight metabolic interrelation exists between the red and white muscle of fish received experimental support from the work of Wittenberg and his co-workers (19, 59). In his interpretation of the function of the red and white muscle of fish, Wittenberg is concerned with carbohydrate metabolism only, and considers that the red muscle ensures essential metabolic conditions for the muscular effort of the white muscle. He suggests that the catabolites of glycolysis, formed in the white muscle, are oxidized completely in the red muscle which, in turn, furnishes phosphorous compounds to the white muscle. In higher vertebrates, this function is accomplished by liver. In fish, which has a very poor capillarization of the white muscle, the oxidation may, in part, take place in the adjacent red muscle.

In view of the high amounts of glycogen in the lateral line muscle, it may also be considered that this tissue acts in fish as a storage site of glycogen. In this connection, it is noteworthy that studies with mammalian muscle have shown that ^{14}C-glucose is incorporated to a much larger extent into glycogen in red fibers than in white ones (60). The enzyme involved in biosynthesis of glycogen was also shown in mammals to have higher activity in the red than in the white muscle (61).

The special function of the red muscle may also be considered in relation to regulation of body temperature in some species of fish. It has been found that tuna fish maintain a body temperature above the ambient water temperature. Since higher temperatures were observed in the red than in the white muscle (62), it appears very likely that the burning of fat in the red muscle may have an important role in heat generating processes in the body of tuna (20).

The use of fish as experimental animal for basic research presents some interesting possibilities for studies on muscle metabolism. The fact that, in contrast to mammals, fish red and white muscle fibers are more confined to separate tissues, provides a better basis for studies on biochemical properties of each type of fiber and it may also allow investigations of the metabolic interaction between the two types of fibers. Due to this differentiation, certain basic physiological processes common to both fish and mammals may appear more evident in fish.

FISH MUSCLE PHOSPHOLIPASES

The enzymic hydrolysis of phospholipids presents an aspect of lipid catabolism in fish muscle which is not related to the utilization of depot fat. This process has been studied in relation to its importance in promoting postmortem changes in muscle. The hydrolysis of muscle phospholipids is known to take place during cold storage of fish (63, 64), and the fish muscle enzymes responsible for these changes were studied. The occurrence of various types of phospholipase activity has been investigated in rainbow trout muscle and it was shown that the main pathway of lecithin catabolism is via glycerylphosphoryl-choline (65, 66). The presence of lysolecithinase (65, 67, 68) and phospholipase A (69) has also been demonstrated in fish muscle.

REFERENCES

1. J. Vague and R. Fenasse, In: HANDBOOK OF PHYSIOLOGY, Section **5**: Adipose tissue (A. E. Renold and G. F. Cahill, Jr., eds.), p. 25, Am. Physiol. Soc., Washington, D. C., 1965.

2. O. R. Braekkan, Report on Technological Research Concerning Norwegian Fish Industries **3**: 1 (1959).

3. L. Tashima and G. F. Cahill, Jr., In: HANDBOOK OF PHYSIOLOGY, Section **5**: Adipose tissue (A. E. Renold and G. F. Cahill, Jr., eds.), p. 55, Am. Physiol. Soc., Washington, D. C., 1965.

4. R. Boddeke, E. J. Slijper, and A. van der Stelt, Proc. Koninkl. Ned. Akad. Wetenschap., C **62**: 576 (1959).

5. A. Barets, Arch. Anat. Microscop. Morphol. Exp. **50**: 91 (1961).

6. D. R. Idler and H. Tsuyuki, Can. J. Biochem. Physiol. **36**: 783 (1958).

7. D. R. Idler and W. A. Clemens, Int. Pac. Salmon Fish. Comm. Progr. Rep., p. 80 (1959).

8. H. L. A. Tarr, Proc. Twent. Biol. Colloq., Oregon State College, Marine Biology, p. 36 (1959).

9. I. B. Fritz, Physiol. Rev. **41**: 52 (1961).

10. G. I. Drummond, Fortschr. Zool. **18**: 359 (1967).

11. E. Bilinski, Can. J. Biochem. Physiol. **41**: 107 (1963).

12. J. C. George and F. D. Bokdawala, J. Animal Morphol. Physiol. **11**: 124 (1964).

13. R. E. E. Jonas and E. Bilinski, J. Fisheries Res. Board Can. **21**: 653 (1964).

14. K. Unemura, Igaku To Seibutsugaku **18**: 108 (1951). Cited by CA **45**: 5327 (1951).

15. K. Unemura, Igaku To Seibutsugaku **18**: 204 (1951). Cited by CA **45**: 6232 (1951).

16. H. Fakuda, Bull. Japan. Soc. Sci. Fisheries **24**: 24 (1958).

17. J. C. George, The Am. Midland Naturalist **68**: 487 (1962).

18. M. Gumbmann and A. L. Tappel, Arch. Biochem. Biophys. **98**: 262 (1962).

19. C. Wittenberg and I. V. Diacius, J. Fisheries Res. Board Can. **22**: 1397 (1965).

20. M. S. Gordon, Science **159**: 87 (1968).

21. F. Mastuura and K. Hashimoto, Bull. Japan. Soc. Sci. Fisheries **20**: 308 (1954).

22. G. Hamoir, Advan. Protein Chem. **10**: 277 (1955).

23. G. Hamoir and S. Konosu, Biochem. J. **96**: 85 (1965).

24. H. Buttkus, J. Fisheries Res. Board Can. **20**: 45 (1963).

25. E. Bilinski and R. E. E. Jonas, Can. J. Biochem. **42**: 345 (1964).

26. W. D. Brown and A. L. Tappel, Arch. Biochem. Biophys. **85**: 149 (1959).

27. J. D. Wood, Can. J. Biochem. Physiol. **37**: 937 (1959).

28. J. D. Wood, J. Fisheries Res. Board Can. **16**: 755 (1959).

29. N. Santa and A. Bottesch, Acad. Rep. Populare Romine Studii Cercetari Biol., Ser. Biol. Animala **13**: 143 (1961). Cited by CA **56**: 3932 (1962).

30. F. D. Bokdawala and J. C. George, J. Histochem. Cytochem. **12**: 768 (1964).

31. E. Bilinski and Y. C. Lau, J. Fisheries Res. Board Can. (submitted for publication).

32. D. P. Wallach, J. Lipid Res. **9**: 200 (1968).

33. J. D. Schnatz, Biochim. Biophys. Acta **116**: 243 (1966).

34. J. D. Schnatz and J. A. Cortner, J. Biol. Chem. **242**: 3850 (1967).

35. Y. Biale, E. Gorin, and E. Shafrir, Biochim. Biophys. Acta **152**: 28 (1968).

36. P. Desnuelle and P. Savary, J. Lipid Res. **4**: 369 (1963).

37. A. E. Renold and G. F. Cahill, Jr., (eds), HANDBOOK OF PHYSIOLOGY, Section **5**: Adipose tissue., p. 824, Am. Physiol. Soc., Washington, D. C., 1965.

38. C. R. Hollett and J. V. Auditore, Arch. Biochem. Biophys. **121**: 423 (1967).

39. E. Gorin and E. Shafrir, Biochim. Biophys. Acta **84**: 24 (1964).

40. M. Wenke, Advan. Lipid Res. **4**: 69 (1966).

41. D. Rudman and M. Digirolamo, Advan. Lipid Res. **5**: 35 (1967).

42. T. Farkas, Magy. Tud. Akad. Tihanyi Biol. Kutatoint. Evkonyve **34**: 129 (1967).

43. L. G. Leibson, E. M. Plisetskaya, and T. I. Mazina, Zh. Evol. Biokhim. Fiziol. **4**: 121 (1968). Cited by CA **69**: 1041 (1968).

44. D. S. Fridrickson and R. S. Gordon, Physiol. Rev. **38**: 585 (1958).

45. B. Shapiro, In: HANDBOOK OF PHYSIOLOGY, Section **5**: Adipose tissue (A. E. Renold and G. F. Cahill, Jr., eds.), p. 217, Am. Physiol. Soc., Washington, D. C., 1965.

46. E. Bilinski and L. J. Gardner, J. Fisheries Res. Board Can. **25**: 1555 (1968).

47. A. M. Phillips, Jr., F. E. Lovelace, D. R. Brockway, and G. C. Baltzer, Jr., Portland Hatchery Rept. 22: 1923 (1953).

48. A. M. Phillips, Jr., H. A. Podoliak, D. L. Livingston, R. F. Dumas and R. W. Thoesen, Portland Hatchery Rept. **28**: 56 (1959).

49. F. W. H. Beamish, Trans. Am. Fisheries Soc. **93**: 127 (1964).

50. Y. Inui and Y. Ohshima, Bull. Japan. Soc. Sci. Fisheries **32**: 492 (1966).

51. R. J. Havel, A. Naimark, and C. F. Borchgrevink, J. Clin. Invest. **42**: 1054 (1963).

52. B. Issekutz, Jr., H. I. Miller, P. Paul, and K. Rodahl, Am. J. Physiol. **207**: 583 (1964).

53. E. J. Masoro, L. B. Rowell, R. M. Donald, and B. Steiert, J. Biol. Chem. **241**: 2626 (1966).

54. P. Andersen, J. K. S. Jansen and Y. Løyning, Acta Physiol. Scand. **57**: 167 (1963).

55. Q. Bone, J. Marine Biol. Assoc. U. K. **46**: 321 (1966).

56. M. D. Rayner and M. J. Keenan, Nature **214**: 392 (1967).

57. G. I. Drummond and E. C. Black, Ann. Rev. Physiol. **22**: 169 (1960).

58. O. R. Braekkan, Nature **178**: 747 (1956).

59. C. Wittenberg, Rev. Roumaine Biol. Zool. **12**: 139 (1967).

60. R. M. Bocek, R. D. Peterson, and C. H. Beatty, Am. J. Physiol. **210**: 1101 (1966).

61. S. S. Stubbs and M. C. Blanchaer, Can. J. Biochem. **43**: 463 (1965).

62. F. G. Carey and J. M. Teal, Proc. Nat. Acad. Sci. U. S. **56**: 1464 (1966).

63. J. A. Lovern and J. Olley, J. Food Sci. **27**: 551 (1962).

64. E. C. Bligh and M. A. Scott, J. Fisheries Res. Board Can. **23**: 1025 (1966).

65. E. Bilinski and R. E. E. Jonas, J. Fisheries Res. Board Can. **23**: 207 (1966).

66. E. Bilinski and R. E. E. Jonas, J. Fisheries Res. Board Can. **23**: 1811 (1966).

67. M. Yurkowski and H. Brockerhoff, J. Fisheries Res. Board Can. **22**: 643 (1965).

68. J. Cohen, M. Hamosh, R. Atia, and B. Shapiro, J. Food Sci. **32**: 179 (1967).

69. R. E. E. Jonas and E. Bilinski, J. Fisheries Res. Board Can. **24**: 2555 (1967).

COMMENTS

DR. GOLDBERG: Have you compared glycolysis in white and red muscle? I would presume that it would be higher in white muscle.

DR. BILINSKI: It is likely to be higher in white than in red muscle. In my work I'm confined to lipid metabolism.

DR. GOLDBERG: A number of people, including ourselves, have found that many fish, particularly the salmonids, have a very powerful set of lactic dehydrogenase isozymes confined to the skeletal muscle. We have not found any differences between white and the red muscle.

DR. BOUCK: I wonder why you incubated those tissues at a temperature that would be more than lethal for the organism.

DR. BILINSKI: These assay temperatures were used in order to increase the activity to a measurable level. In several studies conducted on enzymic degradation of lipids in fish muscle (lipases, phospholipases, fatty acid oxidation) we have found the optimal activity usually to be over 30°, despite the fact that the fish lives at a much lower temperature. It is difficult to correlate this thermal reactivity of the tissue with the normal temperature of the animal.

DR. BOUCK: We have observed an inactivation of salmon isocitric dehydrogenase at temperatures approaching 30° and significant inactivation of lactate dehydrogenase below 25°C.

DR. HILTIBRAN: I think when you work with enzyme systems, you can be trapped by the effect of temperature on the enzyme complex. If you isolate the enzyme complex from an organism which has been maintained at a certain temperature, and then incubate the enzyme at a different temperature, you may see an increase in activity which is strictly the effect of temperature on the enzyme complex. How do you control this effect or have you attempted to control it? Since enzyme complexes basically follow the same law of catalysis, namely, if you increase the temperature by ten degrees, the rate of the reaction is about doubled. This is why I am concerned about data obtained from systems removed from the organism and incubated at a different temperature. How do you correct for this or take into account the effect of temperature?

DR. BILINSKI: In theory you are right, but in practice it is different. If you want to study the effect of some enzymatic activity, you look for the optimum conditions to demonstrate this activity. If you would incubate a tissue at 12° C, I don't think you would find, in the case of some of the fish enzymes, a measurable reaction within a short period of incubation.

DR. HILTIBRAN: This is just the point. The trout seems to fare very well at 12°, but I'm sure that the trout will not fare very well at 35°. I would like to see the temperature differential taken into account as well as the effect of temperature upon the enzyme system *per se*. If you do this, then I would not have any disagreement with you.

DR. PAULSRUD: I have studied fatty acid oxidation and succinate oxidation in the rat and bat as a function of temperature. The Arrhenius plots for each reaction are grossly different for the rat and bat. This would seem to reflect the ability of the bat and the inability of the rat to survive a broad range of body temperatures. The Arrhenius plots of crystalline glyceraldehyde-3-PO_4 dehydrogenase from 4 varieties of fish were also found to be different. Therefore, the best way to conduct comparative enzymatic studies is to perform the reaction at a variety of temperatures and report the data in the form of Arrhenius plots. In practice there is no consistent Q_{10} rule. Indeed, Q_{10} may vary from near 0 to 5 or more. Energy of activation values calculated from Arrhenius plots are more meaningfull.

DR. SCARPELLI: I would heartily agree with what has just been said by Drs. Hiltibran and Paulsrud. Temperature optima for enzyme reactions in cold blooded certainly differ from those in warm blooded ones. We have been studying protein synthesis in trout liver and have found that there is a 3 to 4 fold increase in amino acid incorporation by trout liver microsomes when the experiments are performed at 20-23° C. instead of the 30-35° customarily employed in such studies on homeotherms.

DR. CONTE: We isolated aminoacyl synthetases and not all of them have the same temperature optima. We have looked at a number of these and find that the glutamic acid synthetase is very temperature sensitive even in the mammalian system. Not all proteins have to be thermally denatured. I think one should be concerned about this. We looked at thymidine kinase and thymidylate kinase. You can't explain the temperature effects just on the collision frequency or Q_{10}. There are probably allosteric effects or conformational changes involved in these proteins. There are some enzymes that are going to have thermal stability; there are others that will not.

DR. SIPERSTEIN: If one wishes to be a purist about such things, it should be pointed out that thermal stability depends on a lot of other environmental factors as well as the optimum temperature. Magnesium concentration comes to mind. We don't try to simulate that too accurately *in vitro* and the factor which is least well simulated, I suspect, is pH. Most of us run our analyses at a pH of 7 or so for muscle whereas the pH of muscle, certainly of mammalian muscle, normally is about 5.99 or 6.0[1]. This will probably turn out to be true of fish also. If one wishes to be truly physiological in developing temperature curves, he should pick a more acid pH than most of us employ.

1. N. Carter, F. Rector, D. Campion, and D. Seldin, J. Clin. Invest. **46**: 920 (1967).

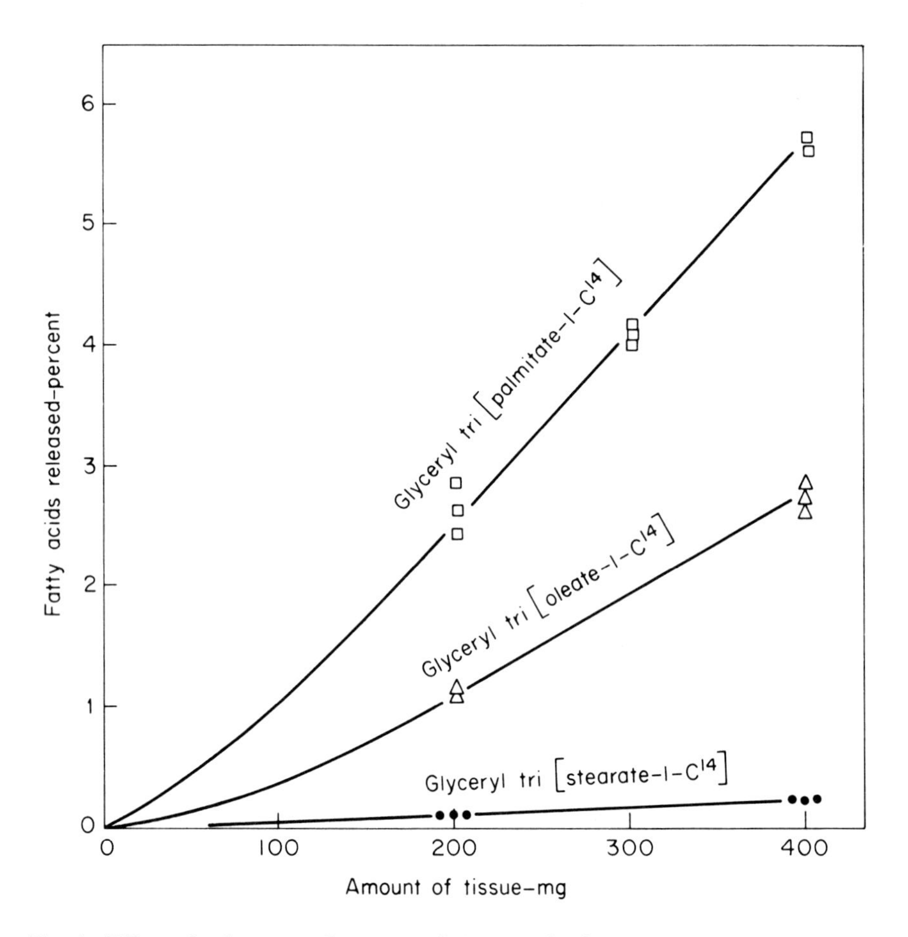

Fig. 1: Effect of substrate and amount of tissue on lipolytic activity in rainbow trout lateral line muscle. Incubation Conditions: 40 mμ moles ^{14}C-triglyceride, 0.8 mg fish muscle phospholipid, 60 min incubation at 35°, NaK phosphate, pH 7.3, 100 mM, total volume 4.0 ml (from Bilinski and Lau (31)).

PRESENTATION OF DR. H. L. A. TARR

Dr. H. L. A. Tarr was sponsored by the Society of the Sigma Xi and was introduced by Dr. James Adams, President of the University of South Dakota Chapter.

Since the major functions of the Society of the Sigma Xi are to stimulate and to honor excellence in scientific research, it is indeed a pleasure to welcome all of you to this Symposium on behalf of the University of South Dakota Chapter of the Society of the Sigma Xi. I am particularly pleased for this Chapter now has the opportunity to help carry out both of its functions by sponsoring a special lecture as part of this Symposium. The Chapter is extremely privileged to honor Dr. H. L. A. Tarr, who will present the Sigma Xi Lecture.

While born in England, Dr. Tarr holds degrees from the University of British Columbia as well as the Ph.D. in Biochemistry from Cambridge. Dr. Tarr is a member of many professional societies as well as a Fellow of the Royal Society of Canada. In the past he has worked as a bacteriologist and is presently the Director of the Vancouver Laboratory of the Fisheries Research Board of Canada. With such qualifications, I am sure his lecture will serve to stimulate us. The title of the presentation is, "The Contrasts Between Fish and Warm Blooded Vertebrates and the Enzyme Systems of Intermediary Metabolisms." I am sure that this lecture will be a most worthwhile addition to the Symposium.

CONTRASTS BETWEEN FISH AND WARM BLOODED VERTEBRATES IN ENZYME SYSTEMS OF INTERMEDIARY METABOLISM

H. L. A. Tarr

INTRODUCTION

Studies of fish intermediary metabolism are indeed incomplete and fragmentary in comparison with those that have been carried out on warm blooded vertebrates. Early studies of fish biochemistry usually suffered from the fact that, because of practical pressures, they were frequently designed to solve immediate and practical problems of the fishing industry. Whereas medical research has been endowed with many and often prodigious sources of funds, fisheries research has been comparatively feebly supported, and biochemical research on fish has received but a small fraction of the whole. Consequently, research of this type has been conducted on a small scale in government or university laboratories, sometimes because a single individual has developed an interest in the subject or simply wishes to use fish as a tool in some phylogenetic or other investigation. However, interest in the subject now appears to be gaining momentum.

Perusal of reviews written up to 1958 indicated the paucity of information then available concerning enzymes and systems of intermediary metabolism in fish (1-5). Some of the more recent advances in this general area will be discussed, particularly those concerned with carbohydrate metabolism, terminal oxidation of metabolites and nucleic acid metabolism. Unfortunately, time will not permit a discussion of amino acid and protein metabolism, and other speakers in this program have covered lipids and certain aspects of endocrinology. Where possible, comparisons will be made with similar systems in warm blooded vertebrates. There has been some work concerned with influence of the environment on systems of intermediary metabolism, and this will also be discussed.

TEMPERATURE ACCLIMATION OF FISH AND WARM BLOODED ANIMALS

The Embden–Myerhof and hexose monophosphate shunt pathways together account for most of the carbohydrate metabolism in warm blooded vertebrates. Research, most of which has been carried out during the past 10 or 15 years, has demonstrated that these pathways also occur in fish. At the outset it must be pointed out that, while warm blooded animals usually maintain a comparatively constant temperature, poikilothermic animals such as fish usually assume the temperature of their environment, and this fact has attracted the attention of a number of physiologists. Some mention must, therefore, be made of this important field of investigation.

The muscle of live healthy resting fish usually contains 0.2-1.0% of glycogen. A very brief period of rapid muscular activity reduces this to a low level. The energy derived from its breakdown *in vivo* appears to be utilized for emergencies, and recovery to normal resting levels is very slow, taking 8 and 24 hours for lactate and glycogen respectively. Obviously, sources of energy other than glycogen must be used during prolonged activity as in spawning migrations, and recent research points to utilization of fat reserves for this purpose.

The knowledge that fish, unlike warm blooded vertebrates, usually adopt the temperature of their environment, which may vary in different instances from about 0 to 30°, has naturally aroused the interest of physiologists and biochemists. Early investigations showed that poikilotherms in general had lower metabolic rates at higher temperatures, and it is now known that the acclimation temperature often influences the use of different metabolic pathways. Thus it was discovered that oxygen consumption by gill homogenates of 10°–acclimated goldfish was 53% inhibited by iodacetate (inhibitor of glyceraldehyde 3–phosphate dehydrogenase), while 77% inhibition resulted when homogenates from 30°–acclimated fish were used (6). This finding indicated greater participation of the Embden–Myerhof glycolytic pathway in the warm-acclimated fish. When *Salmo fontinalis* (brook trout) were acclimated to 4° and 15°, muscle homogenates of the cold-acclimated fish consumed more oxygen at 4°, 10° and 15° than did those from warm-acclimated fish. On the other hand at 20°, homogenates from warm-acclimated fish consumed oxygen while those from cold-acclimated fish did not. This is an interesting finding since it indicates that certain enzymes when formed at lower temperatures may be more unstable. At 4°, muscle of cold-acclimated fish consumed twice as much oxygen as that from warm-acclimated fish and iodoacetate inhibited this more strongly in the latter. When $^{14}CO_2$ was injected into the fish, the glucose obtained from the glycogen synthesized was C-3 and C-4 labeled about ten times more markedly in the warm-acclimated fish. In cold-acclimated fish there was labeling of C-1, 2, 5 and 6 of glucose. These results all suggest a much more marked operation of the Embden–Myerhof pathway in warm-acclimated fish, and an increased use of the hexose monophosphate shunt pathway in cold-acclimated fish (7).

The fact that gluconate-1-^{14}C was incorporated only very slightly into liver glycogen in warm-acclimated fish was taken to indicate that it was metabolized by the hexose monophosphate shunt pathway (8). It is well known that glucose 6-phosphate is oxidized to 6-phosphogluconolactone and thus to 6-phosphogluconic acid by mammalian enzymes, and that both reactions are practically irreversible *in vivo*. Thus the rate of liberation of $^{14}CO_2$ from C-1 labeled 6-phosphogluconic acid should give some indication of the importance of the hexose monophosphate shunt pathway in metabolism of hexoses. Experiments showed that, in rainbow trout acclimated naturally to between 5° and 7°, $^{14}CO_2$ was formed 50 to 75 times as fast from intraperitoneally or intramuscularly injected glucose-U-^{14}C as from gluconic acid-U-^{14}C. Injection of C-6 or uniformly labeled 6-phosphogluconic acid, or of glucose-U-^{14}C, caused labeling of muscle hexose sugars and hexose phosphates, while injections of C-1-^{14}C 6-phosphogluconic acid caused no such labeling (9). All these findings are consistent with a rather slight utilization of the hexose monophosphate shunt pathway in these cold-acclimated fish.

Warm adaptation in trout apparently causes a striking loss of activity in some of the 14 or so lactic dehydrogenase isoenzymes of trout muscle. It has been suggested that in cold-adapted trout there may be an increase in the lactic dehydrogenases based in sub-cellular particles (10).

It must not be concluded that these experiments with fish are unique, for there have been numerous, similar experiments with warm-blooded animals. For example, there was a Federated Societies symposium on cold adaptation in 1963 at which, I believe, fish were not even mentioned. It appears that with warm-blooded animals we have what appears to be an inverse situation. In warm-blooded animals the normal situation would seem to be similar to fish held at higher temperatures, for in both fish and mammals lowering environmental temperature increases oxygen consumption and metabolic rate. With warm-blooded animals, lowering the ambient temperature from say 22° to 3° or 4° causes increased traffic over the metabolic pathways which in turn causes changes in levels of enzymes of intermediary metabolism and chemical intermediates. Thus glucose 6 - phosphate, fructose 6 - phosphate and fructose 1, 6 - diphosphate levels in liver are higher (11). Cold acclimation first causes "shivering thermogenesis," and this is soon replaced by "non-shivering thermogenesis" as the animal adapts to the cold environment and can produce heat without excess muscular activity. There are marked changes in levels of many enzymes almost immediately and these are usually maintained, though not always at the same level. Many of the glycolytic enzymes show greater activity in cold-adapted mammals, and it has been reported that the rate of glycogen synthesis increases, and the ability to oxidize tricarboxylic acid cycle intermediates is accelerated. The dehydrogenases and cytochromes involved in electron transport are more active and there is an increase in transaminase activity and in urea cycle intermediates (12, 13).

ENZYMES INVOLVED IN INTERMEDIARY METABOLISM

Carbohydrate Metabolism

Phosphorylase. Though this enzyme was demonstrated in crude extracts of fish muscle some ten years ago it is only very recently that it has been partially purified and some of its properties studied (14). An aqueous extract of rainbow trout muscle was held 7-8 minutes at 37°, chilled, and denatured protein removed. The soluble fraction was made 0.42 saturated with ammonium sulfate and the precipitated protein was dissolved in, and dialyzed against, 0.05 M tris buffer pH 7.4 containing 0.02 M 2-mercaptoethanol. Portions of the dialyzed extract were adsorbed on a 5 x 1.8 cm potato starch powder column and it was washed with 0.1 M sodium fluoride containing 0.012 M cysteine. pH 6.1, and the enzyme was eluted with the same solution to which 0.4% of glycogen was added, the enzyme being eluted after the main body of protein, presumably as a glycogen–phosphorylase complex. This technique required a careful adjustment of the amount of protein applied to the column so that it was neither over- nor underloaded. Attempts to purify the enzyme further were unsuccessful, presumably due to its marked instability.

The enzyme thus obtained was phosphorylase b and the purification was 70-90 fold. The enzyme was very labile and had to be used almost immediately after elution from starch columns since it lost half its activity in 90 minutes at 0-2° C. All attempts at stabilization failed. The optimum pH was about 6.8, being similar to that found for mammalian muscles. The K_m with glucose 1-phosphate as substrate, in the presence of 1.5 mM 5′-AMP, was 10-15 x 10^{-3} M, and the enzyme was competitively inhibited by glucose. ATP also inhibited the enzyme. The K_m value for 5′-AMP was 0.2-0.4 mM which is similar to that reported for rabbit muscle phosphorylase b. The results indicated the fish muscles, as mammalian muscles, contain phosphorylase b kinase, the enzyme that converts phosphorylase b to a, the form that is active in absence of 5′-AMP. Rabbit phosphorylase b kinase converted fish muscle phosphorylase b to the a form. Fish muscle also contained phosphorylase phosphatase which converted the a to b form.

Phosphoglucomutase, Phosphoribomutase and Phosphoglucoisomerase. These three enzymes were partially purified from lingcod muscle and some of their properties investigated (15). An aqueous extract of the muscle was saturated with ammonium sulfate and the precipitated protein was dialyzed to remove the salt. The dialyzed extract was flash heated to 55° and promptly chilled. The soluble fraction was freeze-dried. It was chromatographed on DEAE cellulose columns, the enzymes being eluted by a linear gradient of 0.01 M tris into 0.25 M tris, pH 7.0. Phosphoglucoisomerase (PGI) was eluted first, there being two "peaks" of activity, the first very small. Presumably these were isoenzymes. Phosphoglucomutase (PGM) and phosphoribomutase (PRM) were fairly well separated from PGI and were eluted together. Both proved to be fairly unstable, losing activity quite rapidly at 0° and being completely destroyed by freezing.

Fish muscle PGM had properties similar to those of mammalian PGM. The dephosphoenzyme was inactive unless glucose 1, 6 - diphosphate was added. Ribose diphosphate and deoxyribose diphosphate were less active. The enzyme required Mg^{++} and cysteine. The ratio of glucose 1-phosphate to glucose 6-phosphate was 5:95. As with other PGM enzymes the PGM and PRM activities could not be separated, the former being about 100 times as active as the latter. PRM required ribose 1, 5 - diphosphate for optimum activity, and this could be replaced, though less effectively, with deoxyribose 1, 5 - diphosphate and glucose 1, 6 - diphosphate. It differed from the PGM activity in that it was strongly inhibited by cysteine and Mg^{++} At equilibrium the ratio of ribose 1-P to ribose 5-P was 15:85. The optimum pH of PGM was 7.5 and of PGI 9.0; the ratio of glucose 6-phosphate to glucose 1-phosphate at equilibrium with PRI was 38:62.

Electrophoresis of aqueous extracts of rainbow trout muscle showed five separate protein zones possessing PGM activity. Three were very unstable, losing all activity in 19 hours at 0°, the other two becoming inactive after 67 hours at 0° (16).

Improved purification of PGM from fish muscles (shark and flounder) has recently been achieved (18). An aqueous extract of the muscle was heated to between 57 and 59° and fractionated with ammonium sulfate (500 g per liter). The precipitated protein was adjusted to pH 7.5 (tris buffer), Mg^{++} was added to protect the enzyme during 0.5-1.0 minute heating at 60°. The soluble fraction was fractionated with ammonium sulfate, desalted using a Sephadex G 25 column, chromatographed on DEAE cellulose and again precipitated with ammonium sulfate. The purification attained was 146 fold for shark and 115 fold for flounder enzyme and the corresponding specific activities were 606 and 630 (μ moles of substrate used per mg of protein per minute), which is similar to that of rabbit PGM. The enzymes were not quite homogeneous (ultracentrifuge), were rather labile and the molecular weights were about 63,000. No isoenzymes were found, but this is not surprising since apparently "pure" enzyme fractions were studied and paper electrophoresis was used.

The enzymes were activated 1.5 to 2.0 fold by incubation with histidine or imidazole buffer containing Mg^{++}, and this activation is lower than that reported for rabbit muscle PGM. The enzyme exhibited a much more marked glucose 1, 6 - diphosphatase activity than did the rabbit enzyme, and this was shown to be due to phosphorylation of the enzyme by glucose 1, 6-diphosphate and spontaneous dephosphorylation of the phosphoenzyme. This reaction does not occur with rabbit PGM. The fish enzymes, unlike the rabbit enzyme, were extremely unstable when held in buffers of low ionic strength. Also the fish enzymes were irreversibly inactivated below pH 5.0, while the rabbit enzyme was activated by exposure to pH 3.0. The flounder enzyme is much more readily denatured by urea than is the rabbit enzyme.

Enolase. Gel electrophoresis showed that 8 different species of salmon all possessed 3 distinct muscle enolases, and the quantitative ratios of these did not alter with changes in extraction time and temperature, so that it was concluded that they were not artefacts induced by the isolation methods (18).

Crystalline enolase was prepared from rainbow trout (19) by fractionating an 0.05% EDTA extract of frozen muscle with acetone at -5°, the protein precipitating between 35 and 40% being retained. The protein was dissolved in 0.05 M imidazole buffer (pH 7.9) containing 0.001 M Mg^{++}, the acetone removed, and the solution heated 3 minutes at -5° and chilled. The soluble fraction was fractionated with ammonium sulfate under nitrogen, the 0.55 - 0.75 saturated fraction being retained and crystallized under nitrogen by cautious addition of saturated ammonium sulfate solution.

The recrystallized enzyme had a specific activity of 390, representing 16-fold purification and 15% recovery. The molecular weight was about 91,000. It contained 3 moles of free SH groups and 3 moles of disulfide, and its activity was accelerated by 0.2 - 0.3 M KC1. The K_m was 4×10^{-5} M, the pH optimum 6.9 and it had an absolute requirement for Mg^{++}. It was fairly stable between pH 7.9 and 11.0, but was very acid sensitive (half life 2 hours at pH 5.0 and a few minutes at pH 4.0). It was inactivated by 2 M urea in 1 hour at 25°. The crystalline enzyme had 3 electrophoretically-distinct forms, and their ratio did not alter during purification.

One striking difference noted between fish and mammalian enolase was the ease of irreversible denaturation of the former, while the latter can be reversible denatured.

Aldolase. Carp muscle aldolase was first prepared in 1958 (20). Aqueous extracts of the muscle were fractionated, the protein precipitating between 1.75 and 2.54 M being dialyzed against dilute KCN and then freeze-dried. The enzyme proved less heat-stable than rabbit muscle aldolase. The specific activity was also lower, being 0.6-2.1 in comparison with 3.9 for the rabbit enzyme. Several years later aldolase was prepared from muscle of 2 species of tuna by fractionation and re-fractionation with ammonium sulfate (21). The specific activity (μ moles fructose diphosphate cleaved per mg protein per minute at pH 9.0) was 7.25. The enzyme had 60% of the activity of rabbit aldolase and was less stable. The activation energy (4.2×10^3 cals.) was 1/3 of that of the rabbit enzyme, while the K_m (2×10^{-3} M)was similar.

Triose phosphate dehydrogenase. An aqueous extract of white muscle of carp was dialyzed and the fraction precipitating at half saturation with ammonium sulfate was dissolved in H_2O, dialyzed and flash heated to 63° The soluble fraction was fractionated at -15° with acetone, the fraction precipitating between 33 and 42% of acetone being retained. The yield was 0.04% of the muscle. The activity was 3 times that of the rabbit enzyme, and it was free from aldolase (22).

This enzyme was crystallized from muscle of halibut and sturgeon and also from warm-blooded vertebrates (23). All preparations had a molecular wt. (M.W.) of about 120,000, and showed a single protein zone by starch gel electrophoresis. However, differences were shown in migration and direction (anode or cathode). The sturgeon enzyme behaved like that from chicken, the halibut enzyme like that from warm-blooded mammals, as far as migration was concerned. The amino acid compositions of halibut and sturgeon were vastly different from those of warm-blooded mammals. The sturgeon enzyme had an amino acid composition similar to that from chicken and turkey and cross reacted in complement fixation tests more strongly with turkey antiserum than did that from halibut. These results suggest that this enzyme has undergone evolutionary changes and that the biochemical findings are consistent with anatomical data which suggest that the sturgeon is a primitive fish which has changed very little from a common ancestor of the bony fishes, while halibut is an advanced species which has undergone marked evolution from a common ancestor.

Lactic dehydrogenase (LDH). The first published work on LDH of fish muscle showed that it was rather unstable even at 25° (24). It was also observed that the amount of lactic acid formed *post-mortem* was much lower than that which would be expected if all muscle glycogen was degraded to this compound.

In 1964 a detailed study of crystalline LDH from various sources was carried out (25). In animals 2 types of this enzyme are recognized namely, H (heart) which is strongly inhibited by pyruvate, and M (muscle) which is not so inhibited. Hybrids containing both H and M forms are known so that 5 types are recognized (H_4, H_3M_1, H_2M_2, H_1M_3 and M_4). Twenty-five M and H LDH's were crystallized, including 2 from bony fishes and 1 from a cartilaginous fish, and all had M.W. of about 140,000. Though the histidine content varied considerably, the quantity in LDH's from fish and warm-blooded mammals was rather similar. Peptide "finger print" patterns of 4 classes of vertebrate enzymes (chicken, bullfrog, halibut and dogfish) differed radically, indicating very diverse amino acid sequences. However, it is of considerable interest that one peptide obtained from the active site gave, in all 4 cases, an identical amino acid sequence. The fish LDH's exhibited temperature stabilities of the same order as those from warm-blooded mammals, namely, 50% inactivation at 60° in 20 minutes. In zone electrophoresis the fish H_4 LDH enzymes moved more slowly than did those from warm-blooded mammals. Immunological results obtained using a very sensitive microcomplement fixation technique showed that fish LDH enzymes, and also triosephosphate dehydrogenase, are very widely separated from the corresponding avian enzymes. When rabbit antisera prepared from animals immunized against comparatively pure fish M_4 LDH and triosephosphate dehydrogenase were studied, it was found that, in approximate order of evolution, mackerel, carp, eels, salmon species, pike, sturgeon, lungfish, dogfish and lamprey enzymes are increasingly removed from halibut and other "flat fishes."

It was also reported (26) that, while mammalian and avian crystalline LDH enzymes have 4 functionally essential thiol groups per mole of enzyme, halibut M_4 does not appear to possess essential thiol groups. Halibut M_4 enzyme bound parahydroxy-mercuribenzoate to 10 moles of non-essential thiol groups before it precipitated, while dogfish M_4 LDH possessed essential thiol groups that were protected from this inhibitor by reduced coenzyme and its acetylpyridine analog.

M_4 LDH's, homogeneous by starch gel electrophoresis, ultracentrifuge patterns and immunological relationships, were prepared from halibut, dogfish and lamprey, and these enzymes had greater turnover numbers than did those from birds and mammals. The dogfish enzyme formed unusually large crystals and contained one reactive non-essential thiol group per enzyme sub-unit, whereas similar enzymes from mammals and birds contain 4 such groups. These results indicate that there are marked differences in LDH and triosephosphate dehydrogenase and that certain of these differences are probably due to evolutionary changes.

Enzymes of the hexomonophosphate shunt (pentose phosphate) pathway. It has already been observed that under certain circumstances fish tissues appear to utilize this pathway, usually to a rather minor extent. Two of these enzymes were purified considerably from lingcod muscle several years ago, and some of their properties were investigated (27).

An aqueous extract of the muscle was saturated with ammonium sulfate, dialyzed and flash heated to 55°. The insoluble protein was removed and the soluble fraction was absorbed on a DEAE cellulose column and eluted with a gradient of 0.25 into 0.02 M tris.

Phosphoriboisomerase (PRI) emerged after most of the protein (86%) had been eluted, and was closely followed by ribulose 5′ phosphate-3′-epimerase (EPIM). Due to its comparative stability and low protein content, PRI was purified about 2000-fold. Its specific activity (μ moles substrate utilized per mg protein per minute at 37°) was 20.4, and the optimum pH 7.0-8.0, with a wide range of activity (pH 4.0-11.0). At equilibrium the ratio of ribose

5-phosphate to ribulose 5-phosphate was 55-60: 45-40%. The enzyme was inhibited 28% by 5 x 10^{-4} and 45% by 5 x 10^{-3} M phosphoribonic acid.

The pH optimum of EPIM was similar to that of PRI and at equilibrium the ratio of xylulose 5-phosphate:ribulose 5-phosphate was 60:40%. The specific activity was almost identical with that of PRI (21.6). In general the properties of these two enzymes were very similar to those which have been isolated from other sources. The specific activity of EPIM was similar to that of the rabbit muscle enzyme for which values of between 8 and 20 were recorded (28). So far other enzymes of this pathway, including transaldolase, transketolase, phosphofructoisomerase and fructokinase do not appear to have been studied in fish tissue.

Post-morten carbohydrate degradation. A discussion of enzymes concerned with carbohydrate metabolism in fish would hardly be complete without some reference to post-mortem carbohydrate degradation in the muscle. In 1964 the whole subject of glycogen metabolism in mammals was ably reviewed (29). In the mammal much is now known regarding both phosphorolytic cleavage of glycogen (by glycogen phosphorylase and enzymes of the Embden–Myerhof pathway) and hydrolytic cleavage. Even now it cannot be said with any certainty whether the enzymes of hydrolytic cleavage fit into the economy of the living cell. In rabbit muscle there appear to be at least two α (alpha) amylases and 3 glucosidases capable of degrading glycogen. The α amylases degrade glycogen to oligosaccharides and the α glucosidases will further degrade these to glucose. Other α glucosidases probably act directly on glycogen degrading it to glucose and, in certain instances, leaving undegraded "limit dextrins."

The position regarding hydrolytic cleavage of glycogen in fish muscles is presently rather vague, though it would appear that the picture is quite complex as with mammals (30). Studies with radioactive glycogen and starch have shown that both phosphorolytic and hydrolytic routes operate in fish muscles, but to a variable extent. Labeling occurs in dextrins, maltose, glucose, hexose phosphates, ribose 5-phosphate, glucose and lactate. None of the hydrolytic pathway enzymes from fish has been purified to any extent. Synthesis of hexose phosphates, glucose, ribose and other intermediates, and of a polymer of quite high molecular weight occurred when radioactive lactate was introduced into muscle blocks post-mortem. From the point of view of the enzymologist, this whole area is practically unexplored.

Tricarboxylic acid cycle (TCA) enzymes. This system, of course, accounts for the major portion of the terminal oxidation of sugars, fats and amino acids, the end products being CO_2 and H_2O. Part of the energy released is stored as ATP. Since most of the oxygen used by respiring organisms is employed to oxidize the various intermediates of the cycle, oxygen consumption is a fairly useful measure of its comparative activity.

Limited studies carried out between 1950 and 1952, using homogenates of fish tissues, indicated activities that could be attributed to the presence of the condensing enzyme, cytochrome oxidase and dehydrogenases for citrate, malate, succinate, glutamate and pyruvate. However, up to the present time only two investigators have made an intelligent study of enzymes of the TCA cycle in fish (31, 32). They used homogenates and mitochondria from muscle or liver of carp or minnows.

The QO_2 values for succinic and malic dehydrogenases of carp muscle were only about one-tenth those reported for rat skeletal muscle. However, with one exception, the fumarase activities were similar to those of mammals. Isocitric dehydrogenase activities were lower. Citric acid was found in liver and muscle, and the fact that it accumulated in liver and muscle when homogenates of these metabolized oxalacetate and pyruvate indicated the presence of condensing enzymes.

The performance of the TCA cycle in carp liver mitochondria was investigated using radioactive pyruvate and C-1 and C-2 labeled radioactive DL-alanine (34). As with mammals, alanine was utilized by oxidative decarboxylation, since, under conditions for oxidative phosphorylation in the presence of α-ketoglutarate, alanine-1-^{14}C yielded 50 to 60 times as much $^{14}CO_2$ as did C-2 labeled alanine. Both routes by which pyruvate normally enters the TCA cycle, namely, by oxidative decarboxylation and attendant formation of acetyl CoA, and CO_2 fixation to form malate and oxalacetate, were demonstrated in fish mitochondria. On the other hand, in contrast to mammalian mitochondria, pyruvate and alanine were oxidized without addition of a TCA intermediate such as fumarate or malate. Carp liver mitochondria may, therefore, retain sufficient endogenous substrate to act as primer to sustain oxalacetate for condensation with acetyl CoA. In the various experiments, the fact that both soluble and mitochondrial enzymes were present was taken to indicate that fish liver mitochondria are readily damaged during extraction procedures.

Though the above 2 papers indicate the presence of the TCA cycle in fish tissues, no attempt has been made to investigate in detail the isolated enzyme systems. Moreover, possible influence of physiological condition of the fish on these enzymes (maturation, stress, etc.) has yet to be investigated. It is interesting to note that fish mitochondrial enzymes exhibit the instability experienced with those of carbohydrate and nucleic acid metabolism. Thus it was noted that addition of glucose and hexokinase to carp liver mitochondria caused a much less marked rise in O_2 consumption than that observed with rat preparations, indicating a feeble respiratory control which could well be attributed to instability (33).

Enzymes of Nucleic Acid Metabolism

Deoxyribonucleases (DNAases). A survey of fish DNAase activity at approximately neutral pH values showed that, of 7 organs studied, exclusive of gonads, kidney was most active. The enzyme was found to be very unstable. It was destroyed by ammonium sulfate fractionation, by brief dialysis and isoelectric precipitation (pH 4.5-5.0). A very crude kidney preparation hydrolyzed fish DNA to the constituent deoxynucleosides but more than one enzyme may have been responsible (34).

Mature salmon testes were shown to possess 3-5 times as much DNAase activity as calf spleen and 6-15 times as much as calf thymus, and yielded a highly purified type II DNAase (acid DNAase) (35). The enzyme was prepared by a laborious extraction procedure involving dilute acid extraction, ammonium sulfate fractionation, heating and ethanol fractionation. The purification was about 20,000-fold; and yield, over 50%. The preparation was free from RNAase and phosphomono- and phosphodiesterase activities. It was most stable between pH 4 and 5, its pH optimum was 4.8, and the temperature optimum was 55°. It was at least 10 times as active as the best preparations described for calf thymus and spleen, and required divalent cations for maximum activity.

Ribonuclease, phosphomonoesterase and phosphodiesterases. It was first shown that free ribose was liberated from crude commercial RNA at 0°C by ground lingcod muscle (36). However, no hydrolysis of the endogenous RNA in lingcod muscle was observed in up to 21 days at 0° (37). The reason for this is not known but may be due to an inability of the fish muscle enzymes to attack native RNA, or because the RNA is linked to protein.

Ribonuclease, phosphomonoesterase and phosphodiesterases have been demonstrated in fish muscle (38). The RNAase was partly purified and contained some phosphodiesterase and a trace of phosphomonoesterase. The RNAase was inhibited by 2×10^{-2} M EDTA while the phosphodiesterase was not affected. The RNAase degraded yeast RNA and, but more slowly, highly polymerized lingcod muscle RNA. The products were the 3′-ribonucleotides of adenine, guanine, cytosine and uracil. A similar enzyme was demonstrated in muscles of 6 other widely different fish species.

The RNAase activity in fish muscle extracts was much lower than that reported for calf spleen and liver extracts. In its lability to heat at acid pH values, the enzyme resembled certain calf spleen, thymus and pancreas RNAases. It resembled most closely a spleen nuclease fraction III (39). The enzyme mixture was later purified by chromatography on a small column of DEAE (free base form (41)). The RNAase fraction hydrolyzed RNA to give ribonucleoside 3′-phosphates and adenosyl 3′-benzyl phosphate to yield adenosine 3′-phosphate. 2′, 3′-Adenosine and uridine phosphates and DNA were not hydrolyzed. Thymidine 3′ and 5′ nitrophenyl phosphates were both hydrolyzed; the 5′ ester, very slowly. Thus both the 3′ links of ribose and deoxyribose were hydrolyzed. Neither p-nitrophenylphosphate nor bis-p-nitrophenylphosphate was hydrolyzed. Thus in nearly all respects the enzyme had the same specificity as the spleen enzyme.

Five distinct phosphomonoesterase activities were also reported in fish muscle using DEAE in the free base form (40). They were distinguished by their pH optima and range of activity, metal ion requirements (Zn^{++} and Mn^{++}), inhibition or stimulation by EDTA, and sensitivity to formaldehyde, tartrate, cystine and fluoride. The substrate specificities were not studied. A recent study has been made of a phosphomonoesterase from cod muscle (41). This enzyme was in the soluble fraction. It was specific for uridylic, inosinic, guanylic and cytidylic acids, and had no significant activity for other substrates. It is inhibited by 2-mercaptoethanol, reduced glutathione, carbonate and EDTA. Only manganese stimulated the enzyme. It was very labile and was partially destroyed by maintaining the temperature at 37° for 1 hour and rapidly inactivated at 42°. It was stable for almost a week at 0° and could withstand freezing and thawing, but was rapidly inactivated when stored at -30°. Though it is well known that inosinic acid is hydrolyzed in mammalian and avian muscles, the enzyme responsible for this has not yet been isolated. The fish enzyme is unlike any other phosphomonoesterase yet described.

Adenosinetriphosphatases. Early work on fish ATPases was carried out with muscle homogenates or whole muscle. In 1960 both particulate (sarcosomal) and soluble ATPases were isolated from carp and rabbit muscle (42). The activity of the carp particulate enzyme was 2-3 times that of the similar rabbit enzyme. The activity of both was accelerated markedly by 1 x 10^{-3} M Mg^{++} and slightly by Ca^{++}. The carp enzyme was rapidly inactivated at 35° while 42° was required to inactivate the rabbit enzyme. A soluble ATPase was also obtained from both muscles.

At about the same time it was found that the actomyosin ATPase of cod muscle was strongly activated by Mg^{++} while the myosin ATPase was suppressed by Mg^{++}; and these results paralleled those obtained with similar mammalian enzymes. Both enzymes were very unstable (43). Another investigator prepared a fish muscle myosin ATPase and digested it very briefly with trypsin. This procedure resulted in formation of a single product with a sedimentation constant similar to that of L meromyosin, while with rabbit myosin both L and H meromyosins were formed (44).

Deaminases. Adenylic acid deaminase was first demonstrated in a crude extract of carp muscle in 1959 (45); the specific activity was 6 μ moles/mg protein/hr at 30°. In 1964 a partly purified deaminase preparation was obtained from lingcod muscle (46) by dialyzing an 0.8 saturated ammonium sulfate fraction from lingcod muscle against H_2O, applying the extract to a small DEAE column, washing it with tris buffers of less than 0.3 M concentration, and finally eluting it with 0.3 M tris. Preparations thus obtained had 0.25 mg protein per ml, were very unstable and were used within a few hours of preparation. All attempts to purify the preparation further by use of comparatively mild treatments, such as absorption of alumina-Cγ or calcium phosphate gel failed, and 2-mercaptoethanol did not protect the enzyme.

The preparation was crude and contained both adenylic acid and adenosine plus deoxyadenosine deaminase activities. The former but not the latter was destroyed at 58°. Dexoyadenosine was deaminated about 2300 times as fast as adenine. The preparation also possessed nucleoside phosphorylase activity and the adenosine or deoxyadenosine formed was rapidly deaminated to inosine. The relationships are shown herewith:

$$\begin{array}{ccc}
\text{Adenine + ribose (deoxyribose)} & \overset{a}{\rightleftharpoons} & \text{Adenosine (deoxyadenosine) + Pi}' \\
c\downarrow & & \downarrow d \\
\text{Hypoxanthine + ribose (deoxyribose)} & \underset{b}{\rightleftharpoons} & \text{Inosine (deoxyinosine) + Pi}'
\end{array}$$

a and b = strong nucleoside phosphorylases; c = very weak deaminase; d = very strong deaminase.

It has been known for many years that free guanine increases in the skin of young salmon during smolting and a crude enzyme was prepared from the skin of roach (*Leucisus rutilus*) which deaminated guanine. It was stimulated by EDTA, strongly inhibited by Zn^{++} and Cu^{++}, and not inhibited by fluoride or iodoacetate (47).

A guanine deaminase was prepared from lingcod muscle (48). A 0.4-0.6 saturated ammonium sulfate fraction of an aqueous extract was heated 1 minute at 58° and dialyzed 4 hours. The crude unheated extract possessed 2 guanine deaminase activities, one with a pH 5.6 optimum and the other with a pH 8.5 optimum. Heating removed the "pH 8.5" activity. The "pH 5.6" fraction was purified by ammonium sulfate fractionation (0.4-0.6 saturated fraction retained), 4 hour dialysis, absorption on a DEAE cellulose column and elution with a pH 6.0 buffer containing 22 mM citrate and 6.4 mM phosphate. The most active fraction was purified 280-fold and had a specific activity of about 1950 (millimicromoles of guanine deaminated per minute at 22°). The K_m was 3.3×10^{-5} M. The enzyme was strongly inhibited by cyanide, fluoride, parahydroxymercuribenzoate, but not by Cu^{++} or iodoacetate. The purified enzyme was quite stable at 0°, but freezing was very destructive. In only one of several other species of fish examined was the deaminase present. In a subsequent study (49) the specificity of the enzyme was thoroughly investigated. The results of these studies showed that only guanine and 8-azaguanine were deaminated; guanylic acid, guanosine, 2-aminopurine, and 2, 6 - diaminopurine were not deaminated. The enzyme removed Cl from 2-chloro-6-hydroxypurine, hydrazine from 2-hydrazine-6-hydroxypurine and methylamine from 2-methylamino-6-hydroxypurine. The K_m and V_{max} values with these substrates varied considerably. The results supported the view that the lactam tautomeric form of the substrate is necessary to support activity.

Nucleoside hydrolases. A nucleoside hydrolase enzyme was prepared from lingcod muscle (50) by making a 0.6 M NaCl extract, adjusting this to pH 4.6, centrifuging, fractionating the supernate with ammonium sulfate, retaining the 0.4-0.6 saturated fraction, dialyzing it, absorbing the active fraction on XE-64 cation exchange resin and eluting the enzyme with 2.5 M NaCl. The enzyme was purified about 1500-fold, and was stable for weeks at 37° when freeze-dried. The preparation had two pH-activity peaks, one at about pH 8.6 and the other at pH 5.5. At pH 8.6 only inosine was hydrolyzed, and its hydrolysis was inhibited by adenosine. At pH 5.5 adenosine, guanosine and inosine were hydrolyzed rapidly, xanthosine and cytidine slowly, while uridine was not hydrolyzed. Ribose 1-phosphate and ribose 5-phosphate were not hydrolyzed. Subsequent work has shown that this enzyme does not hydrolyze deoxynucleosides. This enzyme occurs in muscle of other fish, and there is indirect evidence that it occurs in mammalian muscle but it has apparently not been isolated from this source.

Nucleoside phosphorylases. Fish muscle nucleoside phosphorylase was first slightly purified from lingcod muscle 10 years ago (51), and has recently been purified employing milder fractionation methods now available (52). An 0-0.6 saturated ammonium sulfate fraction of an aqueous extract of lingcod muscle was dialyzed overnight against 0.01 M phosphate buffer pH 7.2. At this stage the enzyme was quite stable, but all subsequent purification steps had to be carried out in 25 mM 2-mercaptoethanol. The dialyzed enzyme was absorbed on a DEAE cellulose column and eluted with a concave increasing gradient of 0.5 M into 0.01 M phosphate buffer. Three different "peaks" of nucleoside phosphorylase activity were obtained, together with a very weak peak of pyrimidine nucleoside phosphorylase activity that could only be detected by use of radioactive tracer technique. The most active fraction was further purified (about 120-fold) by absorption on alumina-Cγ and elution. The specific activity (μmoles inosine phosphorylyzed per hour at 30°) was 22.

The enzyme utilized hypoxanthine, 6-mercaptopurine, guanine, 8-azaguanine, xanthine, adenine, 2, 6 - diaminopurine and 6-methylpurine in the presence of ribose 1- or deoxyribose-1-phosphate. Several other purines tested were neither substrates nor did they inhibit the enzyme. The K_m with inosine in the presence of excess orthophosphate which would facilitate ribose-1-phosphate formation was 3.2×10^{-5} M. The pyrimidine nucleoside phosphorylase activity was very weak being about one-thousandth that of the most active purine nucleoside phosphorylase, and it utilized uridine and thymidine, deoxyuridine feebly, and not cytidine or deoxycytidine. The enzyme was inactivated by freezing and lost activity rapidly in the absence of 2-mercaptoethanol. Orthophosphate was required for maximum stability.

In the earlier work referred to, the enzyme was used to prepare ribose-1-phosphate (R1-P) and deoxyribose 1-phosphate (dR1-P) and to prove that chemically synthetic R1-P had the α and not the β configuration. It was, and perhaps still is, used by one biochemical company to prepare dR1-P.

Recently a salmon liver nucleoside phosphorylase, which also possessed deoxyribosyl transferase activity, was prepared from salmon liver (53). The reactions involved are:

1. Deoxyriboside (e.g. thymidine) + $P_i \rightleftharpoons$ dR1-P + pyrimidine base

2. Thymine deoxyriboside + uracil $\rightleftharpoons$ uridine deoxyriboside + thymine

This enzyme also required 2-mercaptoethanol for stabilization. A 100,000xg supernate of salmon livers was made 0.5 saturated with ammonium sulfate, and the precipitate, in a small volume of 0.01 M phosphate buffer, was passed through a large Sephadex G200 column. The enzyme fraction, freed from ammonium sulfate and much protein, was then adsorbed on a DEAE cellulose column which was eluted with a gradient of 0.3 M into 0.01 M phosphate buffer. The enzyme was eluted as a sharp peak, and this was concentrated on a Diaflo semi-permeable membrane under 50 lb. pressure (nitrogen) to a small volume, and this was repeated until the concentration of phosphate was about 1 mM (pH 6.8). The enzyme thus obtained was adsorbed on an hydroxyapatite column, and was eluted at 0.3-0.35 M phosphate buffer concentration either by stepwise or gradient elution.

The specific activity varied from 70-145 (μ moles uracil formed from deoxyuridine per hour by 1 mg protein at 30°) and the purification was

68-140 fold. The phosphorylase was specific for deoxyuridine, thymidine, 5-iododeoxyuridine, 5-bromodeoxyuridine, and 5-fluorodeoxyuridine. Uridine, deoxycytidine, cytidine, deoxyinosine, deoxyadenosine and deoxyguanosine were not utilized. The K_m with deoxyuridine as substrate was 1.7 x 10^{-3} M and with thymidine 1 x 10^{-2} M.

Under conditions where equilibrium was strongly in favor of deoxynucleoside formation (low orthophosphate concentrations) marked deoxyribosyl transfer occurred between 5-substituted deoxynucleosides and pyrimidine bases. Thus, deoxyuridine and thymine (^{14}C) gave thymidine ^{14}C plus uracil, the same substrates being active as with the nucleoside phosphorylase activity. When xanthine was substrate the transfer reaction and the phosphorylase reaction both yielded 3-deoxyribosyl xanthine. The ratio of transferase to phosphorylase remained approximately constant at 4:1 during purification. This ratio differs from ratios reported for transferase: phosphorylase in similar systems from mammalian or other sources. The pH-activity curves for phosphorylase and transferase were quite different, and in this respect were comparable to certain similar enzymes isolated from other sources.

Enzymes involved in deoxyribonucleic acid formation. So far as fish are concerned, this work has centered largely on DNA biosynthesis in male testes which synthesize comparatively large amounts of DNA. Only with the *de novo* pathways of uridylic and inosinic acid biosynthesis has liver also been used.

It was found that both immature testes and liver of salmon are capable of forming purines and pyrimidines by the recognized biosynthetic pathways. However, it was found necessary to employ comparatively large concentrations of radioactivity in order to occasion labeling. $KH^{14}CO_3$, L-aspartate-U-^{14}C, glycine-U-^{14}C and $H^{14}COONa$ were employed as substrates, and, since uridylic and inosinic acids were degraded in the tissues during incubation, uracil and hypoxanthine were isolated using the usual carrier methods. The results (54) showed the carbonate entered C-6 of hypoxanthine and glycine C-4 and C-5 while aspartate was not utilized. Formate also caused labeling of hypoxanthine. Uracil was labeled in C-4 and C-5 by aspartate. In contrast to previous work with mammalian tissues, carbonate did not label the ureide (C-2) carbon of uracil exclusively but also labeled C-4, 5, and 6. This is believed to be because radioactive carbonate was used to form radioactive aspartate by the fish tissue preparation. This could readily occur by combined activity of pyruvate carboxylase which would occasion oxalacetate formation, and the amination of oxalacetate by a transaminase to give aspartate.

Experience showed that in studies of the function of DNA in salmon testes it is imperative, or at least highly desirable, that immature testes be used. These can be selected when young salmon are killed for food purposes, or their formation may be encouraged by pituitary hormone injection.

Preliminary results (55) showed that specific labeling of the purine or pyrimidine bases in testes DNA could be effected by injecting the gonads in live fish, or whole excised gonads, with radioactive adenine, guanine and thymidine. In this work the testes used were somewhat too mature, and more recently slices of immature testes were used with much greater success (56). About 0.2 g of slices was incubated 16-20 hours at 5° with a very small amount of high specific activity radioactive purine or pyrimidine base, nucleotide or nucleoside. DNA was isolated in crude form and hydrolyzed to the constituent bases with 70% perchloric acid to determine the specificity of the label. In these experiments it is interesting to note that very slight homogenization of the testes, or even their disruption by application of vacuum, largely prevented labeling of the bases of DNA.

The results showed the purine and pyrimidine bases, nucleosides, nucleotides and especially their deoxy-derivatives were incorporated into DNA in a manner which suggested that they were incorporated by recognized biosynthetic pathways. Also, labeling of the 2-deoxyribose moiety of DNA occurred when C-1, C-6 or U-^{14}C glucose, 5′-AMP, deoxyuridylic acid, R5-P or 5-phosphoribosylpyrophosphate (PRPP) were substrates.

The occurrence of pyrophosphokinase in male gonads was also demonstrated (57). Soluble extracts of the gonads were prepared and the nucleic acids removed with protamine sulfate. These extracts were held 4 hours at 35° with tritiated PRPP plus ^{14}C labeled purine or pyrimidine base, or orotic acid. Appropriate 5′ nucleotides were added as carriers, and these were purified by ion exchange resin columns and repeated paper chromatography until products of constant specific activity resulted. In this way UMP (from both uracil and orotic acid), CMP, TMP, AMP and GMP were prepared. Acid hydrolysis of the purine nucleotides yielded tritiated D-ribose and the appropriate ^{14}C labeled free base.

A cell-free extract of testes was also found to form nucleosides, deoxynucleosides, and the corresponding mononucleotides (58). Thus, when tritiated R1-P or dR1-P were incubated with ^{14}C labeled purine or pyrimidine bases for 2 hours at 35°, the corresponding nucleosides or deoxynucleosides in which the base was ^{14}C labeled and the deoxyribose ^{3}H labeled were formed. If ATP were included, the mononucleotides were formed from adenosine, guanosine, uridine, cytidine, thymidine and deoxyuridine. In these experiments thymine riboside was formed as well as thymine deoxyriboside.

It was also demonstrated that immature salmon testes possessed enzymes which transfer phosphoryl groups, namely (where N=nucleoside):

1. N-triphosphate + N-monophosphate $\rightleftharpoons$ N-diphosphate + N-diphosphate

2. N-triphosphate + N-diphosphate $\rightleftharpoons$ N-triphosphate + N-diphosphate

These activities are generally known as nucleoside monophosphokinase and nucleoside diphosphokinase respectively. A crude enzyme mixture containing such activities was prepared from immature testes by isoelectric precipitation of extracts at pH 4.6-4.8, and adjustment of the soluble fraction to pH 7.0 (59). The enzyme was very unstable, losing 65% of its activity in 1 hour at 30° and 85% in 2 days at 0°. The simple purification procedure nearly doubled the specific activity but eliminated some of the undesirable nucleoside triphosphatase activity. The preparation required

ATP, or another nucleoside triphosphate, and was most active at pH 7.4-8.0. Its activity was greatly enhanced by Mg^{++} and Ca^{++}, less by Mn^{++}, Co^{++} and KCN, and Ni^{++}, Hg^{++}, Ag^{++}, Zn^{++}, iodoacetate, fluoride, EDTA and p-hydroxyparamercuribenzoate were strongly inhibitory. 2-Mercaptoethanol and reduced glutathione were without effect or slightly inhibitory. The best substrates were 5′-CMP, AMP, UMP and d-AMP; GMP and d-CMP were poorer substrates and IMP and TMP were only feebly utilized. When CDP, UDP or ADP were employed as substrates in absence of added nucleoside triphosphate, an equilibrium mixture of mono-, di- and tri-phosphates resulted in which roughly half the activity was in the di-phosphate fraction. Evidently immature salmon testes possess very diverse monophosphokinase, and also possibly diphosphokinase, activity.

Other enzymes. Uridinediphosphate glucose (UDPG), uridinediphosphate galactose (UDP Gal) and uridinediphosphate acetylglucosamine have been isolated from fish livers (60). UDPG dehydrogenase and UDP Gal-4′-epimerase were isolated from salmon liver. Nicotinamide hydrolase (NADase) and NADP hydrolase, enzymes that cause hydrolysis at the N-ribosyl linkage, were demonstrated in carp liver (61) and the muscle of cod, (62) and some of their properties were studied. The general reaction is:

$$\text{Adenine-ribose-P-O-P-ribose nicotinamide} \rightleftharpoons \text{ADP-ribose + nicotinamide.}$$

REFERENCES

1. R. T. Williams (ed.) "The Biochemistry of Fish." Biochem. Soc. Symp. 6, (Cambridge, Eng.) p. 105. University Press, 1951.

2. E. Geiger, Fortschr. Chem. Org. Naturstoffe. **5**: 267 (1948).

3. G. Hamoir, Adv. Protein Chem. **10**: 227 (1955).

4. H. L. A. Tarr, Ann. Rev. Biochem. **27**: 223 (1958).

5. M. Gubmann, W. D. Brown, and A. L. Tappel, Intermediary metabolism of fishes and other aquatic animals. U. S. Fish and Wildlife Serv., Spec. Sci. Rep. Fisheries, **288**: 51 (1958).

6. D. R. Ekberg, Biol. Bull. **114**: 308 (1958).

7. P. W. Hochachka and F. R. Hayes, Can. J. Zool. **40**: 261 (1962).

8. P. W. Hochachka, Gen. Comp. Endocrinol. **2**: 499 (1962).

9. H. L. A. Tarr, Can. J. Biochem. **41**: 313 (1963).

10. P. W. Hochachka, In: MOLECULAR MECHANISMS OF TEMPERATURE ADAPTATION, p. 177, Am. Assoc. Adv. Sci., 1967.

11. H. B. Stoner, Federation Proc. **22**: 851 (1963).

12. J. P. Hannon, Federation Proc. **22**: 856 (1963).

13. G. J. Klain and D. A. Vaughan, Federation Proc. **22**: 862 (1963).

14. M. Yamamoto, Can. J. Biochem. **46**: 423 (1968).

15. G. B. Martin and H. L. A. Tarr, Can. J. Biochem. **39**: 297 (1961).

16. E. Roberts and H. Tsuyuki, Biochim. Biophys. Acta.**73**: 673 (1963).

17. T. Hashimoto and P. Handler, J. Biol. Chem. **241**: 3940 (1966).

18. H. Tsuyuki and F. Wold, Science, **146**: 535 (1964).

19. R. P. Cory and F. Wold, Biochemistry **5**: 3131 (1966).

20. T. Shibata, Bull. Fac. Fisheries, Hokkaido Univ. **9**: 218 (1958).

21. T. W. Kwon and H. S. Olcott, Comp. Biochem. Physiol. **15**: 7 (1965).

22. L. Ludovicy-Bungert, Arch. Intern. Physiol. Biochim. **69**: 265 (1961).

23. W. B. Allison and N. O. Kaplan, J. Biol. Chem. **239**: 2140 (1964).

24. F. Nagayama, Bull. Japan Soc. Sci. Fisheries **27**: 1961 (1961).

25. A. C. Wilson, N. O. Kaplan, L. Levine, A. Pesce, M. Reichlin and W. S. Allison, Federation Proc. **23**: 1258 (1964).

26. T. P. Fondy, J. Everse, G. A. Driscoll, F. Castillo, F. E. Stolzenbach and N. O. Kaplan, J. Biol. Chem. **240**: 4219 (1965).

27. H. L. A. Tarr, Can. J. Biochem. **37**: 961 (1959).

28. M. Tabachnick, P. A. Sere, J. Cooper and E. Roche, Arch. Biochem. Biophys. **74**: 315 (1958).

29. W. J. Wheelan and M. P. Cameron (ed.), "Control of Glycogen Metabolism." Ciba Foundation Symposium, J. and A. Churchill, London (1965).

30. H. L. A. Tarr, J. Fisheries Res. Board Can. **25**: 1539 (1968).

31. M. Gubmann and A. L. Tappel, Arch. Biochem. Biophys. **98**: 262 (1962).

32. M. Gubmann and A. L. Tappel, Arch. Biochem. Biophys. **98**: 502

(1962).

33. W. D. Brown and A. L. Tappel, Arch. Biochem. Biophys. **85**: 149 (1959).

34. A. Mannan and H. L. A. Tarr, J. Fisheries Res. Board Can. **18**: 349 (1961).

35. M. R. McDonald, J. Gen. Physiol. (Pt. 2, Suppl.) **45**: 77 (1962).

36. H. L. A. Tarr, Food Technol. **8**: 15 (1954).

37. N. Tomlinson and V. Creelman, J. Fisheries Res. Board Can. **17**: 603 (1960).

38. N. Tomlinson, Can. J. Biochem. Physiol. **36**: 633 (1958).

39. L. A. Hepple, R. Markham and R. J. Hilmoe, Nature, **171**: 1152 (1953).

40. N. Tomlinson and R. A. J. Warren, Can. J. Biochem. Physiol. **38**: 605 (1960).

41. H. L. A. Tarr, L. J. Gardner and P. Ingram, (In preparation).

42. A. Reuter, Arch. Intern. Physiol. Biochim. **68**: 339 (1960).

43. J. R. Dingle and J. A. Hines, J. Fisheries Res. Board Can. **38**: 1437 (1960).

44. A. Kover, M. Szabolcs and K. Benko, Acta. Physiol. Acad. Sci. Hung. **23**: 229 (1963).

45. S. Nara and T. Saito, Bull. Fac. Fisheries, Hokkaido Univ. **10**: 68 (1959).

46. H. L. A. Tarr and A. G. Comer, Can. J. Biochem. **42**: 1527 (1964).

47. P. E. Lindahl and P. O. Srard, Acta. Chem. Scand. **11**: 846 (1957).

48. J. E. Roy, Can. J. Biochem. **44**: 1093 (1966).

49. J. E. Roy and K. L. Roy, Can. J. Biochem. **45**: 1263 (1967).

50. H. L. A. Tarr, Biochem. J. **59**: 386 (1955).

51. H. L. A. Tarr, Can. J. Biochem. Physiol. **36**: 517 (1958).

52. H. L. A. Tarr and J. E. Roy, Can. J. Biochem. **45**: 409 (1967).

53. H. L. A. Tarr, J. Roy and M. Yamamoto, Can. J. Biochem. **46**: 407 (1968).

54. H. L. A. Tarr, Can. J. Biochem., In press (1969).

55. H. L. A. Tarr, Can. J. Biochem. **42**: 51 (1964).

56. H. L. A. Tarr and J. Roy, Can. J. Biochem. **44**: 1435 (1966).

57. H. L. A. Tarr, Can. J. Biochem. **42**: 575 (1964).

58. H. L. A. Tarr, Can. J. Biochem. **42**: 1535 (1964).

59. H. L. A. Tarr and J. Roy, Can. J. Biochem. **44**: 197 (1966).

60. R. J. Forrest and R. G. Hansen, Can. J. Biochem. Physiol. **37**: 751 (1959).

61. K. Raczynska-Bojanowska and I. Gasiorowska, Acta. Biochim. Polon. **10**: 117 (1963).

62. N. R. Jones and J. Murray, Bull. Japan Soc. Sci. Fisheries, **32**: 197 (1966).

COMMENTS

DR. MANOHAR: You said that there are 2 enzymes to break down the nucleosides. One is a phosphorylase and the other a hydrolase. Are these similar to the enzymes in mammalian tissues? As far as I know, the phosphorylase is more active in mammalian tissues. Do you consider the hydrolase to be equally active in the fish muscle?

DR. TARR: They are both present in fish muscle. I don't know about mammalian muscle other than that you do get free ribose in meats, so they must have a riboside hydrolase. I have never heard anybody describe the nucleoside phosphorylase or the deoxynucleoside phosphorylase of mammalian muscle; it is probably there also, but I don't think that it has ever been isolated, at least as far as I know.

TOPICS IN GENETICS

E. Goldberg, Chairman

QUANTITATIVE INHERITANCE AND ENVIRONMENTAL RESPONSE OF RAINBOW TROUT

Graham A. E. Gall

INTRODUCTION

A suitable population of fish maintained for the purposes of research will have to be one which can be maintained on a small population size in terms of the effective number of parents. It would be desirable for this population to remain stable genetically. Small population size is necessary because maintenance on a large population base will result in greater genetic variability among the individuals. Such variability will, in general, be disadvantageous to most research programs except those involved in studying the heritance of particular characteristics.

The environment in which the population is to be maintained is also critical. The environment will have a direct effect upon the phenotypic expression of the character or characters of interest in the sense that the environment can shift the mean of the population. Secondly, the environment can affect the normal developmental pattern of the organism. For example, excessively warm water during the maturation of rainbow trout eggs can result in very high embryonic abnormalities. Other stresses, such as handling and noise, can also cause embryonic death. An additional effect of the environment is one resulting from an interaction between the genotypes in the population and the environments to which they are exposed commonly referred to as a genotype by environment interaction.

FACTORS INFLUENCING A RESEARCH STOCK

Wild-type stock. The first considerations in establishing a research stock of fish is to ascertain the nature of the population desired. In some instances the researcher may desire a wild-type stock, that is, one that is representative of fish in nature. One could only maintain such a stock by repeat sampling of parents from the wild habitat. This type of population would probably only be of interest to someone carrying out research on the nature of fish populations *per se* rather than for other research purposes. A most likely population would be a domesticated stock. The stock could be developed through any one or a combination of three methods: a random bred stock, an inbred stock, or a selected stock.

Random-bred stock. The random-bred stock is the most unlikely choice. We desire a stable population structure which means that the population is not under the influence of selection, mutation or genetic drift. The maintenance of these conditions, if it is at all possible, would require an enormous population size. The size would in general be beyond the limits of most facilities and would only be maintained where a genetic control population was needed. The principal use of a random-bred stock is to control all genetic and environmental effects other than the one of interest to the researcher. This implies that the researcher is intending to change one of the characteristics and is not what is generally required of a research population in general. Another disadvantage of random-bred population is that there would be a natural temptation, because of the fecundity of fish, for institutions to maintain the population on the basis of a few parents, a practice which would result in very rapid genetic drift.

Selection and inbreeding. We are left then with establishing our population on the basis of either selection or inbreeding. We have noted the fact that the population will likely be maintained on a relatively small population size. This immediately implies inbreeding either intentionally or coincidentally. The population could be developed on the basis of inbreeding alone, but such a process would require tremendous effort, ambition and persistence. One would be required to start with a very large number of family lines, 100 to 200 as a minimum. The philosophy is that we are assuming that some of the families will survive through an inbreeding depression. The rapid deterioration of many of the lines could be due to the exposure of deleterious genes carried in the stock or from the general disruption of the genetic and physiological homeostasis of the individuals.

The most likely approach would be one which combined selection and inbreeding. Such a procedure would allow intense selection for reproductive capacity, the most likely characteristics to show an inbreeding depression. One must acknowledge, however, that the small population sizes will result in genetic drift and that many characteristics of the population will be lost.

It may, however, be possible to select the stock for the particular purpose that the researcher has in mind. This again will be a very extensive and expensive procedure and one would need to be very sure of his end point before attempting such a procedure.

CONSIDERATIONS AFFECTING POPULATION

Regardless of the research purpose of the stock, there are a number of characteristics which must be considered and kept at a maximum level. These include characteristics of fecundity, in particular egg number, egg size, and hatchability of the eggs. The genetic and phenotypic relationships between these characters is also of importance although not well known. For example, egg number may not be of great importance since the number of eggs is normally large. However, egg size may be important since there are indications that the size of the egg exerts a maternal effect on the performance of the fingerlings and since there may be a negative genetic correlation between egg size and egg number. Other performance characteristics of importance would include the growth and survival of the fingerlings and the age and size at sexual maturity. The latter two are of importance from the standpoint of reducing the expense of maintaining the population and also the size at maturity may have an effect on fecundity.

OBSERVATIONS OF FOUR DOMESTIC LINES OF RAINBOW TROUT

The Inland Fisheries Branch of the California Department of Fish and Game maintains four domestic lines of rainbow trout in addition to a wild stock. The egg number and egg size of the four domestic stocks are outlined in Table I. The first important observation is that these four stocks differ in the number of eggs spawned per female. At two years of age, RTH[1] produces the largest number of eggs while RTV[1] produces approximately 30% fewer eggs. Two of the stocks, RTW[1] and RTH show a substantial increase in number of eggs at three and four years of age over that spawned at two years of age; the RTV and RTS[1] stocks do not show this response with age. The stocks also differ in the size of eggs, RTV having the smallest egg size and RTS having the largest egg size at two years of age. All of the stocks show an improvement in egg size with increasing age; however, the age difference is again more pronounced for RTW and RTH. These data also shed some light on the effects of inbreeding since both RTV and RTH have undergone a considerable but unknown amount of inbreeding. In addition, RTV has been under selection for a much shorter time than RTH, and therefore the selection applied has been much less. The contrast between these two stocks would indicate that RTH has shown little inbreeding depression or that selection has been able to counteract any inbreeding depression since this stock has a superior egg number and egg size. It does however have a slightly slower growth rate than an RTV.

1. RTH, RTV, RTS, RTW: domestic lines of Rainbow Trout, H, V, S, and W respectively.

TABLE I

Spawning Performance of Four Domestic Rainbow Trout Broodstocks.

1A Stock	Number of Eggs per Female Age (years)		
	2	3	4
RTS	2825	2525	2663
RTV	2467	2785	2738
RTW	2894	---	4429
RTH	3673	4032	---

1B Stock	Egg Size (eggs/oz.) Age (years)		
	2	3	4
RTS	346	239	222
RTV	438	392	394
RTW	368	---	261
RTH	392	275	---

TABLE II.

Body Weight of Domestic (RTW) and Wild (RTK) Rainbow Trout Lines and Their F_1 Cross

	Weight (oz./100 fish)			
Number Weeks	RTW	RTK	F_1*	$F_1 - \bar{P}$*
13	4.4	2.5	5.7	2.3
24	34.5	8.3	20.0	-1.4
34	88.9	14.7	55.5	3.7
41	133.3	26.3	83.3	3.5
48	228.6	66.7	133.5	-14.3

* F_1 - RTW x RTK cross
$\bar{P}$ - average of RTW and RTK

One method of evaluating the possible extent of inbreeding depression is to carry out crosses between existing lines and evaluate their response in terms of heterosis. An example of such a cross involving a domestic stock, RTW, with a wild rainbow trout, designated RTK, is shown in Table II. This experiment involved an evaluation of growth rate as indicated by the body weights of the fish from 13-48 weeks of age. The wild trout has a much lower growth rate than the domesticated stock. The performance of the first cross (F_1) in the last column of Table III is indicated as the difference between the F_1 value and the average of the two parent stocks. Heterosis is defined as existing if the F_1 performance exceeds that of the mid-parent value. This experiment showed little indication of heterosis for growth rate. Heterosis is expected to result from an increase in heterozygosity in the population. The absence of a definite heterotic affect in the cross between these two stocks may be a result of the wild stock being excessively heterozygous in its own right, and the degree of heterozygosity in the F_1 is little different. It may in fact be decreased.

TABLE III.

Mean Calculated Weights of Two Lines of Domestic Rainbow Trout and Their F_1 Cross

Mean calculated 140-day weights (oz/100 fish) observed in a 2-way factorial experiment involving 3 lines of Rainbow trout (RTS, RTW and F_1 cross) and 3 stocking densities (500, 1000 and 1500 fish/bin).*

Line	Mean Weight	Density	Mean Weight
RTS	13.4[a]	500	15.4[a]
RTW	13.3[a]	1000	14.0[b]
F_1	14.6[b]	1500	11.9[c]

*mean values with same superscript are not significantly different (P< 0.05).

In a second experiment (1) a cross between two domestic lines was evaluated in the presence of three different environments namely three densities of stocking rate in rearing troughs. Table III gives the mean calculated 140-day body weight for the three lines and for the three densities. Contrary to the results of the experiment involving the wild stock heterosis was observed in the cross of the two domestic lines to the extent that the average of the cross exceeded that of either parent. This suggests that a slight inbreeding depression might be expected for growth rate. The effect of density may not be important in itself, but it does demonstrate the environment can shift the mean performance of the population, the population in this case being all three lines. There may be a threshold level for density since there was a much greater decrease in performance between the densities of 1000 and 1500 than between 500 and 1000.

This experiment also allows an evaluation of genotype by environmental interaction. We note from Table III that the two parent stocks RTS and RTW are equal when averaged over all three densities. However, a plot of each line performance against density (Fig. 1) suggests that the two stocks are not equal at all densities. RTS is superior at low density, RTS and RTW are approximately equal at intermediate density, and RTW is superior at high density. This result is a genotype by environmental interaction of the simplest kind. The hybrid performance is superior at all three densities but is somewhat more superior at a 1000 density than at the other two densities. The major point made here is that stocks do not always rank the same from one environment to the next.

SUMMARY

We have looked at only one environment, but similar results are likely to hold for other environments such as water temperature, water quality and nutritional level, etc. It indicates that fish are much more susceptible to the influence of the environment than we usually expect for mammals. This property of genotype by environmental interaction could be advantageous in maintaining a research stock for a particular purpose. It may be possible to produce stress conditions which will separate genetic lines or which will vary the response of a particular line. In such cases, the researcher would be able to pick the stress and the stock which would produce an exaggerated or abnormal phenotypic response. He would then have research material available in which the differences were magnified and controllable.

ACKNOWLEDGMENT

The author acknowledges the Inland Fisheries Branch of the California Department of Fish and Game which kindly supplied the data.

REFERENCES

1. Calaprice, J. R., An Experimental Evaluation of the Relative Role of Environmental and Genetic Factors Influencing Qualitative Changes in Salmonid Populations, Ph.D. dissertation. University of California, Santa Barbara, 1966.

COMMENTS

DR. GOLDBERG: I would like to ask a question regarding the inbred strains for research purposes. There is evidence accumulating that salmonids and many other fish are tetraploids and that there has been extensive gene duplication. Now what are the consequences of this as far as your hopes to achieve isogenicity?

DR. GALL: First impressions would suggest that it would be more difficult. However, if the salmonids are autotetraploid but show diploid segregation, as many plants do, the rate at which isogenicity would be reached would be the same as that for a diploid organism.

DR. IDLER: We are presumably trying to get animals which have reproducible characteristics. In view of the experiences with the laboratory rat it might be a mixed blessing to suddenly find large numbers of fisheries scientists all devoting their attention to a single strain of a single species simply because it is available on request. I know it certainly does not solve all problems, but one possibility might be to take advantage of the very good reproducibility of some of the natural runs. Scientists at the Vancouver Laboratory, for example, investigated migrating sockeye salmon in the Fraser River over many years. These fish are extremely reproducible as to timing, sexual development and so forth, and also in regards to their biochemical components. A continuous supply of such fish and eggs taken from the natural environment might have advantages over highly inbred strains with the problems attending this approach.

DR. GALL: Rainbow trout are very sensitive in their response to the environment. Applying some environmental stress provides a method of producing exaggerated differences. The same effect can be achieved in this way much faster than by selective breeding.

DR. IDLER: Even within species you can have quite unique characteristics. For example, when we were studying various species of skates we found an interesting characteristic in one which we though was worthy of further investigation. We did a comparative study and found that only the one species had the characteristic. So there is a risk in too narrow a selection of experiment material.

DR. GALL: Oh, yes, there is. However, the purpose in producing unique lines of fish is to have material which will give reproducible results; for example, in the production of a neoplasia. Many people have alluded to the desirability of having lines of fish comparable to the mouse in which the histocompatibility loci are identified and specified.

DR. HUNN: Even if you do have a selected test animal, the conditions under which you test the animal are crucial. One often reads of studies in which the movement of certain drugs across the gills are determined in fish placed in distilled water. It is common to find in the literature the results of studies in which the investigator neglects to tell you the characteristics of the water, for example, the pH or any of the other factors which quite dramatically influence the system. If the environment isn't properly controlled, the experiment is meaningless.

DR. GALL: That is true.

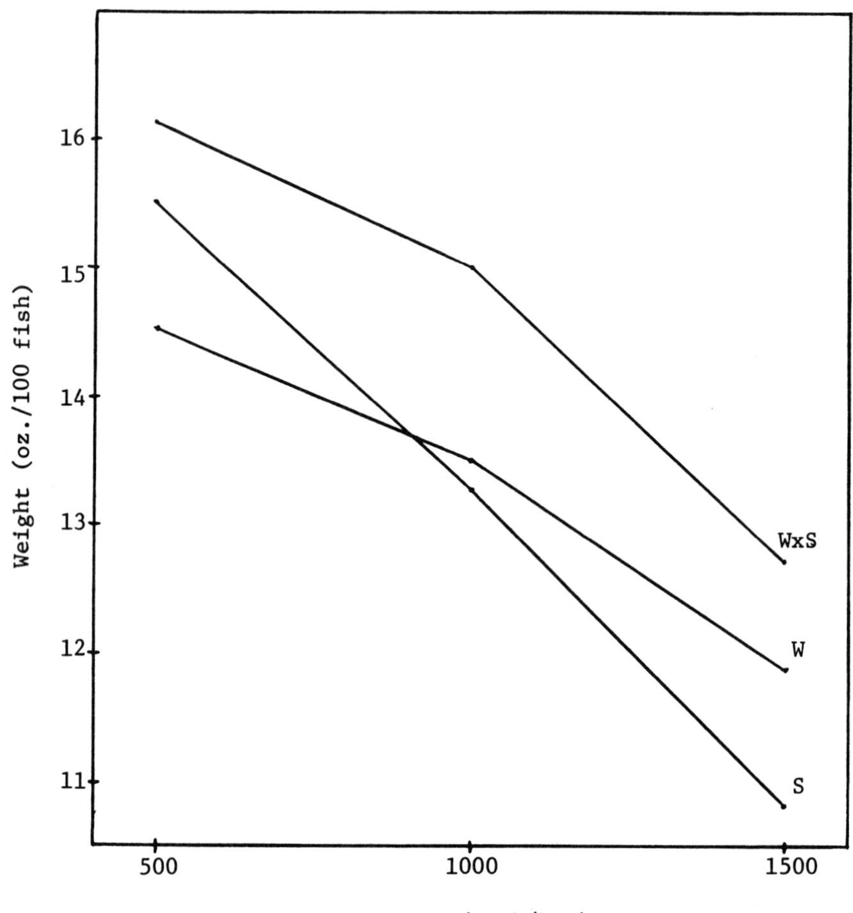

Fig. 1. Performance of 2 domestic rainbow trout stocks and their F_1 cross plotted against density. S – RTS parent stock; W – RTW parent stock; W x S – the RTW x RTS cross.

BLOOD GROUPS IN SALMONID FISHES

George J. Ridgway

INTRODUCTION

As an introduction to this paper, I consider two major reasons for studying blood groups in salmonid fishes. First, in any study of comparative biochemistry and immunogenetics one would expect to find basic principles that would help provide answers to problems in human biology and medicine. For example, Clarke (1) described how an interest in the genetics of butterflies led him to discover an effective preventive measure for the hemolytic disease of newborn babies caused by Rh incompatability. Biochemistry and genetics are unifying disciplines in modern biology, and principles of these sciences discovered in work with any species are widely applicable.

A second reason for studying the blood groups of fishes is the importance to fishery management of recognizing the racial or subpopulation structure of economically important fish stocks. This subject has been treated by several authors including Marr (2), Ridgway (3), Marr and Sprague (4), and Cushing (5).

In this presentation I review the published work on blood groups in salmon and trout, consider some of the patterns my associates and I have found in blood groups of salmon and trout, illustrate how blood groups have been used to distinguish geographic races of salmon, and point out opportunities for research in this field.

BLOOD-TYPING STUDIES ON SALMON AND TROUT

Table I summarizes studies on trout and salmon in which blood types have been demonstrated. The same sort of naturally occurring and experimentally induced agglutinins have been used as have been used for typing the blood of man and other animals.

THE PRODUCTION OF ISOIMMUNE SERA IN RAINBOW TROUT

Early in our studies (6), we found that immunization of rabbits and chickens did not produce the number of reagents with rather narrow ranges of specificity that were required for investigating complex population problems. We turned to isoimmunization, particularly with rainbow trout, and the results were much more productive. The reason for our greater success in finding individual differences with isoimmune sera was probably due to what Landsteiner (7) has called "immunological perspective." Rabbits and chickens are antigenically too distant from fish to be able to distinguish the finer differences among individuals. Conversely, when a rainbow trout was exposed to the blood cells of another trout, the only antigenic differences recognized and responded to were intraspecific differences.

We isoimmunized several hundred trout with a variety of antigen doses, and used various injection schedules, and injection routes. Our most successful results were obtained by isoimmunizing healthy fish not over 3 years old and weighing 500 grams or more. We anesthetized the fish by immersing them in a solution of 1:20,000 tricaine methane sulfonate when bleeding or injecting them. Most of the sera were produced by using pairs of fish that served as both donor and recipient. From each member of each pair, 0.5 ml of blood was taken by heart puncture into 0.1 ml Alsever's solution and injected intraperitoneally into the other member. Injections were made once a week for about 9 weeks and usually booster injections were given later at monthly intervals. Isoagglutinins were detected after 3 weeks in a few animals and most had responded after 9 weeks. After detectable agglutinins had formed, 10 to 12 ml were bled from the trout every 3 or 4 weeks. After a number of bleedings, the sera were tested against a panel of red cells from 30 trout and those bleedings from an individual trout exhibiting the same specificity were pooled. Many of these pools were absorbed with red cells from salmon or trout to produce reagents with unit specificities.

TABLE I

BLOOD TYPING STUDIES ON SALMON AND TROUT

Species	Typing reagents	Reference
Sockeye salmon (*Oncorhynchus nerka*)	Normal pig serum	Ridgway, Cushing and Durall (8)
Sockeye salmon, chinook salmon (*O. tshawytscha*) chum salmon (*O. keta*), pink salmon (*O. gorbuscha*)	Immune rabbit and chicken sera, isoimmune sockeye salmon sera	Ridgway and Klontz (6)
Rainbow trout (*Salmo gairdneri*)	Normal pig serum	Balakhnin (9)
Rainbow trout	Isoimmune rainbow trout sera	Ridgway (10)
Rainbow trout, brown trout (*Salmo trutta*)	Immune rabbit sera	Sanders and Wright (11)
Cutthroat trout (*Salmo clarki*)	Isoimmune sera, normal bighorn sheep serum	Bingham (12)
Rainbow trout, brook trout (*Salvelinus fontinalis*), lake trout (*Salvelinus namaycush*)	Immune rabbit and chicken sera	Wright Sklenarik, and James (13)
Rainbow trout, golden trout (*Salmo aguabonita*)	Immune rabbit sera	Calaprice and Cushing (14)
Pacific salmon (*Oncorhynchus*), several species	Plant extracts	Utter, Ridgway and Hodgins (15)
Rainbow trout, sockeye salmon	Rainbow trout isoimmune sera	Ridgway (16)
Atlantic salmon (*Salmo salar*)	Immune rabbit sera	Alabaster and Durbin (17)

SEROLOGICAL PROPERTIES OF RAINBOW TROUT

Isoimmune Sera. The agglutinin titers of the isoimmune sera in rainbow trout ranged from 1:2 to 1:256. Frequently these titers were higher when the sera were diluted in normal trout serum than when diluted in one percent sodium chloride solution, indicating the presence of incomplete antibodies. None of the sera were hemolytic for trout cells, although the same sera often contained hemolytic systems for the red blood cells of rabbits. The isoagglutinating antibodies were active from 4°C to 30°C.

Testing Methods. Some of the agglutination tests were made with the standard tube method by using equal quantities of diluted serum and a 2 percent suspension of washed red cells in the modified Alsever's solution described by Hodgins and Ridgway (18). Most of the tests, however, were conducted with a modification of the capillary agglutination method of Chown and Lewis (19), which required much smaller volumes of serum and was more sensitive. Equal volumes of the serum, usually diluted 1:2, and a 20 percent suspension of washed red cells in modified Alsever's solution were used. Readings were taken after 15 and 30 minute intervals.

EXTENT OF ANTIGENIC DIVERSITY IN SOCKEYE SALMON

About 150 reagents were produced by the methods described above; many were found to detect individual differences in sockeye salmon and 35 were used in extensive tests on populations of this species. In one of the investigations on sockeye salmon from the Copper River in Alaska, blood samples from 462 fish were tested with 15 reagents. We attempted to use automatic data processing methods to find recurrent combinations that might be diagnostic of certain racial groups in the tributaries of the Copper River. The incidences of combinations of the reactions with these 15 reagents were machine listed; most combinations occurred only once, a few twice, and the most frequent combination (no reaction with any of the reagents) occurred only 16 times. On careful consideration, this result is not surprising; if the reactions are independent there are 2^{15} or 32,768 possible combinations. We had nearly demonstrated, with just 15 reagents, the antigenic individuality of the 462 fish tested.

Actually, the reactions of the various reagents are not entirely independent; some form systems or patterns of interrelationship similar to the patterns in the Rh blood group system of man, and the B blood group systems of cattle and chickens.

THE B-SYSTEM OF BLOOD GROUPS IN SALMON AND TROUT

I have called the most complex system of blood groups found in sockeye salmon and rainbow trout the B-system. The pattern of statistical interrelationships among the reactions of 11 rainbow trout isoimmune sera in this system with the blood cells of 1500 sockeye salmon from the Copper

River in Alaska is illustrated in Fig. 1. Reagents connected by lines show significant lack of independence in their reactions. The statistical association of the reactions with these antisera, as well as the results of absorption analyses, indicate that these reagents possess related, overlapping specificities and that they distinguish sets of related antigens. The interrelated antigens detected by this group of antisera are what I have called the B-system. Following the convention of Stormont (20), I have called the related reoccurring groups of reactions in this system phenogroups. A few of the common phenogroups found in Copper River sockeye salmon are ABHJ, HJM, FHJLM, CHJM, and CHJ. Such groups of reactions are considered to designate single antigens.

GENETICS OF THE B-SYSTEM OF BLOOD GROUPS

We have obtained some first-hand knowledge of the genetics of the B-system from experimental matings of rainbow trout. The data available concern only a single generation and consist of tests made on a dozen or so individuals in each of several sibstrips. The reagents which show statistical relationships in population data also delineate groups of reactions which appear to be inherited as units. Examples of these groups in rainbow trout are ABHJM, CHJ, CJ, and HJ.

BLOOD GROUP FREQUENCY DIFFERENCES AMONG GEOGRAPHIC RACES OF SOCKEYE SALMON

Differences in the frequencies of blood groups among geographic races of man have been demonstrated (21, 22).

My associates and I have found similar differences in the frequencies of the blood groups of sockeye salmon among geographic races or subpopulations (8, 16, 23, 24). An example is from data on subpopulations of sockeye salmon that spawn in tributaries of the Naknek River, Alaska. The question whether these tributary areas are occupied by separate reproductively isolated population units – perhaps differing in year class composition and time of migration – or by a single homogeneous interbreeding stock is important to the management of this fishery. This question has been examined by Hartman and Raleigh (25) and by Straty (26), by tagging and with age composition data. Their papers contain evidence for independence of the stocks occupying these tributaries. Despite the small sample sizes, our blood group data provide confirmatory evidence. Table II lists the frequencies of reactions with 4 of the B-system reagents with samples of sockeye salmon from 4 tributary areas. These frequencies are significantly different with 3 of the 4 reagents used. In Fig. 2 the phenogroup frequencies are plotted. These phenogroups probably represent the characters which are inherited as units. The frequencies of some of these phenogroups are different in the different subpopulations.

OPPORTUNITIES FOR FURTHER RESEARCH

Certain methodological problems deserve attention in future research.

TABLE II

FREQUENCY DIFFERENCES IN REACTIONS WITH FOUR B-SYSTEM REAGENTS AMONG POPULATIONS OF SOCKEYE SALMON IN TRIBUTARIES OF THE NAKNEK RIVER.

		REAGENT							
		A		B		D		F	
Area	Number tested	Number positive	Percentage positive	Number positive	Percentage positive	Number positive	Percentage positive	Number positive	Percentage positive
American River	27	5	19	2	7	11	41	6	22
Hardscrabble Creek	31	2	7	5	16	8	26	9	29
Brooks River	24	7	29	7	29	6	25	2	8
Hidden Creek	26	13	50	7	27	16	62	12	46
Chi-square		15.3		4.98		10.3		9.27	

Such sensitive typing methods as hemolytic and antiglobulin tests have been tried, but serious problems remain to be solved before these tests can be applied to the typing of fish blood. The study of the chemical nature of the blood group substances in fishes is an open field for research. An understanding of the magnitude and causes of blood group polymorphism in salmon and trout would be important to studies of evolution and genetics. Our present knowledge of blood groups in trout can be expected to have wide applicability in studies of tumor and tissue transplantation immunogenetics.

ACKNOWLEDGMENTS

The contributions of my former associates Warren Ames, Robert Carter, Harold Hodgins, George Klontz, Edward Larson, Robert Smith, and Fred Utter were essential to many of the studies I have described here.

REFERENCES

1. C. A. Clarke, Sci. Am. **219**: 46 (1968).

2. J. C. Marr, U. S. Fish Wildlife Serv. Spec. Sci. Rep. Fish. no. 208: 1 (1957).

3. G. J. Ridgway, U. S. Fish Wildlife Serv. Spec. Sci. Rep. Fish. no. 208: 39 (1957).

4. J. C. Marr and L. M. Sprague, Int. Comm. Northwest Atl. Fish. Spec. Publ. 4: 308 (1963).

5. J. E. Cushing, Advan. Mar. Biol. **2**: 85 (1964).

6. G. J. Ridgway and G. W. Klontz, U. S. Fish Wildlife Serv. Spec. Sci. Rep. Fish. no. 324: 1 (1960).

7. K. Landsteiner, "The Specificity of Serological Reactions." Revised ed. p. 330, Dover Publ., New York, 1962.

8. G. J. Ridgway, J. E. Cushing and G. L. Durall, U. S. Fish Wildlife Serv. Spec. Sci. Rep. Fish. no. 257: 1 (1958).

9. I. A. Balakhnin, Dokl. Akad. Nauk SSSR (English translation) **141**: 500 (1961).

10. G. J. Ridgway, Ann. N. Y. Acad. Sci. **97**: 111 (1962).

11. B. G. Sanders and J. E. Wright, Ann. N. Y. Acad. Sci. **97**: 116, (1962).

12. D. A. Bingham, Proc. West. Ass. State Game Fish Comm., July 10-12, 1963, pp. 224-233.

13. J. E. Wright, R. Sklenarik, and S. M. James, Proc. of the XIth International Congress of Genetics, Vol. I, The Netherlands, 1963, Pergamon Press, Oxford, 1963.

14. J. R. Calaprice and J. E. Cushing, Calif. Fish Game **50**: 152 (1964).

15. F. M. Utter, G. J. Ridgway, and H. O. Hodgins, U. S. Fish Wildlife Serv. Spec. Sci. Rep. Fish. no. 472: 1 (1964).

16. G. J. Ridgway, Xe Congres europeen sur les groupes sanguins et le polymorphisme biochemique des animaux. Institut National de la Recherche Agronomique, Paris, 1966.

17. J. S. Alabaster and F. J. Durbin, Salmon Res. Trust Ireland, Rep. Statement Accounts, 1964, pp. 38-39.

18. H. O. Hodgins and G. J. Ridgway, Nature **210**: 1336 (1964).

19. B. Chown and M. Lewis, Can. Med. Ass. J. **55**: 66 (1946).

20. C. J. Stormont, Ann. N. Y. Acad. Sci. **97**: 251 (1962).

21. W. C. Boyd, "Genetics and the Races of Man." p. 453, Little, Brown and Company, Boston, 1950.

22. A. E. Mourant, p. 438, C. C. Thomas Publ. Springfield, Ill., 1954.

23. G. J. Ridgway and F. M. Utter, Int. N. Pac. Fish. Comm. Annu. Rep. **1963**: 149 (1964).

24. F. M. Utter and W. E. Ames, Int. N. Pac. Fish. Comm. Annu. Rep. **1964**: 111 (1966).

25. W. L. Hartman and R. F. Raleigh, J. Fish. Res. Board Can. **21**: 485 (1964).

26. R. R. Straty, U. S. Fish Wildlife Serv. Fish. Bull. **65**: 461 (1966).

COMMENTS

MR. VON LIMBACH: In developing the titer and isoimmune reaction did you use artificially elevated temperatures or just the normally appearing 15° C. water?

DR. RIDGWAY: We used only the 15° water available at Hagerman. This is about optimal for rainbow trout.

MR. VON LIMBACH: You couldn't take it up much higher, I don't think, with any safety?

DR. RIDGWAY: No, 15° is optimal for the production of antibodies and for growth; however, it is not optimal for breeding. This is another important problem that we encountered.

DR. GOLDBERG: What differences in antigenicity did you find in spawning-fish versus non-spawning or feeding-fish versus non-feeding? What are the effects of these parameters, if any?

DR. RIDGWAY: I don't think there is any effect on the antigenic makeup of the animal with respect to these factors. We have recently found a complicated situation in herring which is similar to a phenomenon also found in humans. Individuals who were anemic had a large proportion of immature red cells in their circulation and this resulted in some differences in the results of blood typing. In herring we found whole populations that were subjected to physiological stresses which changed not the antigenicity, but the way that this antigenicity is expressed in terms of an agglutination reaction.

DR. TARR: As you know, there's a very considerable amount of work being done on polyacrylamide and starch gel electrophoresis of muscle myogens. Would you say that these methods are as sensitive as immunological methods? We have never been able to distinguish races of salmon in the Fraser River by these techniques. You can take 5-600 sockeye salmon from different areas; they all have the same electrophoretic patterns. When you study the races of these fishes in various rivers, these differences don't show up. I think the only hope from the point of view of the chemist is isolating the individual proteins and making peptide maps and possibly isolating individual peptides and then determining amino acid sequences which is very laborious. Would you say the immunological methods are more sensitive?

DR. RIDGWAY: I think where you're using muscle myogen patterns, that the problem is the great diversity of different kinds of proteins involved. A paper has been submitted to the Journal of the Fisheries Research Board of Canada, on lactic dehydrogenase isozymes of sockeye salmon and their variations in populations. H. O. Hodgins, W. E. Ames, and F. M. Utter, J. Fish. Res. Bd. Canada **26**: 15 (1969). I expect that such variations in enzymes will reveal differences among races of sockeye salmon.

DR. GOLDBERG: One of the problems, I think, in blood typing or tissue typing in terms of protein patterns is that you may get electrophoretic identities which do not necessarily mean that you have the same proteins in these different races. Now I think to increase the sensitivity all one has to do, aside from your 50 year study of protein purification and sequencing, is to combine your immunological testing with electrophoresis. This will give you another order of magnitude of sensitivity.

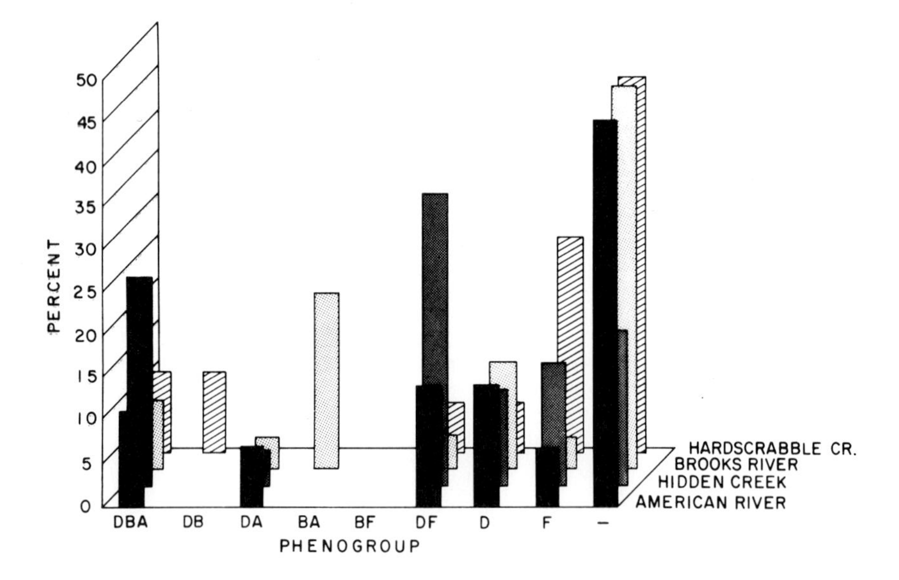

Fig. 1: Statistical relationships among the reactions of 11 rainbow trout isoimmune sera with blood cells from 1500 sockeye salmon from the Copper River, Alaska.

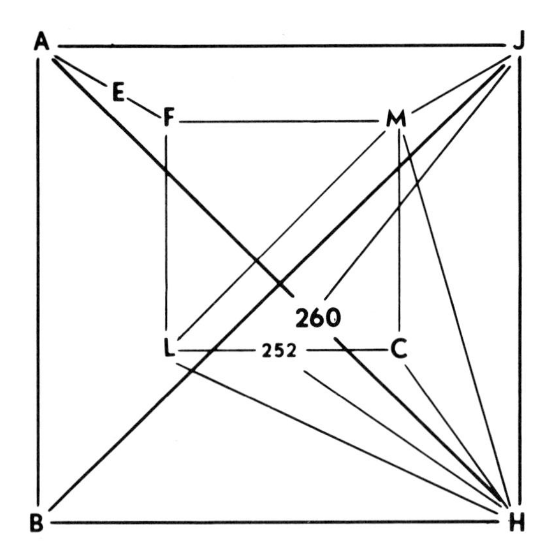

Fig. 2: Frequencies of B-system phenogroups in sockeye salmon of tributary areas of the Naknek River, Alaska.

ONTOGENY OF LACTATE DEHYDROGENASE ISOZYMES IN TROUT

Erwin Goldberg

INTRODUCTION

A central problem in biology today is the elucidation of the molecular nature of the mechanism(s) for regulating gene function. Within a complex metazoan there are many different cell types, each of which represents the product of differentiative mechanisms transforming the zygote into the adult organism. The characteristic features of adult cells are ultimately attributable to their protein and more specifically, enzyme complement in terms of types of enzymes present and in the molecular heterogeneity of a particular enzyme. Since genes for all of the proteins found in an organism are present in each of the cell types, according to current biological dogma, it is of considerable interest to study their differential expression to gain insight into the factors controlling this expression (or cell phenotype).

PROBLEMS OF GENE EXPRESSION

A meaningful way to approach the overall problem of gene expression is to examine qualitative and quantitative aspects of a specific protein during the development of an organism. The enzyme lactate dehydrogenase (EC 1.1.1.27; L-lactate:NAD oxidoreductase) is particularly useful for such a study, because it exists in multiple molecular forms, or isozymes. These isozymes generally are found in species and tissue-specific patterns. In addition, the relative proportions of the isozymes change in accordance with the state of differentiation of the tissue, the type of adult tissue, and the tissue environment. This form of differential gene expression may endow an organism with a precise and refined control over metabolic events, as enzymatic requirements of different tissues fluctuate under varying metabolic stresses. It seems probable that both genetic and epigenetic factors regulate the expression of this enzyme in terms of its isozyme complement.

LACTATE DEHYDROGENASE

Isozymes. The molecular basis for LDH isozymes is now well established. LDH is a tetramer and in most vertebrate species there are generally five forms of this enzyme which arise from combinations of two distinct subunit types. The polypeptide subunits represent the primary products of a pair of co-dominant alleles at each of two loci and combine as a tetramer to form the enzymatically active protein. Characteristically, adult heart muscle synthesizes predominantly the B (or H) subunit and contains, therefore, an abundance of LDH-1, while most skeletal muscles produce primarily the A (or M) subunit resulting in a preponderance of LDH-5.

When the subunits differ in net charge it is possible to separate, by electrophoretic techniques, five forms of LDH which have the following compositions: B_4, B_3A, B_2A_2, BA_3, A_4. Most of the evidence on LDH tetramer formation, especially in mammals, supports a mechanism involving random association of the subunits produced as a consequence of the activity of two genetic loci. Therefore, a cell which synthesizes A and B polypeptides in approximately equal amounts will be expected to contain the LDH isozymes in the binomial proportions of 1:4:6:4:1 (1). Deviations from this result can often be explained in terms of greater activity at the A or B locus depending upon whether the LDH-5 or LDH-1 end of the isozyme spectrum predominates. On the other hand, there is a growing body of evidence to indicate that in fish (2) and amphibians (3-5) restrictions on tetramer assembly may occur.

LDH of Fish. The LDH isozymes in fish have proven to be particularly interesting and useful objects to study. Markert and Faulhaber (2) in their investigation of some 30 species of fish found one major isozyme system of A-B subunit-containing tetramers in all fish, as well as two minor systems restricted to eyes and to gonads in some species. Subsequent work by a number of investigators has revealed that many species of fish have more than two loci coding for electrophoretically distinct types of LDH.

GENETIC STUDIES USING SPLAKE TROUT

Synthesis of LDH in brook trout (*Salvelinus fontinalis*) and lake trout (*S. namaycush*) is controlled by at least five structural gene loci (6, 7). Two of these cistrons are responsible for the production of A and B polypeptide subunits (6) which combine to form the tetrameric LDH isozymes (1, 8, 9). Lake trout and brook trout may be readily crossed to produce the fertile hybrid, the splake trout (10, 11). The A subunit from lake trout migrates more rapidly toward the anode during electrophoresis than that of the brook trout while the B subunits are not distinguishable by net charge. Therefore, the LDH genotype for lake trout may be designated $\frac{\text{LDA LDB}}{\text{LDA LDB}}$ and for brook trout $\frac{\text{LDA}' \text{LDB}}{\text{LDA}' \text{LDB}}$. Since transmission of the A and B genes is presumably autosomal (6, 7), F_1 splake have the LDH genotype, $\frac{\text{LDA LDB}}{\text{LDA}' \text{LDB}}$, with A and A$'$ acting as co-dominant alleles. These hybrids contain, in effect, three classes of electrophoretically distinguishable polypeptide subunits coded at the A and B loci (i.e., A, A$'$, and B), so that there are 15 isozymes which can be formed in this system (Fig. 1). Only 9 isozymes are demonstrable in the splake because LDH-1, 3, and 5 of lake trout have the same electrophoretic mobility as LDH-1, 2, and 3, respectively, of brook trout (6).

Since A and A$'$ subunits are distinguishable in splake trout and represent products of structural gene loci, a comparison of LDH isozyme patterns during the ontogeny of reciprocal hybrids should provide information on both differential gene activation and differential gene activity (as reflected in isozyme proportions). A number of studies have revealed that the relative proportions of A and B subunits change during development. In the mouse, for example, most embryonic tissues initially contain principally LDH-5; as development proceeds there is a gradual transposition toward the LDH-1 end of the spectrum (12). Conversely, in the chick embryo, LDH-1 predominates, with the shift toward LDH-5 occurring in most organs during development (13, 14). The adult pattern finally achieved, of course, is tissue- and organ-specific.

ONTOGENIC STUDIES USING LDH ISOZYMES

Parental influences on LDH in early development of amphibians were studied by Wright and Moyer (3, 4) in hybrid embryos containing variant genes for this enzyme. Hitzeroth *et al.* (15) were able to show that there was an asynchronous activation of the paternal genome coding for certain LDH subunits as well as for alcohol dehydrogenase in rainbow trout x brown trout larvae.

In this article I will briefly outline the ontogeny of LDH isozyme patterns involving A and B subunits of lake trout and brook trout hybrids. Reciprocal crosses were used to relate paternal gene activation with developmental stage in the hybrid embryos. Embryonic stages in lake trout and brook trout have been described according to external anatomical features evident in preserved embryos (16, 17). This description was followed in correlating isozyme changes with embryonic development.

Brook Trout. In the unfertilized egg of the brook trout and immediately after fertilization, LDH-1, 2, and 3 predominated and were of approximately equal relative activity in terms of amount of formazan deposition on electropherograms. LDH-4 and 5 were barely detectable on gels as very faint bands. Densitometric analysis confirmed this. The relative proportions of A and B subunits were calculated by assuming a tetrameric structure for LDH as depicted in Fig. 1. At this stage of development the ratio of B subunits to A subunits was approximately 3:1. Therefore, A subunits were available in relatively small quantity for tetramer assembly. By day 16 post-fertilization (stage 6, three quarters epiboly in which the germ ring has overgrown three-quarters of yolk; somites differentiating) the ratio has decreased to 2:1. This is reflected by an increase in LDH-4 and 5. On day 21 after fertilization (stage 8, blastopore closed; segmenting trunk ending in node) the 5 isozyme bands appeared to be of approximately equal activity.

Lake Trout. LDH extracts from lake trout ova contained LDH-1, 2, 3, and 4 with a ratio of B to A subunits of approximately 2:1. This pattern changed only slightly by day 24 after fertilization (stage 9, trunk free of sac; segmenting trunk has grown beyond and over sac, beginning to coil) at which time 5 formazan bands could be detected. The LDH-5 zone, however, was relatively weak. The lake trout (LT) x brook trout (BT) reciprocal hybrids had LDH isozyme compositions which initially reflected the typical pattern expected from the above descriptions of the maternal parental types.

Splake. In the LT ♀ x BT ♂ cross the A′ allele was contributed to the genome of the offspring by the male parent. This subunit of LDH was first detected at 29 days of development (Fig. 2) in embryos obtained from 2 separate sources in each of two different breeding seasons. According to Garside's table of development (16), these embryos are between stages 9 (day 25) and 10 (day 33 and referred to as fin-fold in which the fin-fold is evident in the mid-dorsal line). Examination of preserved embryos from the same group indicated that they had not reached stage 10 at day 26. The isozyme pattern on day 26 indicated that LDH-1, 2 and 3 predominate and are of approximately equal activity. There is somewhat less LDH 4, and the LDH-5 band is just detectable on the gels, indicating that B subunit synthesis predominates in early embryonic stages. Synthesis of the A′ allele was revealed by the appearance of LDH isozymes which migrated more slowly toward the anode during electrophoresis than lake trout LDH-5. On day 29, lake trout LDH-1, 2 and 3 were still the most abundant forms of the enzyme. The typical hybrid pattern, however, became more obvious as development progressed (Fig. 3).

Brook trout embryos incubated at about 5°C reach stage 9 in 25 days and stage 10 in 32 days (17). In these hybrids, *i.e.*, BT ♀ x LT ♂, the splake LDH isozyme pattern appeared on day 32 (Fig. 3) and examination of preserved embryos revealed that they were indeed at stage 10.

The brook trout maternal isozyme pattern predominated in these reciprocal hybrids at the time of paternal gene activation as indicated by wide-spaced heavy formazan bands with intercalated lighter zones of enzyme activity. Even though these bands increase in intensity the pattern of alternating dark and light formazan zones persist at least through day 60 post-fertilization (Fig. 3). By hatching, however, isozyme patterns in reciprocal hybrids are indistinguishable.

The LDH isozyme patterns suggest that both A and B subunits of maternal origin are involved in tetramer assembly during early development of hybrid trout embryos. Whether synthesis of these subunits during embryogenesis is mediated through transcription of the genome or translation of stable mRNA is not known. However, it is probable that chemical events during embryonic development in these organisms is comparable to that in sea urchins and amphibians: species in which it has been demonstrated that protein synthesis during cleavage stages until gastrulation is a result of the activation of pre-existing mRNA molecules rather than *de novo* synthesis of new messages (cf. 18). The change in maternal LDH isozyme patterns after gastrulation has begun in trout embryos (stage 3) may reflect a specificity indicative of transcription rather than translation. The LDH isozyme composition of unfertilized and cleavage embryos reveals that more B subunits than A subunits have been synthesized prior to or during these stages. It is possible that this synthesis is mediated by preformed (during oogenesis?) stable mRNA. Then, at gastrulation, the increase in proportion of LDH-5 suggests that new mRNA synthesis takes place providing more A subunits for tetramer assembly at a time when many biosyntheses are accelerated in embryos. This pattern of predominance of B subunits in early embryonic stages of lake and brook trout, with a gradual increase in the activity of A-containing isozymes as development proceeds is similar to LDH ontogeny in frog (3, 4, 19) and chick embryos (14). The reverse pattern of isozyme formation in the mouse embryo has been described by Markert and Ursprung (12).

At a specific and relatively late stage during morphogenesis the paternal contribution to the LDH subunit pool becomes apparent, at least as far as the A (or A$'$) polypeptide is concerned. The possibility that synthesis of the paternal B subunit is activated at an earlier or later stage cannot be excluded. However, this question can be examined in intraspecific crosses of brook trout that exhibit polymorphism at this locus (6). In any case, production of both A and B subunits of paternal origin, assuming that the only contribution to the embryo by the male is nuclear, must involve synthesis of new messages rather than merely unmasking of pre-existing mRNA.

The present results showing specific timing of paternal gene activation are in accord with the findings of Wright and Moyer (3, 4) who studied LDH ontogeny in reciprocal hybrids of *Rana sp*. as well as in crosses between *R. p. pipiens* x *R. p. sphenocephala*, the latter with a variant B subunit. These investigators showed that the maternal LDH pattern was maintained until stage 19 (heart beat) and in addition, contributed most of the enzyme activity at least through stage 23 (hatching at stage 20). In addition, their results (4) showed that LDH synthesized during oogenesis persisted for at least 11 days after larvae began feeding and 15 days after the embryo began to synthesize its own LDH. The present findings also suggest the occurrence of such a stable store of "spare enzyme" in trout embryos, especially in the BT ♀ x LT♂ hybrids which show a dominance of the maternal pattern after paternal A subunit synthesis begins.

TOTAL LDH ACTIVITY IN HYBRIDS

It was of interest to determine if synthesis of the A or A′ LDH subunit would be reflected in a significant increase in LDH activity. Therefore the total LDH concentration, as indicated by enzyme specific activity, was measured in embryos from fertilization through hatching. These data (Fig. 4) reveal that B x L hybrids maintain a virtually constant level of LDH for approximately 48 days of embryogenesis. The reciprocal hybrids (L x B), on the other hand, show a distinct and continuous increase in enzyme concentration commencing about day 23 (stage 9, trunk free of sac, segmenting trunk has grown beyond and over sac, beginning to coil). An increase in LDH specific activity of B x L embryos is evident by stage 15 (complete eye pigment). No clearly defined relationship between onset of paternal LDH synthesis and total LDH activity is apparent from these data. Of considerable interest, however, is the marked difference in both initial enzyme concentration and rate of increase of specific activity in reciprocal hybrids.

REGULATION OF GENE OUTPUT

By definition, reciprocal hybrids should have identical genomes. Therefore, the obviously significant difference in LDH specific activity suggests a maternal effect in the regulation of gene output. This suggests that phenotypic differences are a consequence of cytoplasmic factors acting on the nucleus. It may be argued that the LDH level of embryos reflects translation of maternally transcribed mRNA during oogenesis and different amounts are available depending upon the species of the maternal parent. If this were the case, one would expect a cytoplasmic effect to disappear during morphogenesis with activation of the paternal genome as well as with the development of well-defined organ systems that, in the adult at least, will maintain their own characteristic levels of this enzyme (20).

If a cytoplasmic component does indeed regulate gene output, the fact that this regulation persists during development suggests that either the regulator must be replicated during cell division or that it becomes distributed throughout the organism during embryogenesis. The latter possibility seems most likely, with the yolk as the most probable source of this regulator. In fact, from the data in Fig. 4, it is apparent that in newly hatched, *i.e.*, yolk sac larvae, the maternal effect is present and then disappears when the yolk sac is resorbed. It is probable that by this stage of development, especially with onset of feeding by the larvae, any maternal effect would be obscured in the analysis of homogenates of whole fish.

SUMMARY

LDH ontogeny in trout embryos, then, involves a dual control mechanism. First, there must be a highly allele specific process associated with switching on the appropriate gene. Second, a cytoplasmic-mediated modulation of amount of gene product occurs. Further study of this system in trout may provide information on the important biological question of whether control is achieved through regulator genes of nuclear-cytoplasmic feedback, similar to bacterial systems.

ACKNOWLEDGMENTS

This work was supported by Grant GB 7271 from the National Science Foundation. The cooperation of J. P. Cuerrier and J. C. Ward, Canadian Wildlife Service, is sincerely appreciated.

REFERENCES

1. C. L. Markert, Science **140**: 1329 (1963).
2. C. L. Markert and I. Faulhaber, J. Exp. Zool. **159**: 319 (1965).
3. D. A. Wright and F. H. Moyer, J. Exp. Zool. **163**: 215 (1966).
4. D. A. Wright and F. H. Moyer, J. Exp. Zool. **167**: 197 (1968).
5. E. Goldberg and T. Wuntch, J. Exp. Zool. **165**: 101 (1967).
6. E. Goldberg, Science **151**: 1091 (1966).
7. W. J. Morrison and J. E. Wright, J. Exp. Zool. **163**: 259 (1966).
8. C. L. Markert, *In* "Cytodifferentiation and Macromolecular Synthesis" (Michael Locke, ed.), p. 65, Academic Press, New York, 1963.
9. C. Shaw and E. Barto, Proc. Nat. Acad. Sci. U. S. **50**: 211 (1963).
10. J. E. Stenton, Can. Fish. Cult. no. 6: 2 (1950).
11. K. Buss and J. E. Wright, Trans. Amer. Fish. Soc. **87**: 172 (1957).
12. C. L. Markert and H. Ursprung, Devel. Biol. **5**: 363 (1962).
13. R. D. Cahn, N. O. Kaplan, L. Levine, and E. Zwilling, Science **136**: 962 (1962).
14. D. T. Lindsay, J. Exp. Zool. **152**: 75 (1963).
15. H. Hitzeroth, J. Klose, S. Ohno and U. Wolf, Biochem. Genetics **1**: 287

(1968).

16. E. T. Garside, Can. J. Zool. **37**: 689 (1959).

17. E. T. Garside, J. Fisheries Res. Board Can. **23**: 1121 (1966).

18. A. Tyler, *In* "Control Mechanisms in Developmental Processes" (Michael Locke, ed.), p. 170, Academic Press, New York 1967.

19. F. H. Moyer, Connie B. Speaker and David A. Wright, Ann. N. Y. Acad. Sci. **151**: 650 (1968).

20. E. Goldberg, unpublished data.

COMMENTS

DR. MALINS: Studies with goldfish showed that the LDH components are very much dependent upon environmental temperature; I note that you didn't give this factor much credence. Perhaps you are not taking into account that the work of Hochachka who found significant differences in the isozyme patterns of the liver of warm- and cold-adapted goldfish. Wouldn't this factor be critical in your studies?

DR. GOLDBERG: Some work has been done on temperature effects on LDH in goldfish; we have also done some using trout. Temperature leads to differences in enzyme levels but not in the isozyme pattern. We have studied gene activation and enzyme levels in embryos developing at 3, 8, 12, and even some at 15°C. The only thing that changes is the time at which the gene becomes activated, but this correlates with the developmental stage. At the lower temperatures, development is slowed. If you look specifically at the embryology and correlate this with isozyme composition, you find that you are at the same stage.

DR. IDLER: Dr. Tarr referred to the difficulties that had been experienced at his laboratory with identifying salmonids by means of protein electrophoretic patterns (presumably myogens). Dr. Odense of my laboratory has found that lactate dehydrogenase isozymes, for example, work well in herring but not in mackerel. I wonder if you might like to comment on what particular tissue or isozyme is particularly useful for salmonids.

DR. GOLDBERG: First of all with reference to LDH, you must check all tissues. For example, there is a tissue specific locus in the retina, more precisely the neural retina. You also have to look at skeletal muscle because in some salmonids, there are two LDH loci that are active only in skeletal muscle. Therefore, you must look at different tissues in a given species to find out how many loci you're dealing with. If one finds only two, things are much less complex until we know more about other tissue specific loci. Then you can worry about only two kinds of subunits, two genes, a possibility of two variants, one at each of these loci. Brain tissue is the best for the study of the two (A and B) genes of LDH in trout. This tissue gives the best resolution and the levels of activity are just about right with the least interference in electrophoretic migration.

DR. HUNN: Studies have been conducted in Sweden on the esterases in salmonids and salmonid crosses. Extensive work has also been carried out on the esterases of various tissues and I think that these are a very nice complement to the LDH work.

DR. BOUCK: We have studied the LDH isozymes in trout blood by a different electrophoretic technique than that used by Dr. Goldberg, in an attempt to capitalize on work that is usually done in clinical chemistry. Our hope was to show tissue damage and to pinpoint whatever tissue or organ was involved. In our studies we have observed as many as 15 LDH isozymes. These patterns were obtained in brook, brown, lake, and rainbow trout as well as with crosses of brown-brook trout and lake-brook trout (splake). With the splake we found at least 28 or 29 isozymes in the LDH pattern. These results appear to abolish the value of using the isozyme pattern in fish blood to diagnose specific tissue damage. In another situation we also studied a variety of tissues from rainbow trout. In muscle, we found 10 isozymes, 5 had a very intense activity and 5 had low activity. This was true for each tissue tested; always 5 of intense and 5 of minor activity.

DR. GOLDBERG: As I've already noted, there may be tissue specific LDH subunit loci; perhaps as many as eight. This presumably is a consequence of tetraploid ancestry in the evolution of the salmonids. I discussed the 5 "ubiquitous" isozymes, i.e., common to all tissues and perhaps homologous to mammalian LDH. There are an additional 5 forms of LDH restricted to skeletal muscle. The neural retina has at least one unique LDH, as does kidney. Whether or not the subunits of these isozymes represent primary gene products remains to be established. In any case, one could resolve from 5 to 20, or more, forms of LDH in trout depending upon the combination of tissues studied. Then you may get epigenetic modification of the subunits.

DR. BOUCK: My point is that we must resolve the question of isozyme multiplicity among trout. Can we build an accurate picture of LDH genetics when trout have 5 LDH isozymes in one case and 15 LDH isozymes in another case?

DR. GOLDBERG: Yes, we must identify and characterize each particular set of isozymes in order to determine the number of gene loci with which we are dealing. Then, we have to take into account the possibility of variant subunits at each locus.

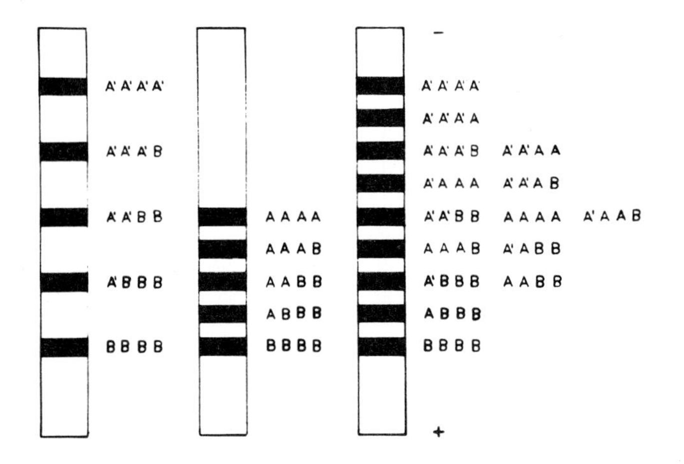

Fig. 1: Diagrammatic representation of electropherograms showing LDH isozymes in Brook Trout, Lake Trout, and Splake (from left to right). The coincident mobility of certain subunit combinations is readily apparent.

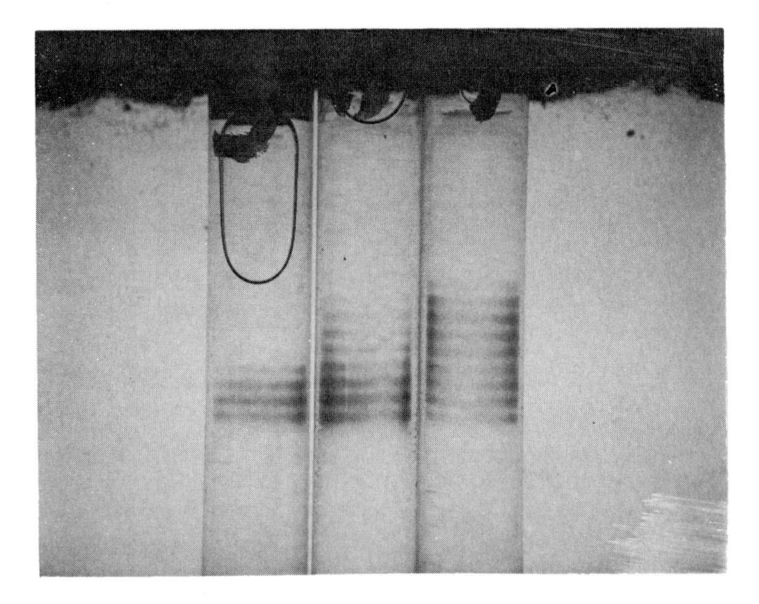

Fig. 2: Electropherogram showing ontogeny of LDH isozymes in LT ♀ x BT ♂ hybrids. From left to right, the gels are from embryos on days 26, 29, and 60, respectively, of development. Orientation of gels as in Fig. 1.

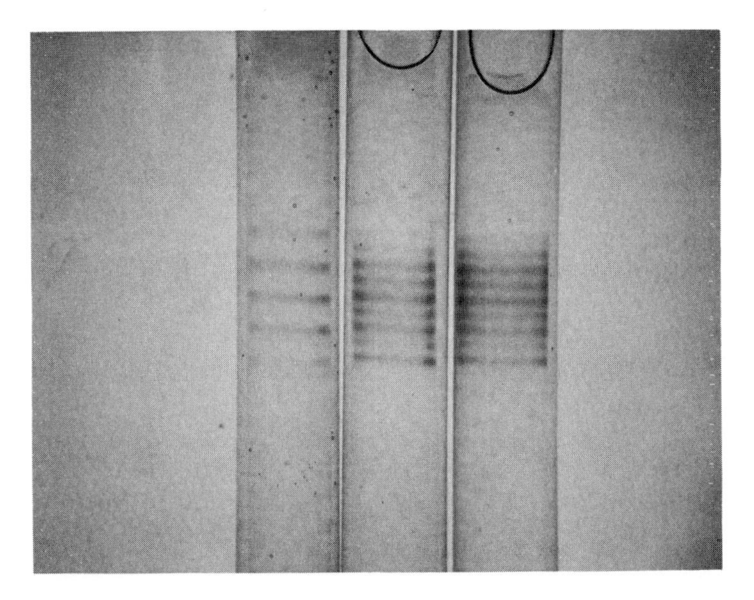

Fig. 3: Electropherograms showing ontogeny of LDH isozymes in BT ♀ x LT ♂ hybrids. From left to right, the gels are from embryos on days 26, 32, and 57 of development. Orientation of gels as in Fig. 1.

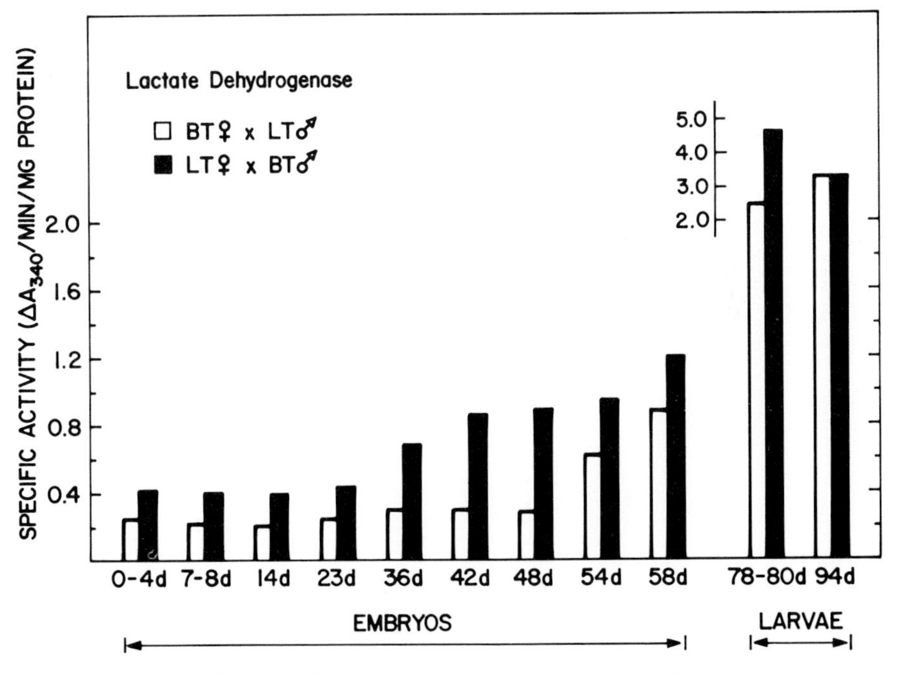

Fig. 4: Amount of lactate dehydrogenase in reciprocal hybrids of LT x BT at several stages during embryonic development through hatching.

TOPICS IN NUTRITION

J. E. Halver, Chairman

VITAMIN REQUIREMENTS

John E. Halver

INTRODUCTION

The foundation for vitamin research with fish was an adequate test diet with positive experimental control over those basic nutrients to be tested. The pursuit of this test diet lasted for nearly 25 years and included scientists from leading research centers throughout the United States. Serious research with fish as the experimental animal began in the mid-twenties at Cornell University, University of Wisconsin and the University of Washington. McCay and Dilly from Cornell worked to identify the anti-anemic factor H present in fresh meat which was necessary for the diet of trout whenever fish were held more than 8-12 weeks on the best test diets then available in animal nutrition (1). Fish became anemic and died forcing termination of nutritional experiments and confining investigations to short term studies to test and describe deficiency syndromes which would appear within a few weeks on particular diet treatments. The first laboratory constructed and programed for fundamental fish nutrition research was built in 1932 and the record shows that A. V. Tunison began the first diet trial on August 24 in a cooperative effort by Cornell University, the New York State Conservation Department and the U. S. Bureau of Fisheries (2). The reports on nutritional requirements of trout issued by this research laboratory have formed the principal source of information for the past 30 years on qualitative and quantitative vitamin requirements of brook trout (*Salvelinus fontinalis*), brown trout (*Salmo trutta*) and rainbow trout (*Salmo gairdneri*).

VITAMIN TEST DIETS

One of the first reports of a specific vitamin deficiency syndrome in fish occurred in 1941 when Schneberger reported paralysis in rainbow trout fed carp at the Thunder River Hatchery in Wisconsin (3). The symptoms could be reduced by injecting crystalline thiamine into individual animals or by feeding dried brewers yeast to the population. Louis Wolf at the Rome, New York, Experimental Hatchery also reported in 1942 on the occurrence of *fish diet disease* which confirmed the fact that thiaminase present in fresh fish tissue would hydrolyze thiamine in test or commercial diet meat-meal mixtures and fish fed these rations would shortly develop a thiamine deficiency syndrome (4). These reports stimulated activity for quantitative requirement studies for the water-soluble vitamins at the fish nutrition laboratory at Cortland, New York. Tunison, Brockway and McCay in 1942 measured levels of thiamine, riboflavin and niacin in the liver, pyloric caeca, kidney and muscle to determine a base line for subsequent experiments to measure requirements for these vitamins (5). Microbiological assay techniques revealed differences between groups of fish fed different diet treatments and over the next five years tentative requirements of rainbow, brook and brown trout for thiamine, riboflavin, pyridoxine, pantothenic acid, biotin and folic acid were determined by feeding groups of young trout different meat-meal mixtures which showed by assay different levels of the particular test vitamin present, and then, after different test time, the livers of each group on each diet treatment were assayed to determine point of maximum, minimum and intermediate vitamin level stored in the organ (6-10). Barbara McLaren and others working with Conrad Elvehjem at Wisconsin used a slightly different approach and developed test diets containing crab meal or dried liver as the source of the anti-anemic factor (11) and were able to report tentative quantitative requirements for thiamine, riboflavin, pyridoxine, pantothenic acid, inositol, biotin, folic acid, choline and niacin based on growth response, food conversion and assay of level stored in the liver at termination of the diet trial (12). No adequate vitamin test diet had yet been developed by this date however, and obviously this missing tool needed to be formed before specific roles of these vitamins in metabolism and growth could be defined.

Xanthopterin was first thought to be the fish anti-anemic factor on the basis of work in 1941 by Simmons and Norris with young chinook salmon (*Oncorhynchus tshawytscha*) held at the University of Washington Fisheries Laboratory (13). Two years later Tunison and McCay thought riboflavin, pyridoxine and pantothenic acid would regenerate anemic brook trout and could be part of McCay's factor H (14). Three years later in 1946, however, Phillips and Brockway could not repeat the riboflavin, pyridoxine, pantothenic acid response even when folic acid was added to the vitamin mixture (8). They could only cure the anemia when dehydrated liver powder was added to the diet (15). In 1949, in unpublished work carried out at the University of Washington, Norris and Halver tested xanthopterin again as the anti-anemic factor for young chinook salmon and injected this material alone, and in combination with folic acid and vitamin B_{12}, into 24 week anemic chinook salmon fingerlings. Some response was obtained from xanthopterin, and some with folic acid, but a dramatic stimulation of erythropoiesis was obtained when the combination of folic acid and vitamin B_{12} was used. The following year these preliminary experiments were confirmed and vitamin B_{12} plus folic acid in a ratio of approximately 1:100 was obviously one factor H for young chinook salmon. That winter Louis Wolf developed a test diet for rainbow trout which contained commercial casein, gelatin, potato starch, Crisco, *alpha*-cellulose flour, minerals, cod liver oil and crystalline vitamins which could be used to induce certain specific vitamin deficiency syndromes in these trout (16). During the summer of 1951 this crude test diet was improved by changing to vitamin-free casein, purified gelatin, white dextrin for potato starch, corn oil for Crisco and a simplified mineral mix. Chinook salmon grew as well on this diet as on Wolf's test diet (17). Later this original complete test diet was improved, Table I, and used for short term feeding studies in coho salmon (*O. kisutch*)

TABLE I

WATER SOLUBLE VITAMIN TEST DIET*

Complete Test Diet		Vitamin Mix		Mineral Mix	
Vitamin-free casein	38 g	Thiamine·HCl	5 mg	USP XII No. 2	plus
Gelatin	12	Riboflavin	20	$AlCl_3$	15 mg
Corn oil	7	Pyridoxine·HCl	5	$AnSO_4$	300
Cod liver oil	2	Choline chloride	500	CuCl	10
White dextrin	28	Nicotinic acid	75	$MnSO_4$	80
Alpha-cellulose mixture	9	Ca-pantothenate	50	KI	15
Alpha-cellulose 8		Inositol	200	$CoCl_2$	100
Vitamins 1		Biotin	0.5	per 100 g of	
9		Folic acid	1.5	salt mixture	
Mineral mix	4	L-ascorbic acid	100		
Water	200	Vitamin B_{12}	0.01		
Total diet as fed	300	Menadione (K)	4		
		Alpha-tocopherol acetate (E)	40		

Note: Delete 2 parts alpha-cellulose and add 2 parts CMC for preliminary feeding.
Dissolve alpha-tocopherol in oil mix.
Add vitamin B_{12} in water during final mixing.

Diet preparation: Dissolve gelatin in cold water. Heat with stirring on water bath to 80°C. Remove from heat. Add with stirring–dextrin, casein, minerals, oils and vitamins as temperature decreases. Mix well to 40°C. Pour into containers; move to refrigerator to harden; remove from trays and store in sealed containers in refrigerator until used. Consistency of diet adjusted by amount of water in final mix and length and strength of beating.

*Modified from (20, 21, 24).

and sockeye salmon (*O. nerka*) and for long term feeding studies through at least three reproductive cycles with rainbow trout in 8°, 10°, 15°, or 17°C water systems (18-20). The tool to test for specific qualitative and quantitative vitamin requirements of salmonids had been formed. Application was rapid with dramatic results. Specific deficiency syndromes occurred whenever one of the required vitamins was deleted from the complete test diet.

THIAMINE

Typical thiamine deficiency symptoms appeared in chinook salmon four to six weeks after start of feeding on the deficient diet, Fig. 1. Body thiamine stores soon became exhausted and fish began to die at a rapid rate after eight weeks on test. When 20% of the population had expired (each exhibiting a paralysis type symptom) the remaining population was split and thiamine was replaced in the diet for half the fish while the other half was allowed to continue on the deficient diet, Table II. Mortality soon ceased in

TABLE II

SPECIFIC VITAMIN DEFICIENCY SYNDROMES

Vitamin	Salmonids, Carp, Catfish	Other Animals[1,2]
Thiamine	Poor appetite; muscle atrophy; convulsions; instability and loss of equilibrium; edema; poor growth.	Poor appetite; muscle atrophy; convulsions; generalized edema; instability and loss of equilibrium; brain lesions; degeneration of peripheral nerve fibers; impaired carbohydrate metabolism.
Riboflavin	Corneal vascularization; cloudy lens, hemorrhagic eyes, photophobia; dim vision; incoordination; abnormal pigmentation of iris; striated constrictions of abdominal wall; dark coloration; poor appetite; anemia; poor growth.	Corneal vascularization; cloudy lens; hemorrhagic eyes; cataracts photophobia; dim vision; incoordination; abnormal pigmentation of iris; scleral congestion; cheilosis and angular stomatitis; impaired erythrocyte formation.
Pyridoxine	Nervous disorders; epileptiform fits; hyperirritability; ataxia; anemia; loss of appetite; edema of peritoneal cavity; colorless serous fluid; rapid post-mortem rigor mortis; rapid and gasping breathing; flexing of opercles.	Nervous disorders; epileptiform fits; hyperirritability; ataxia; anemia; loss of appetite; edema; symmetrical dermatosis; dermal lesions; iron deposition in liver; rise in serum iron.
Pantothenic acid	Clubbed gills; prostration; loss of appetite; necrosis and scarring; cellular atrophy; gill exudate; sluggishness; poor	Prostration; loss of appetite; necrosis and scarring; cellular atrophy; spectacle alopecia; generalized scaling and dermatitis;

TABLE II – continued.

Pantothenic acid (cont.)	growth	diarrhea; hyperemia of intestine; myelin degeneration of nerves.
Inositol	Poor growth; distended stomach; increased gastric emptying time; skin lesions.	Poor growth; increased gastric emptying time; alopecia; nutritional encephalomalacia.
Biotin	Loss of appetite; lesions in colon; coloration; muscle atrophy; spastic convulsions; fragmentation of erythrocytes; skin lesions; poor growth.	Loss of appetite; poor growth; muscle atrophy; seborrheic skin disease; generalized erythema; spastic gait; necrosis of fibers; increased sarcolemma.
Folic acid	Poor growth; lethargy; fragility of caudal fin; dark coloration; macrocytic anemia.	Poor growth; lethargy interposed with convulsions; megaloblastic erythropoesis; nutritional cytopenia; infarction of spleen.
Choline	Poor growth; poor food conversion; hemorrhagic kidney and intestine.	Poor growth; poor food conversion; aversion to food; fatty infiltration of liver; necrosis and scarring; reddish kidney; hemorrhagic eyes.
Nicotinic acid	Loss of appetite; lesions in colon; jerky or difficult motion; weakness; edema of stomach and colon; muscle spasms while resting; poor growth.	Loss of appetite; lesions in colon; jerky or spastic gait; weakness; pain; skin eruptions; edema of stomach and colon; anorexia; diarrhea; neurological lesions; fatty livers, vessel dilation.
Vitamin B_{12}	Poor appetite; low hemoglobin; fragmentation of erythrocytes; macrocytic anemia.	Pernicious anemia; macrocytic anemia; sprue; low or poor growth in young.
Ascorbic acid	Scoliosis; lordosis; impaired collagen formation; altered cartilage; eye lesions; hemorrhagic skin, liver, kidney, intestine, muscle.	Scurvy; capillary fragility; immature fibroblastic formation; impaired collagen and cartilage; hemorrhagic skin, intestine, muscle; eye lesions.
p-Aminobenzoic acid	No abnormal indication in growth appetite, mortality	Alopecia in rat and poor growth in chick.

[1]West and Todd, Textbook of Biochemistry, 3rd ed. Macmillan, New York, 1962.

[2]Hein, Heinz Nutritional Data, 5th ed, Heinz, Pittsburg, 1964.

that group on the recovery diet (complete test diet) and, after four weeks all survivors appeared normal. In contrast, all of the group continued on the deficient ration had expired (21). Recent work has shown injection of thiamine pyrophosphate into thiamine deficient chinook salmon, coho salmon or rainbow trout promotes recovery from gross clinical symptoms of the deficiency and erythrocyte transketolase activity is rapidly restored to near normal or control levels (22). The erythrocyte transketolase activity can thus be used to assay for thiamine status and thiamine intake of these animals, and the relationship of thiamine function in carbohydrate metabolism could thus be studied in trout or salmon held in a number of different environmental water temperatures and exposed to different diet treatments.

RIBOFLAVIN

The response of chinook salmon to the test diet deficient in riboflavin was somewhat different, Fig. 2. Growth differences became apparent after four to six weeks on test and by ten weeks on treatment, mortality began to appear (21). Monolateral and bilateral cataracts appeared in the majority of fish on test for 10-12 weeks. Although some fish survived the 16 week test period, all those on the deficient diet were severely emaciated and were blind. When the missing vitamin was replaced in the ration, recovery occurred except in those fish with cataracts where no cataract repair was noted. Fish with monolateral cataracts on recovery diet began to feed shortly after replacement of the missing vitamin in the ration and continued to grow; but at the end of the 16 week experimental period, the lens of the eye was still opaque and the fish was blind on this side. The role of riboflavin in respiration of poorly vascularized tissues is challenging and relatively unexplored. This same symptom appeared in coho salmon, sockeye salmon and rainbow trout held for 10-14 weeks on riboflavin deficient diets (18, 20, 23) and thus, a number of species of fish might be used to advantage to study the role of riboflavin in metabolism of animals held at different water temperatures and living at different metabolic rates.

PYRIDOXINE

Salmon fed diets devoid of pyridoxine rapidly developed a deficiency syndrome which included severe nervous disorders, muscle incoordination, epileptic type fits and rapid post mortem rigor mortis, Fig. 3. Recovery was equally dramatic and prompt with all fish apparently normal two weeks after replacement of the missing vitamin in the ration (21). Salmonids are tremendous protein synthesizers with dietary requirements in 10°C water of from 40-50% of protein in the ration (24, 25). Since nearly half of the ration should be protein, amino acid metabolizing enzyme systems must be very active. Preliminary experiments have shown intense activity of erythrocyte transaminase in these and other test salmonids (26). Differences in pyridoxine intake have been reflected in both erythrocyte and serum transaminase systems; hence, fish may again be used to advantage to study some of the basic enzyme systems involved in protein synthesis and metabolism, and should furnish new clues to the effect of temperature on reaction rate of tissue components.

FOLIC ACID

Folic acid anemia soon occurred in fish fed diets deficient in folic acid but adequate in vitamin B_{12}, Fig. 4. Gross anemia occurred after 10-12 weeks on the deficient ration and could be detected by observing blood circulating through the gill membranes of normal and deficient salmon or trout, Fig. 5. Abnormal erythrocytes appeared with many distorted nuclei in the population of mature and old cells, Fig. 6. Kidney imprint showed few erythroid megaloblasts and typical normochromic macrocytic anemia occurred, Fig. 7. Shortly after replacement of folic acid in the ration, the erythrocytic precursors in head kidney imprint showed more rapid division into daughter cells and many immature erythrocytes appeared in the blood stream, Fig. 8. Within six to eight weeks the anemia had disappeared and blood smears showed a normal distribution of cell type (27), Fig. 9.

ASCORBIC ACID

Early work with ascorbic acid deficient diets and chinook salmon fingerlings failed to disclose any discrete deficiency syndrome during a 20 week feeding experiment (17, 21). Kitamura *et al.* in Japan later reported lordosis and scoliosis appeared in rainbow trout, carp (*Cyprinus carpio*) and guppies (*Poecilia reticulata*) reared in fresh water for long feeding periods on ascorbic acid deficient diets (28). Nakagawa also sent photographs and roentgenograms of similar conditions in cherry salmon fed low C diets. Tunison had described a similar condition in brook trout fed formalin preserved meat for one year or longer (29). Kitamura *et al.* described twisted tails in carp raised in Japan in pure water (28) and Ikeda and Sato showed that when ^{14}C glucose was ingested by larger carp, ^{14}C ascorbic acid could be isolated from the blood and effluent waters in the aquarium in which the carp were held (30, 31) but evidently insufficient vitamin C was synthesized to satisfy the requirements of young carp held in 20°C water. Poston at Cortland was able to induce spinal deformities in brook trout reared on diets low in ascorbic acid and currently more work on vitamin C requirements for trout is underway there (32).

Recently long term experiments were initiated at the Western Fish Nutrition Laboratory with ascorbic acid deficient diets and young coho salmon fingerlings to determine the nature and extent of the specific vitamin C deficiency syndrome in salmon. After 24-30 weeks on test, the salmon fingerlings began to show a specific deficiency symptom, Fig. 10. Acute lordosis and scoliosis appeared in many of the fish fed low C diet treatments. X-rays of the spinal column of these fish indicated profound acute damage had occurred when fish were continued on this ration for more than 30 weeks, Fig. 11. An Hematoxylin and Eosin stained section of the spinal column showed vertebral displacement up to nearly 120° and indicated atrophy of the spinal cord and hemorrhage in the area of acute lordosis, Fig. 12. Mason's stain of sections of a normal gill arch showed extensive deposit of normal cartilage and supporting tissues, Fig. 13. In contrast, however, Giemsa stain of gill cartilages of vitamin C deficient coho salmon showed proliferation of some cartilage elements and deformed abnormal cartilage deposits, Fig. 14. Close examination of the tip of gill filaments of deficient fish showed twisted cartilage in many of the filaments examined, Fig. 15. This particular symptom could be detected with a hand lens and became apparent in later experiments after 12-14 weeks on test with vitamin C deficient diets, long before acute damage began to be seen in the spinal column, in the gill cover or in the eye support cartilage. The geometry of the eye was obviously affected when fish were held on low ascorbic diets for periods in excess of 30 weeks. Normal eye support cartilage is uniformly and symmetrically deposited; whereas, in vitamin C deficient fish, eye support cartilage showed hyperplasia of nuclei, areas of thickened abnormal cartilage deposit and distortion of the eye ball. At termination of the experiment, ascorbic acid content of circulating blood was determined by pooling five fish samples from each diet treatment, Table III. Replicate assays showed

TABLE III

GROWTH AND ASCORBATE CONCENTRATIONS IN BLOOD AND HEAD KIDNEY TISSUE OF TROUT AND SALMON ON VITAMIN C TEST DIETS

C Diet Treatment	Trout			Salmon		
	Avg Wt at 24 weeks	Ascorbate Conc[1]		Avg Wt at 24 weeks	Ascorbate Conc[1]	
		Blood	Kidney		Blood	Kidney
mg/100 g	g	μg/g	μg/g	g	μg/g	μg/g
0	2.4	--[2]	--[2]	5.0	22.3±2.2	89
5	9.6	34.4±2.9	125	6.0	30.5±1.2	132
10	10.6	34.6±1.3	137	5.7	35.8±1.6	265
20	10.1	38.8±3.3	132	6.1	34.2±2.3	183
40	10.2	46.8±6.2	162	6.3	33.7±2.0	225
100	10.8	51.0±4.6	247	6.0	37.8±2.3	321

(From: Halver, Ashley and Smith (33)).

[1] Average of five samples for blood (± S. D.) and two for head kidney tissue, see text.
[2] No fish available for assay.

minimum level of circulating ascorbic acid for normal growth and survival to be about 35 μg/g of blood for rainbow trout, and about 50 μg/g of blood in fish fed the complete test diet containing 100 mg of ascorbic acid/100 g of dry ration ingredients. Coho salmon in contrast, had about 35 μg of ascorbic acid/g of blood on all diets containing 10 mg or more of ascorbic acid per

100 g ration and lower levels of 20 and 30 μg/g were determined when 0 or 5 mg of ascorbic acid per 100 g diet was fed (33). Thus, the blood reflected the levels of ascorbic acid fed, similar to the different circulating levels of ascorbic acid reported in guinea pigs, monkey and man fed different levels of ascorbic acid. Fish have adrenal cortical tissue present in the head kidney and assay of levels of ascorbic acid stored in this area were determined, Table III. Rainbow trout held on rations with different levels of ascorbic acid showed differences in head kidney storage of the vitamin which varied from approximately 125 μg/g of kidney tissue in those fish fed 5 mg of ascorbic acid in the diet up to twice that level or about 250 μg/g of wet tissue in those receiving the complete test positive control diet containing 100 mg of ascorbic acid/100 g of dry ration (33). In coho salmon kidney tissues, slightly higher levels of ascorbic acid were detected. Fish receiving no ascorbic acid had only 90 μg of vitamin C/g of wet kidney tissue and step wise increases occurred on diets of 5, 10, 20 and 40 mg C/100 g diet up to over 300 μg of ascorbic acid/g of wet kidney in the control complete test diet fed fish. The data on levels of ascorbic acid in circulating blood and ascorbic acid in kidney tissue after 20 weeks on test supported and confirmed the growth data, and indicated that rainbow trout needed at least 10 mg of ascorbic acid/100 g of dry diet ingredients for normal growth in 15°C water. Coho salmon did not, in 20 weeks, develop acute deficiency syndromes in those groups receiving 5 mg or more of ascorbic acid/100 g diet, but tissue assay indicated at least 5 mg of the vitamin per 100 g ration would be needed under these experimental conditions for normal tissue reserves and absence of biochemical deficiency symptoms (33).

Wound repair experiments were initiated at the end of the 20 week feeding trial and results supported the above requirement data. Fish repaired wounds except for those receiving diets devoid of ascorbic acid. Rainbow trout receiving rations containing only 5 or 10 mg of the vitamin per 100 g diet showed slow repair of both abdominal and intramuscular wounds; whereas, slow or intermediate repair of similar wounds in coho salmon was observed on diets of 0 or 5 mg of ascorbic acid. The general rate of wound repair was proportional to the ascorbic acid intake, was more rapid in salmon than in trout for the same dietary titre of L-ascorbic acid, and supports the conclusion that the requirement for ascorbic acid in rainbow trout is approximately twice that amount needed for coho salmon (33). About 10 mg of L-ascorbic acid per 100 g dry diet ingredients would be necessary to maintain normal growth, diet efficiency and tissue reserves for either group of fishes. Proline hydroxylase activity was not measured but obviously impaired formation of the collagen base membrane necessary for normal cartilage development was involved and thus salmonids offer an interesting and challenging model system for exploration of wound repair, cartilage and bone repair and other basic growth processes in young animals.

OTHER VITAMINS

Space does not permit more review or discussion on challenging problems possible for exploration in the area of choline deficiency for impaired digestion, food utilization and fat metabolism studies; in the use of niacin deficient fish to study carbohydrate metabolism, hydrogen transfer systems and muscle contraction; or the role of inositol required by salmonids and necessary for cell membrane metabolism and membrane permeability; for pantothenic acid deficiency and its manifestation of hyperplasia of the gill filament membrane and death by respiratory edema or asphyxiation; or in the study with fish of metabolism of the required fat soluble vitamins A, E and K in vision, cell membrane integrity, polyunsaturated fatty acid oxidation or blood coagulation processes. This brief review of five vitamins with emphasis on only three for salmonids should stimulate interest and demonstrate how fish and fish tissue systems may be used with particular advantage to solve specific problems in basic animal nutrition in the field of vitamin requirements, and function of these compounds in metabolism and physiology of different animals of various age groups held under widely different diet or environmental treatments.

REFERENCES

1. C. M. McCay and W. E. Dilley, Trans. Am. Fisheries Soc. **57**: 250 (1927).

2. C. M. McCay, N. Y. Cons. Dept. Fish. Res. Bull. **1**: 13 (1933).

3. E. Schneberger, Progressive Fish Culturist **56**: 14 (1941).

4. L. W. Wolf, N. Y. Cons. Dept. Fish. Res. Bull. **2**: (1942).

5. A. V. Tunison, D. R. Brockway, J. M. Maxwell, A. L. Dorr and C. M. McCay, N. Y. Cons. Dept. Fish. Res. Bull. **4**: 52pp (1942).

6. A. V. Tunison, A. M. Phillips, H. B. Shaffer, J. M. Maxwell, D. R. Brockway and C. M. McCay, N. Y. Cons. Dept. Fish. Res. Bull. **6**: 21pp (1944).

7. A. M. Phillips, A. V. Tunison, H. B. Shaffer, G. K. White, M. W. Sullivan, C. Vincent, D. R. Brockway and C. M. McCay, N. Y. Cons. Dept. Fish. Res. Bull. **8**: 31pp (1945).

8. A. M. Phillips, D. R. Brockway, E. O. Rodgers, M. W. Sullivan, B. Cook and J. R. Chipman, N. Y. Cons. Dept. Fish. Res. Bull. **9**: 21pp (1946).

9. A. M. Phillips, D. R. Brockway and E. O. Rodgers, N. Y. Cons. Dept. Fish. Res. Bull. **12**: 31pp (1949).

10. A. M. Phillips, D. R. Brockway, A. J. J. Kolb and D. M. Maxwell, N. Y. Cons. Dept. Fish. Res. Bull. **13**: 30pp (1949).

11. B. A. McLaren, E. Keller, D. J. O'Donnel and C. A. Elvehjem, Arch. Biochem. Biophys. **15**: 179 (1947).

12. B. A. McLaren, E. Keller, D. J. O'Donnel and C. A. Elvehjem, Arch. Biochem. Biophys. **15**: 169 (1947).

13. R. W. Simmons and E. R. Norris, J. Biol. Chem. **140**: 679 (1941).

14. A. V. Tunison, D. R. Brockway, H. B. Shaffer, J. M. Maxwell, C. M. McCay, C. E. Palm and D. A. Webster, N. Y. Cons. Dept. Fish. Res. Bull. **5**: 26pp (1943).

15. A. M. Phillips, D. R. Brockway, E. O. Rodgers, R. L. Robertson, H. Goodsell, J. A. Thompson and H. Willoughby, N. Y. Cons. Dept. Fish. Res. Bull. **10**: 35pp (1947).

16. L. E. Wolf, Progressive Fish Culturist **13**: 17 (1951).

17. J. E. Halver, Trans. Am. Fisheries Soc. **83**: 254 (1953).

18. J. A. Coates and J. E. Halver, Spec. Sci. Rep. (Fish) **281**: 1 (1958).

19. J. E. Halver and J. A. Coates, Progressive Fish Culturist **19**: 112 (1957).

20. J. E. Halver, EIFAC. **66**: SC II-3 (1966).

21. J. E. Halver, J. Nutr. **62**: 225 (1957).

22. J. E. Halver, Administrative report, Fish and Wildl. Circular **178**: 35 (1964).

23. J. E. Halver, Proc. Vth Int. Cong. Nutr. 191 (1960).

24. D. C. DeLong, J. E. Halver and E. T. Mertz, J. Nutr. **65**: 589 (1958).

25. J. E. Halver, L. S. Bates and E. T. Mertz, Federation Proc. **23**: 1778 (1964).

26. M. Brin, C. E. Smith and J. E. Halver, Federation Proc. **26**: 880 (1967).

27. C. E. Smith, J. Fisheries Res. Board Can. **25**: 151 (1968).

28. S. Kitamura, S. Ohara, T. Suwa and K. Nakagawa, Bull. Japan. Soc. Sci. Fisheries **31**: 818 (1965).

29. C. M. McCay and A. V. Tunison, N. Y. Cons. Dept. Fish. Res. Bull. **5**: 18pp (1937).

30. S. Ikeda and M. Sato, Bull. Japan. Soc. Sci. Fisheries **30**: 365 (1964).

31. S. Ikeda and M. Sato, Bull. Japan. Soc. Sci. Fisheries **31**: 814 (1965).

32. H. A. Poston, N. Y. Cons. Dept. Fish. Res. Bull. **30**: 46 (1967).

33. J. E. Halver, L. M. Ashley and R. R. Smith, Trans. Am. Fisheries Soc. (In press).

COMMENTS

DR. BUHLER: The comments on ascorbic acid move me to mention another aspect of nutrition of fish. In studies that we did for the Fish-Pesticide Research Laboratory, we studied the influence of diet on lethality of various pesticides to fishes and found that variation in the composition of the diet had a marked influence on lethality. Increasing the lipid content protected the fish against toxic influence of DDT. Also, rather surprisingly, increasing protein content or leaving out several of the water soluble vitamins (including ascorbic acid) protected the fish against lethality for several weeks. Eventually, combination of the pesticide and omission of the vitamins from the diet started to influence the fish and mortality then occurred.

PROF. SINNHUBER: On the ascorbic acid requirement in your earlier work, you failed to show ascorbic acid deficiency on your test diet and several others have also.

DR. HALVER: We put 100 mg of ascorbic acid in 100 g of our test ration and this ration worked for vitamin studies. The original experiments were for 16-20 weeks with chinook salmon and later with some other salmonids. The profound deficiency syndrome appears in 24-30 weeks. Some of the early symptoms might have appeared in gill support cartilage at 10-12 weeks. In later experiments, we also improved the water supply, removed any salad from the water that fish might have eaten by running the water through a large beer filter, and then kept temperature constant. Maybe all these things contributed to a more definitive specific deficiency syndrome. We had more experimental control over variables and therefore, were able to induce the deficiency in more reproducible fashion.

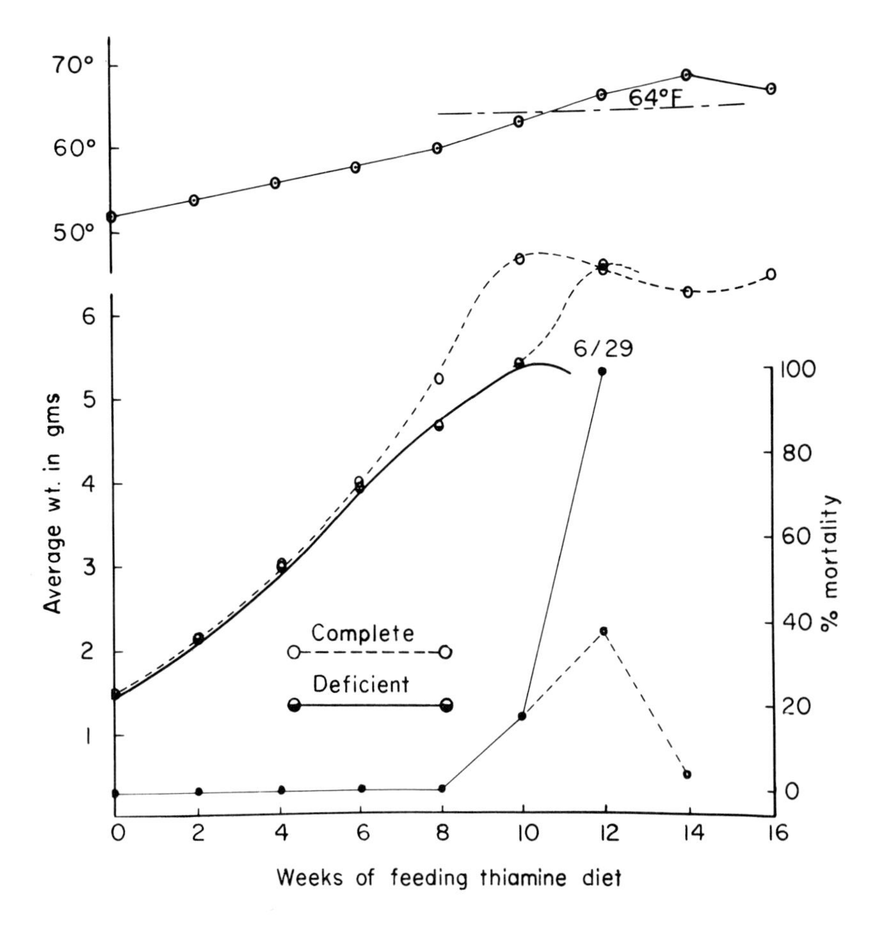

Fig. 1: Growth and mortality of chinook salmon fed complete and thiamine deficient diets. Upper curve shows average water temperature; center curves, growth; and lower curve, biweekly mortality of each group. Junction points in growth and mortality curves represent division of the deficient group into two sublots after deficiency symptoms were apparent in most of the test population. Note growth ceased in control lot when water temperature exceeded 64°F and began near end of experiment when water temperature dropped. (From: Halver, J. Nutr. 62: 230, 1957).

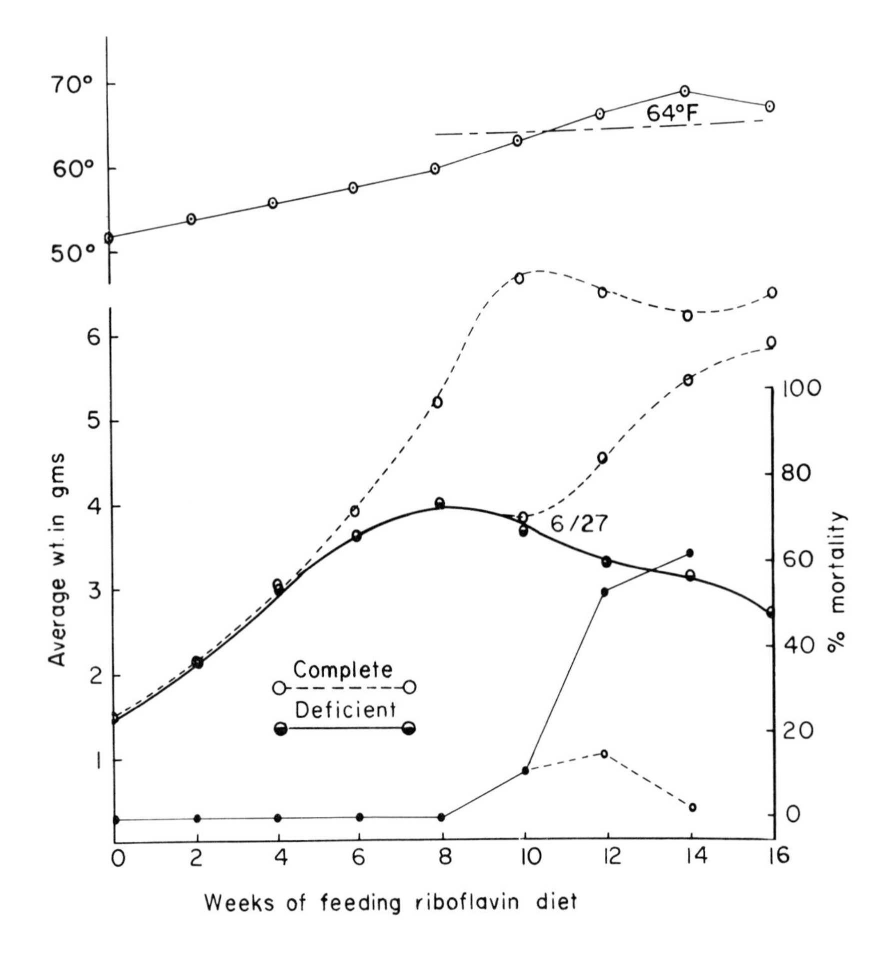

Fig. 2: Growth and mortality of chinook salmon fed complete and riboflavin deficient diets. For details of curves see legend of Figure 1. (From: Halver, J. Nutr. 62: 230, 1957).

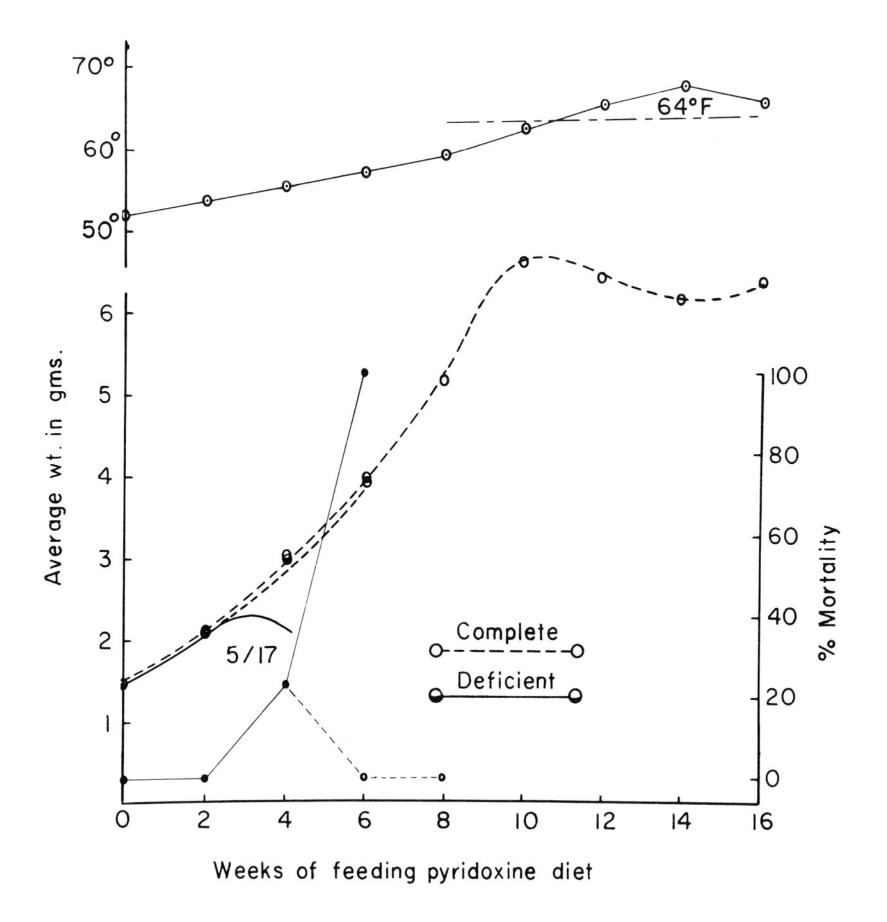

Fig. 3: Growth and mortality of chinook salmon fed complete and pyridoxine deficient diets. For details of curves see legend of Figure 1. (From: Halver, J. Nutr. 62: 230, 1957).

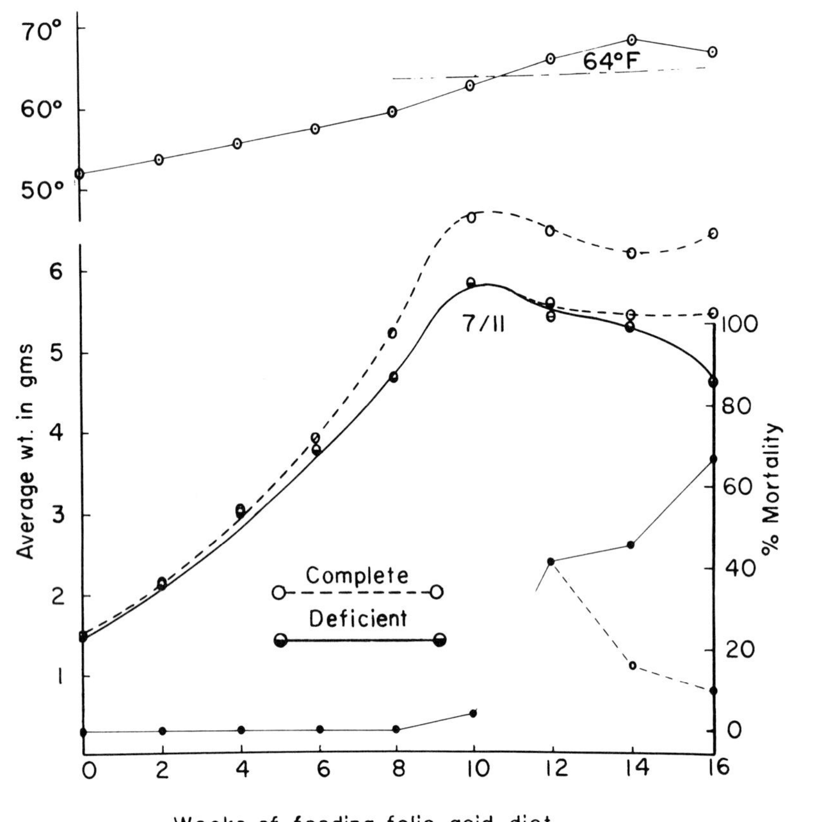

Fig. 4: Growth and mortality of chinook salmon fed complete and folic acid deficient diets. For details of curves see legend of Figure 1. (From: Halver, J. Nutr. 62: 239, 1957).

Fig. 5: Coho salmon fingerlings fed complete and folic acid deficient diets. Upper and lower salmon developed pale gills after 14 weeks on diet devoid of folic acid. Note pale gills and smaller size of deficient fish compared to dark gills and normal size and shape of center, complete test diet fed fish. Courtesy, C. E. Smith, Western Fish Nutrition Laboratory, Cook Washington.

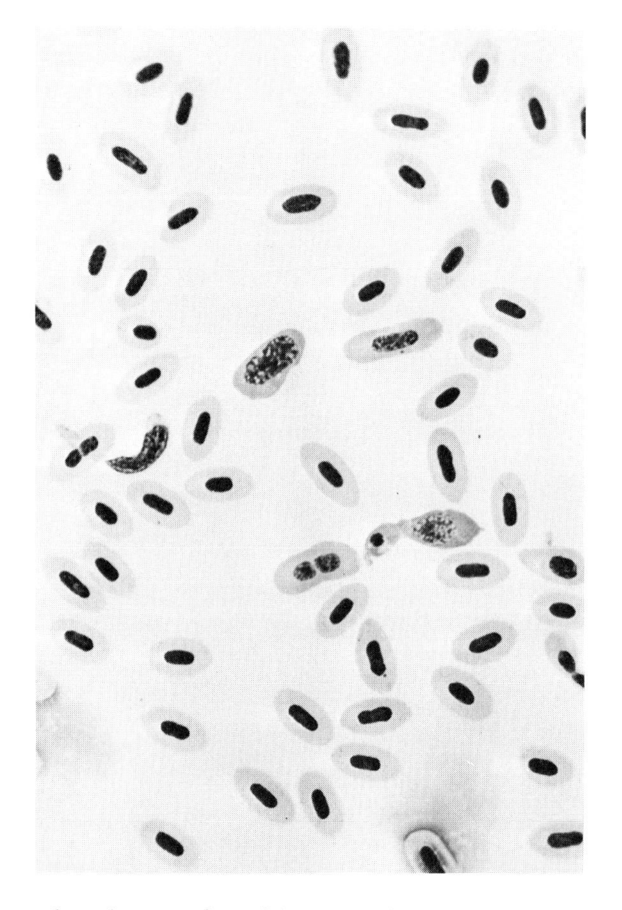

Fig. 6: Abnormal erythrocytes from folic acid deficient fish. Note anisocytosis and nuclear segmentation of erythrocytes with mostly mature and senile cells present. Courtesy, C. E. Smith, Western Fish Nutrition Laboratory, Cook, Washington.

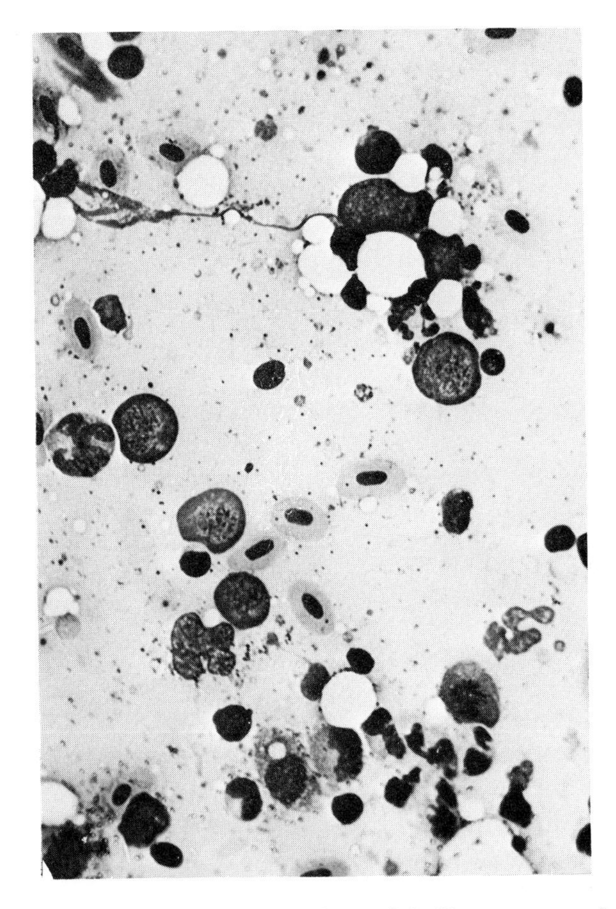

Fig. 7: Kidney imprint from folic acid deficient fish. Note presence of only mature erythrocytes and absence of developing erythroid precursors. Courtesy, C. E. Smith, Western Fish Nutrition Laboratory, Cook, Washington.

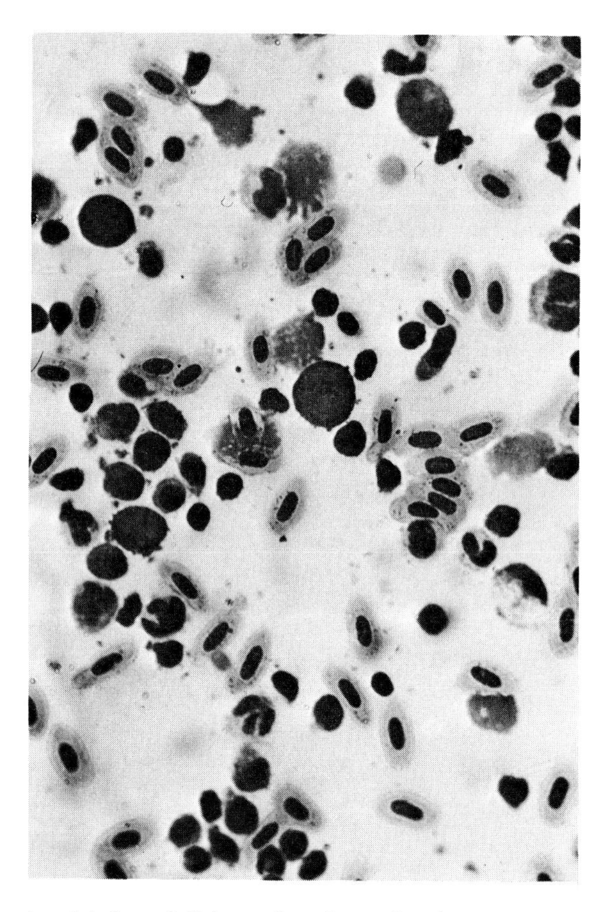

Fig. 8: Kidney imprint from deficient coho salmon placed on recovery diet containing 1.5 mg of folic acid/100 g dry ration for six weeks. Note presence of immature erythrocytes and reduced megaloblastic erythroid precursor cells and reduced erythroid hyperplasia. Courtesy, C. E. Smith, Western Fish Nutrition Laboratory, Cook, Washington.

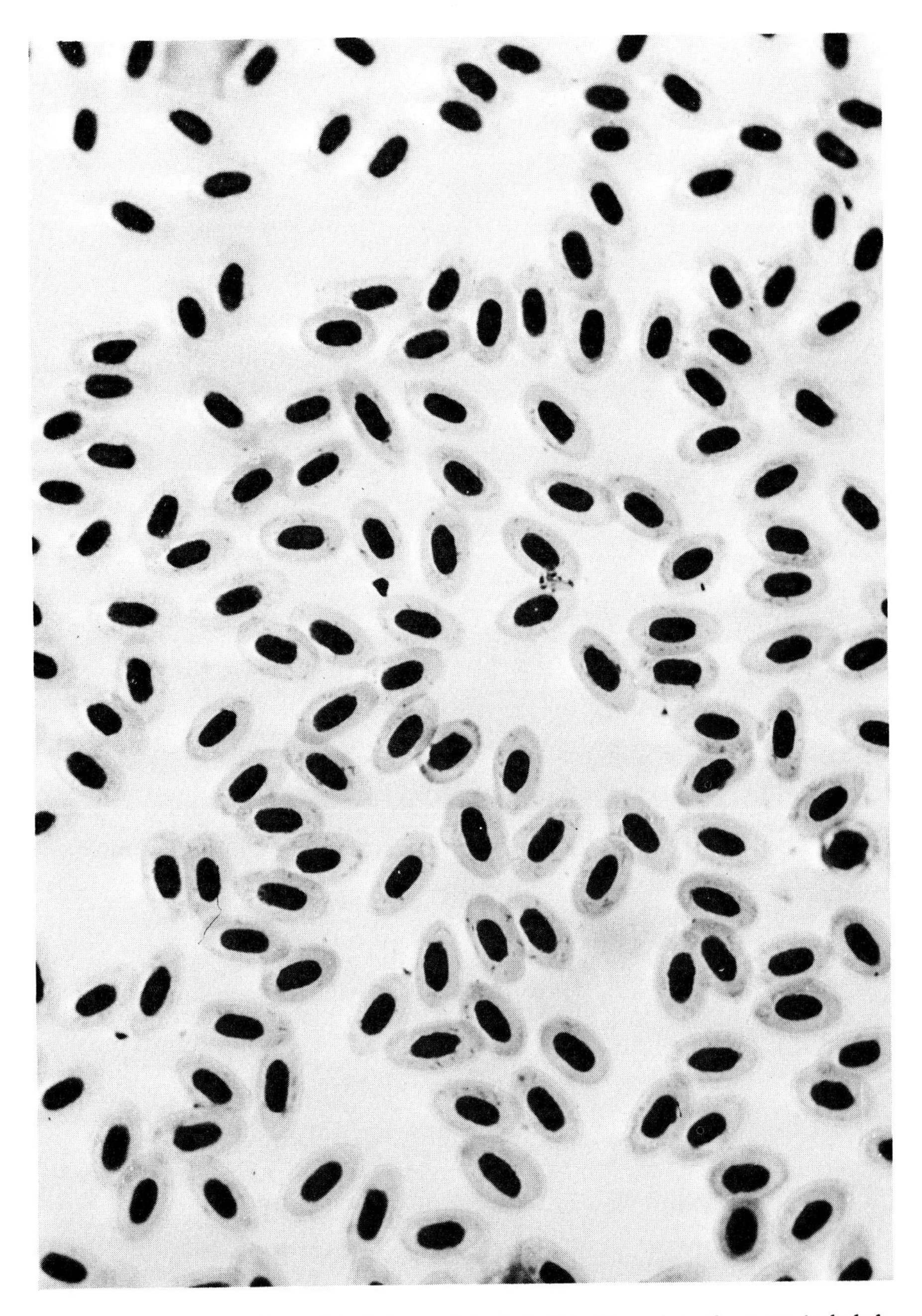

Fig. 9: Blood smear from fish fed complete test diet. Normal erythrocytes included mostly mature cells plus some immature forms with normal nuclei and less cytoplasm. (From: Smith and Halver, J. Fisheries Res. Board Can., 26: 112, 1969).

Fig. 10: Scoliosis and lordosis in fish fed diets devoid of ascorbic acid. Upper fish shows spinal curvature typical of scoliosis. Middle fish apparently normal fed complete test diet containing 100 mg of vitamin C/100 g of dry ration. Bottom fish fed scorbutic diet and shows lordosis (From: Halver, Ashley and Smith, (33)).

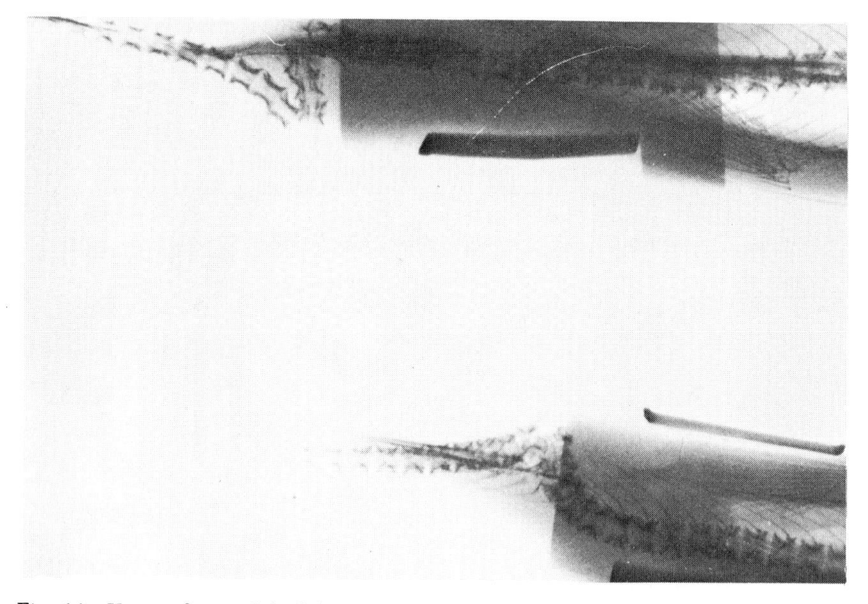

Fig. 11: X-ray of two fish fed ascorbic acid deficient diet for 24-30 weeks. Severe dislocation of vertebrae in spinal column is apparent. (From: Halver, Ashley and Smith (33)).

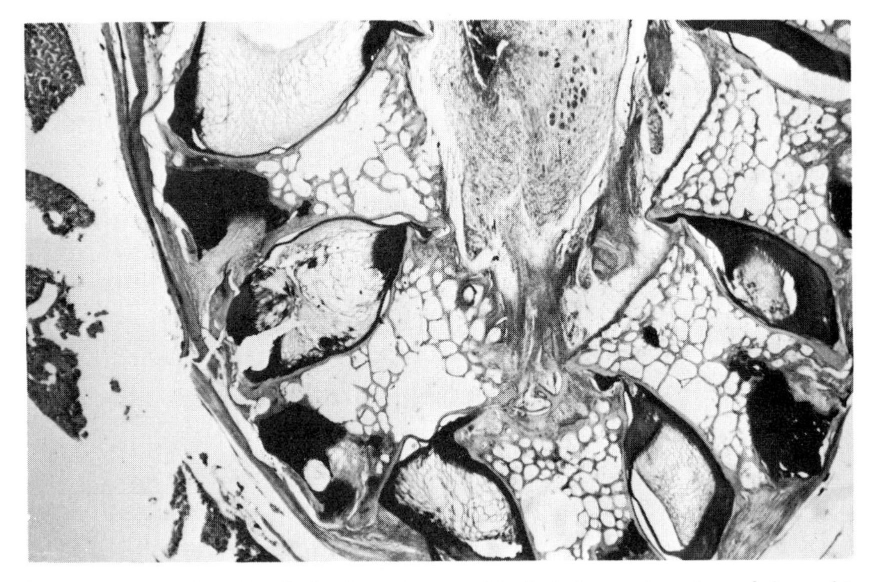

Fig. 12: Section through spinal column in area of lordosis from coho salmon fed ascorbic acid deficient diet for 24 weeks. Hour-glass shape structures with internal spongy bone and peripheral compact bone show slipped vertebrae which bends back on itself Note spinal cord tissue is pinched and reduced to a filament with some hemorrhagic areas present below the filamentous portion. Spinal ligaments are obviously deficient in collagen (From: Halver, Ashley and Smith (33)).

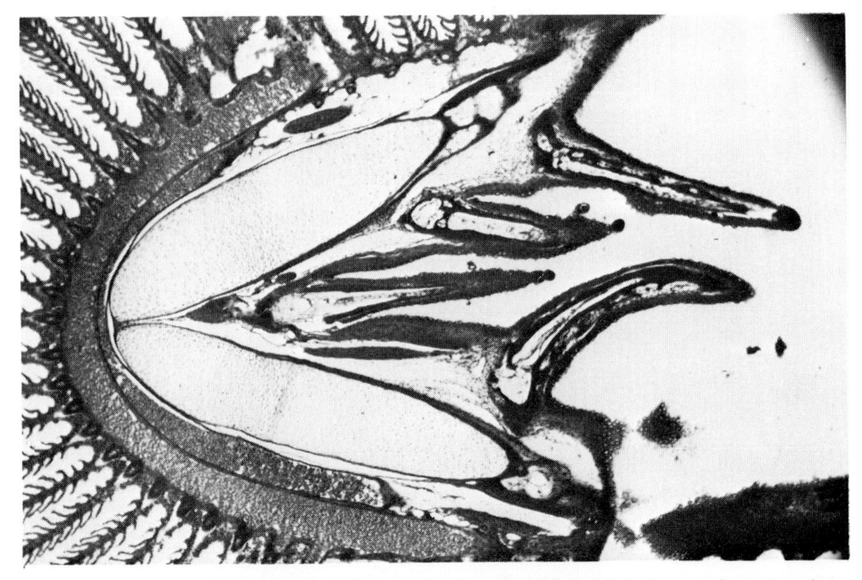

Fig. 13: Mason's trichrome stain of normal gill arch of fish. Note bony rakers, two large gill arch cartilages and arching blood vessels branching into each gill filament from which gill lamellae extend. (From: Halver, Ashley and Smith (33)).

Fig. 14: Section of gill filaments of ascorbic acid deficient fish. Note twisted, spiraled and deformed cartilagenous supporting tissue, apparently the result of impaired collagen formation in young coho salmon fed diets devoid of ascorbic acid for 20 weeks. Courtesy, L. M. Ashley, Western Fish Nutrition Laboratory, Cook, Washington.

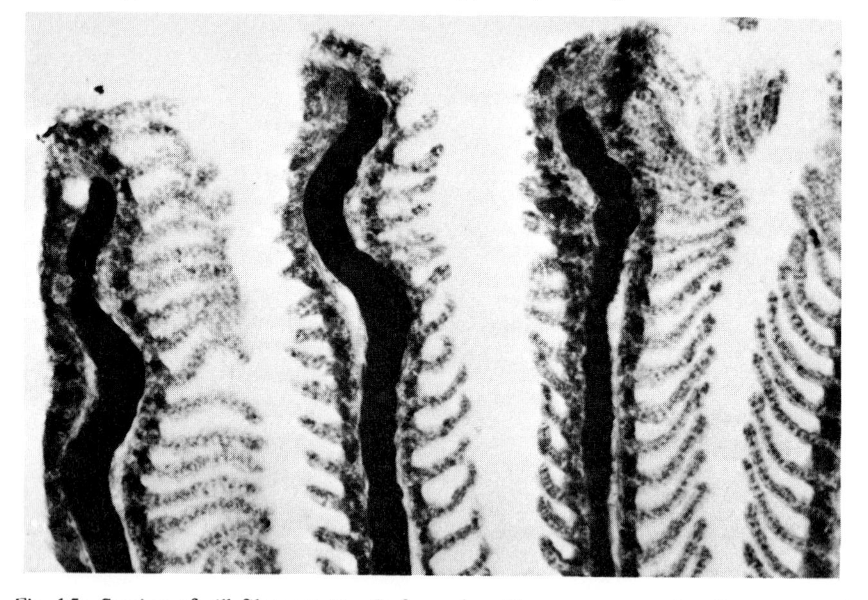

Fig. 15: Section of gill filament tips. Deformed cartilagenous supporting tissue occurred in fish fed diets low in vitamin C for 12-14 weeks. (From: Halver, in MARINE AQUICULTURE, OSU, Corvallis, Ore., 1969).

AMINO ACID AND PROTEIN REQUIREMENTS OF FISH

Edwin T. Mertz

QUALITATIVE AMINO ACID REQUIREMENTS

The first successful feeding test to show clearly the qualitative amino acid requirements of a fish species was reported 11 years ago by Halver, DeLong and Mertz (1). The test diet used (Table I) contained on a dry basis 70% of a mixture of pure L-amino acids with the proportions found in casein, plus 5%

TABLE I

COMPONENTS OF TEST DIET

Component	Amount Used (grams)
Amino acid mix (18A.A.)	70.0
White dextrin	6.0
Corn oil	5.0
Mineral mix	4.0
Vitamin mix	3.0
Cod liver oil	2.0
Water	100.0
Carboxymethylcellulose	10.0
	200.0

(Modified from: Halver, DeLong and Mertz (1)).

corn oil, 2% cod liver oil, 6% white dextrin, 4% mineral mix, and 3% vitamin, α-cellulose flour mix. To 90 parts of a mixture of these ingredients were added 100 parts of water and 10 parts of carboxymethyl-cellulose. By beating air into the mixture, the final diet had the consistency of bread dough and was fed by squeezing through a garlic press. In diets deficient in one amino acid, α-cellulose flour replaced on an equal weight basis the amino acid dropped from the test diet. The test animal was the chinook salmon (*Onchorhynchus tshawytscha*), average individual weight, approximately 2 grams, maintained at a water temperature of 47°F (8°C).

Table II shows that arginine, histidine, isoleucine, leucine, lysine, methionine, phenylalanine, threonine, tryptophan and valine are indispensable for growth of this species of fish. On diets deficient in any one of these amino acids, curbed intake of food was noted in ten days. The fish would swim slowly to the surface, take a piece of food, mouth it, and then spit it out.

TABLE II

AVERAGE INDIVIDUAL WEIGHT GAINS

Missing Amino Acid	Lot Number*	Average Gain, Ten Weeks (gm.)	Control Diet Last Four Weeks (gm.)
None	1	1.42	--
None	2	1.42	--
Arginine	3	–0.17	0.65
Histidine	4	–0.08	0.73
Isoleucine	5	–0.29	0.53
Leucine	6	–0.49	0.43
Lysine	7	–0.06	0.64
Methionine	8	–0.06	0.48
Phenylalanine	9	–0.09	0.64
Threonine	10	0.07	0.70
Tryptophan	11	–0.15	0.55
Valine	12	–0.15	0.41
One Dispensable	13-20	1.33-1.59	--

*At the end of 6 weeks, lots 3-12 inclusive were divided into 2 equal sub-lots, one to be continued as before (Column 3), and the other placed on the control diet (Column 4).

At the end of the sixth week, the deficient fish in each of the lots shown in Table II were split into two sub-groups. Sub-group one was continued on the same amino acid-deficient ration, and sub-group two was placed on the basal diet (Table I) containing all 18 amino acids. By the seventh day, fish in sub-group two were feeding actively. These fish showed an immediate and substantial growth response (Table II). The fish in sub-group one (deficient diet) continued to show no weight gain.

Fig. 1 compares the average growth curve of fish on the arginine-deficient diet with the growth curve of the controls. Similar results were obtained with the other nine essential amino acids. The remainder of the 18 amino acids tested (hydroxyproline was not included) gave growth curves similar to that shown for cystine in Fig. 2, and may be considered dispensable. Pictures comparing the appearance of typical fish selected from the glutamic acid, arginine and leucine-deficient lots with fish from the control lots are shown in Fig. 3 (top, center and bottom rows, respectively).

The qualitative requirements of the fingerling salmon are therefore similar to those of warm blooded animals. The complete inability to synthesize arginine places the salmon fingerling closer to the bird than to the mammal, however. The bird cannot synthesize arginine (2), whereas the weanling rat (3) and the weanling pig (4) can synthesize arginine at about two-thirds of the rate needed for maximum growth.

The same ten amino acids essential for growth in chinook salmon are required by three other species: the sockeye salmon (*O. nerka*) (5), the rainbow trout (*Salmo gairdneri*) (6), and the channel catfish (*Ictalurus punctatus*) (7).

QUANTITATIVE AMINO ACID REQUIREMENTS

Comparison of the quantitative amino acid requirements of different species would be more accurate if all investigators carried out the quantitation studies at, or near, the minimum protein requirement level. Bressani and Mertz (8) found that the minimum requirement level of protein for weanling rats on a balanced corn gluten-amino acid diet was 16%. The lysine requirement at this level of protein was 5.2% of the protein. It dropped to 2.2% of the protein when the diet contained 40% protein. Excess protein in the diet thus tends to depress the calculated requirement value.

DeLong, Halver and Mertz (9) determined the minimum protein requirements of chinook salmon at two water temperatures prior to carrying out amino acid quantitation studies. The diet contained a mixture of L-amino acids, casein and gelatin with an amino acid balance similar to that of whole egg protein (Table III). They found that the minimum protein level for chinook salmon is dependent upon the water temperature, rising from 40% at 47°F (8°C) (Fig. 4) to 55% at 58°F (15°C) (Fig. 5). The protein requirements for fish are thus twice that of the day-old chick, and three to four times that of young mammals, and unlike warm blooded animals, varies directly with the environmental temperature.

Using the egg protein pattern in a mixture of pure L-amino acids, casein and gelatin, Halver and co-workers determined the minimum requirement for each of the ten essential amino acids in chinook salmon fingerlings. These values are summarized in Table IV. The requirement for threonine was determined by DeLong, Halver and Mertz (10), for lysine and methionine by Halver, DeLong and Mertz (11, 12), for isoleucine, leucine, valine and phenylalanine by Chance, Mertz and Halver (13), for tryptophan by Halver (14), and for arginine and histidine by Klein, Halver and Mertz (15).

The values in Table IV show minimum requirements as a percent of the dietary protein. In parentheses are recorded fractions in which the numerator is the requirement as percent of the diet, and the denominator is

TABLE III

COMPONENTS OF DIET

Component	Range gm.
Casein	8-40
Gelatin	2-10
Amino acid mix	4.5-22
Dextrin	61.5-4
Corn oil	5
Cod liver oil	2
Minerals	4
Alpha cellulose + vitamins	3
Carboxymethylcellulose	10
	100
Water	100
Total	200

(From: DeLong, Halver and Mertz (9)).

the percent total protein in the diet. Requirement values for the chick, pig and rat are included for comparison.

It can be seen from Table IV that arginine requirements show the widest variation among different species. The arginine requirement of fishes and birds is similar and high (6% of the dietary protein). The arginine requirement of growing mammals is low (1 to 1½%). Fish and birds lack the

TABLE IV

AMINO ACID REQUIREMENTS OF FOUR SPECIES*

Amino Acid	Chinook Salmon Fingerling	Chick+	Young Pig	Rat
Arginine	6(2.4/40)	6.1(1.1/18)	1.5(0.2/13)	1(0.2/19)
Histidine	1.8(0.7/40)	1.7(0.3/18)	1.5(0.2/13)	2.1(0.4/19)
Isoleucine	2.2(0.9/41)	4.4(0.8/18)	4.6(0.6/13)	3.9(0.5/13)
Leucine	3.9(1.6/41)	6.7(1.2/18)	4.6(0.6/13)	4.5(0.9/20)
Lysine	5(2/40)	6.1(1.1/18)	4.7(0.65/13)	5.4(1/19)
Methionine++	4(1.6/40)	4.4(0.8/18)	3(0.6/20)	3(0.6/20)
Phenylalanine+++	5.1(2.1/41)	7.2(1.3/18)	3.6(0.46/13)	5.3(0.9/17)
Threonine	2.2(0.9/40)	3.3(0.6/18)	3(0.4/13)	3.1(0.5/16)
Tryptophan	0.5(0.2/40)	1.1(0.2/18)	0.8(0.2/25)	1(0.2/19)
Valine	3.2(1.3/40)	4.4(0.8/18)	3.1(0.4/13)	3.1(0.4/13)

*Expressed as percent of dietary protein. In parentheses the numerators are requirements as percent of dry diet, the denominators are percent total protein in the diet.
+The chick required glycine or serine; these are interchangeable (8.9% of protein; 1.6/18).
++In the absence of cystine.
+++In the absence of tyrosine.

urea cycle, which in growing mammals can serve as a source of about three-fourths of the arginine needed.

Fish appear to have the lowest tryptophan requirement of the four species (0.5%). Other species appear to require about twice this amount.

Differences in leucine and isoleucine requirements are also noted in Table IV. Salmon have a lower isoleucine requirement than the other species, whereas chicks have an unusually high leucine requirement, possibly due to feathering. Chicks also differ from the other three species in having an absolute requirement for either glycine or serine.

Except for the differences noted above, requirements of all four species are quite uniform when expressed as a percent of the dietary protein, especially when one considers the variation in the levels of total protein in the diets (13 to 41%).

BLOOD AMINO ACID LEVELS IN FISH

It is interesting to speculate on why fish are able to utilize efficiently such high levels of protein in the diet. If one feeds a 40% protein diet to a chick, one-half of the amino acids are deaminated and lost for protein synthesis. Two-thirds of the amino acids in a 40% protein diet are deaminated and lost for protein synthesis in a rapidly growing weanling pig.

Several years ago Chance, Halver and Mertz (16) determined the plasma amino acid levels in salmon and in weanling pigs. Table V shows the plasma

TABLE V

PLASMA AMINO ACID LEVELS IN FISH AND SWINE*

Amino Acid	Young Ocean Salmon (μg/ml)	Spawning Males (μg/ml)	Young Pigs (μg/ml)
Arginine	63	40	21
Isoleucine	73	22	18
Leucine	126	33	26
Lysine	87	83	32
Methionine	33	13	5
Phenylalanine	30	17	14
Tyrosine	40	10	19
Threonine	58	36	16
Valine	119	49	28
Alanine	159	101	54
Aspartic acid	10	4	2
Serine	41	107	21
Taurine	49	338	4
Cystathionine	4	12	0.6
1-methyl histidine	2	75	trace
Total EAA	608	307	205
Total NEAA	378	379	240

*5 lb. actively feeding salmon, avg. of five 25-30 lb. spawning male salmon, avg. of ten 5 week oil weanling pigs on 19.5% protein corn-soy-skim milk diet.

amino acid levels in a 5 pound chinook salmon which presumably had been actively feeding in the ocean just prior to being caught and bled. Plasma amino acid levels in five week old weanling pigs fed a corn-soy-skimmed milk diet (18½% protein) and in 25 to 30 lb. male spawning salmon in the fasting state are also listed in Table V.

These data show that the levels of individual essential amino acids are three to six times higher in the actively feeding salmon than in the actively feeding pigs, and that the total essential amino acids value is three times higher in the salmon than in the pigs. This may explain the higher tolerance of the fish for protein, for the higher plasma amino acid levels should permit a faster rate of protein synthesis in the body cells. It could also indicate a reduced level in fish of those enzymes which are responsible for the deamination of amino acids.

Table V shows that spawning distorts the pattern of plasma amino acids in fish. The total essential amino acids drop to one-half of the level found in the actively feeding salmon, whereas taurine, a product of cystine and methionine metabolism, increased seven-fold, and 1-methyl histidine, a product of histidine metabolism, increased thirty-fold. These changes suggest that with starvation and accelerated spermatogenesis extensive catabolism of body cells occur.

SUMMARY

Fish have the same qualitative amino acid requirements as young mammals. The minimum protein requirements, however, are more than twice that of either young mammals or birds, and vary directly with the water temperature. Based on the fingerling salmon, fish have a high requirement for arginine, which is similar to that of birds when expressed as a percent of the dietary protein. The tryptophan and isoleucine requirement is lower than that of mammals and birds, but other amino acid requirements are similar to that of mammals.

It is suggested that fish can utilize more of the dietary protein for synthesis than other species because of higher plasma amino acid levels. These may exist in conjunction with increased levels of enzymes and nucleic acids involved in the synthesis of protein within the cell. It is also possible that enzymes involved in the breakdown of amino acids may be less active in fish cells than in mammalian and avian cells.

Because of its unusual ability to utilize for cell division and growth very high levels of protein from the diet, the fish is an ideal animal for the study of amino acid balance, imbalance, and pathways of protein synthesis and breakdown. As the most efficient converter of vegetable protein into tasty high quality animal protein, more attention should be paid to fish. Fish farming is an important means for reducing world wide malnutrition, especially in areas of food protein deficiency.

REFERENCES

1. J. E. Halver, D. C. DeLong, and E. T. Mertz, J. Nutr. **63**: 95 (1957).

2. H. J. Almquist and C. R. Grau, J. Nutr. **28**: 325 (1944).

3. A. Borman, T. R. Wood, H. C. Black, E. G. Anderson, M. J. Oesterling, M. Wormack and W. C. Rose, J. Biol. Chem. **166**: 585 (1946).

4. E. T. Mertz, W. M. Beeson, and H. D. Jackson, Arch. Biochem. Biophys. **38**: 121 (1952).

5. J. E. Halver and W. E. Shanks, J. Nutr. **72**: 340 (1960).

6. W. E. Shanks, G. D. Gahimer and J. E. Halver, Progressive Fish Culturist **24**: 68 (1962).

7. H. K. Dupree and J. E. Halver, Tran. Am. Fisheries Soc., in press.

8. R. Bressani and E. T. Mertz, J. Nutr. **65**: 481 (1958).

9. D. C. DeLong, J. E. Halver and E. T. Mertz, J. Nutr. **65**: 589 (1958).

10. Ibid, J. Nutr. **76**: 174 (1962).

11. J. E. Halver, D. C. DeLong, and E. T. Mertz, Federation Proc. **17**: 1873 (1958).

12. Ibid., Federation Proc. **18**: 2076 (1959).

13. R. E. Chance, E. T. Mertz and J. E. Halver, J. Nutr. **83**: 177 (1964).

14. J. E. Halver, J. Nutr. **24**: 229 (1965).

15. R. G. Klein, J. E. Halver and E. T. Mertz, Unpublished data (1968).

16. R. E. Chance, I. Amino Acid Requirements of Chinook Salmon. II. Free Amino Acids and Related Compounds in the Plasma of Chinook Salmon and Swine. Ph.D. Thesis, Purdue University, Lafayette, Indiana, August 1962.

COMMENTS

DR. TARR: I find this need of young salmon for arginine a very interesting thing because in maturing salmon near point of maturity nucleoprotamine synthesis must go on for quite a time after feeding. The gonad increases in size and has a very high content of arginine, and therefore, there must be some interesting metabolic change in these fish at maturation.

DR. MERTZ: I think perhaps that the arginine these fish use in spermatogenesis is probably from tissue breakdown. Thus, histidine is a byproduct and that may be why we find so much of the methylated form appearing in the blood.

DR. CONTE: Your minimum protein requirement studies for fishes have all been on fresh water species and appear to be about 40% of the dry diet. What is the minimum protein requirement for fishes that are in the marine environment, in other words, the marine fishes? We know that the intestinal tract changes remarkably in a physiological sense when these go into the marine environment simply because they are ingesting all the sea water which has a very odd mineral balance. We do know that these are shifting ionic constituents especially in the intestine. You have shown that in terms of plasma levels elevated circulating amino acids occurred over what was present in small fingerlings. Where is the source of elevation of the plasma proteins when there is no tissue degradation because one sees tremendous net gain of weight in the ocean fish? One cannot say it is catabolism that accounts for the circulating levels. Could it be that the protein requirement, the minimum protein requirement has gone up when the environment changes?

DR. MERTZ: We have no idea of what might happen in the ocean. We do know, of course, that in mammals and in avian species as the animal matures there is a remarkable drop in the requirement for protein. We do not know if this happens in fish. This is something that would be a very interesting experiment for someone to do. We have one little suggestion, that this might happen, because in older, larger salmon that were sampled for plasma amino acids, the level of amino acids is lower, and it seems to drop the larger the fish sampled. We do not have much information but this would suggest that perhaps the protein requirement drops or less protein is consumed.

DR. CONTE: Another thing that is happening too is the fact that a lot of the excess amino acids that would be spilling out normally in fresh water fish would not occur in marine fish because of lower filtration. The kidney shuts down a great deal in the marine environment to conserve fluids.

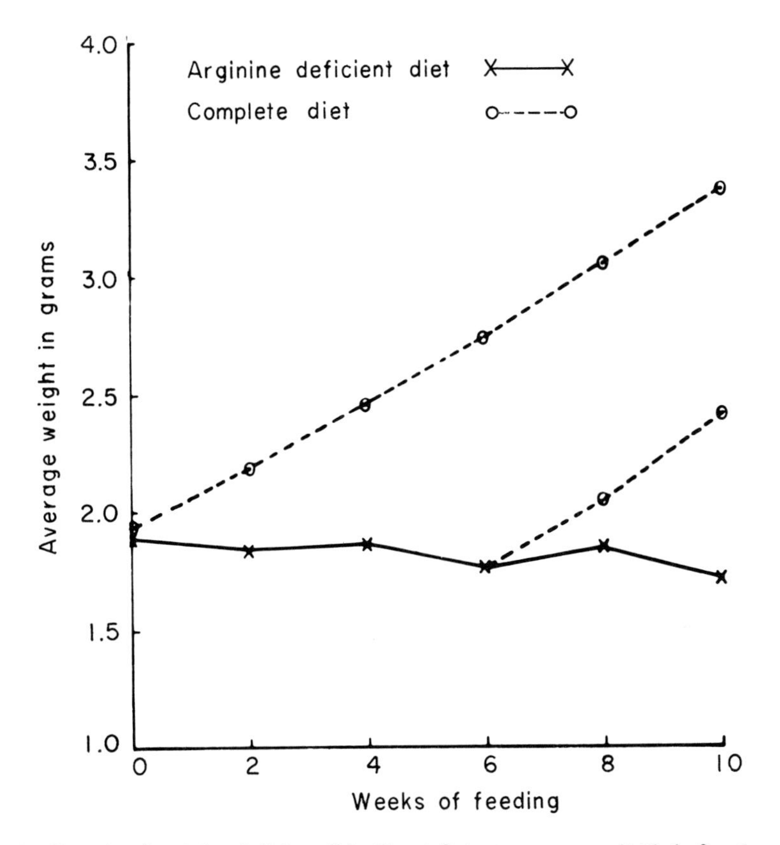

Fig. 1: Growth of arginine deficient fish. The deficient group was divided after 6 weeks on the deficient diet and the missing amino acid was replaced in one of the two sub-lots. (From: Halver, DeLong and Mertz (1)).

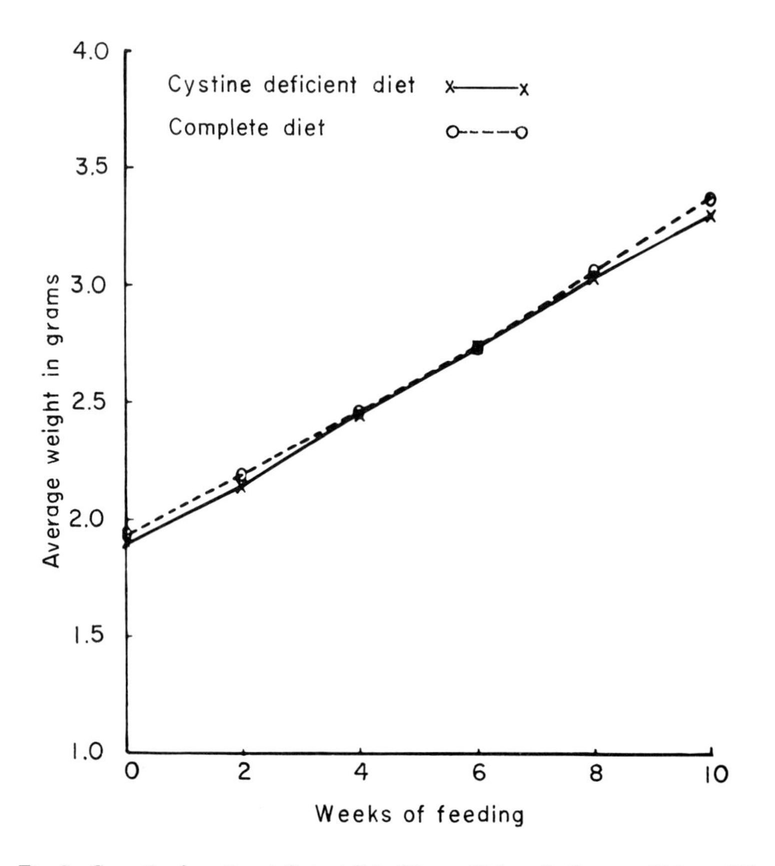

Fig. 2: Growth of cystine deficient fish. (From: Halver, DeLong and Mertz (1)).

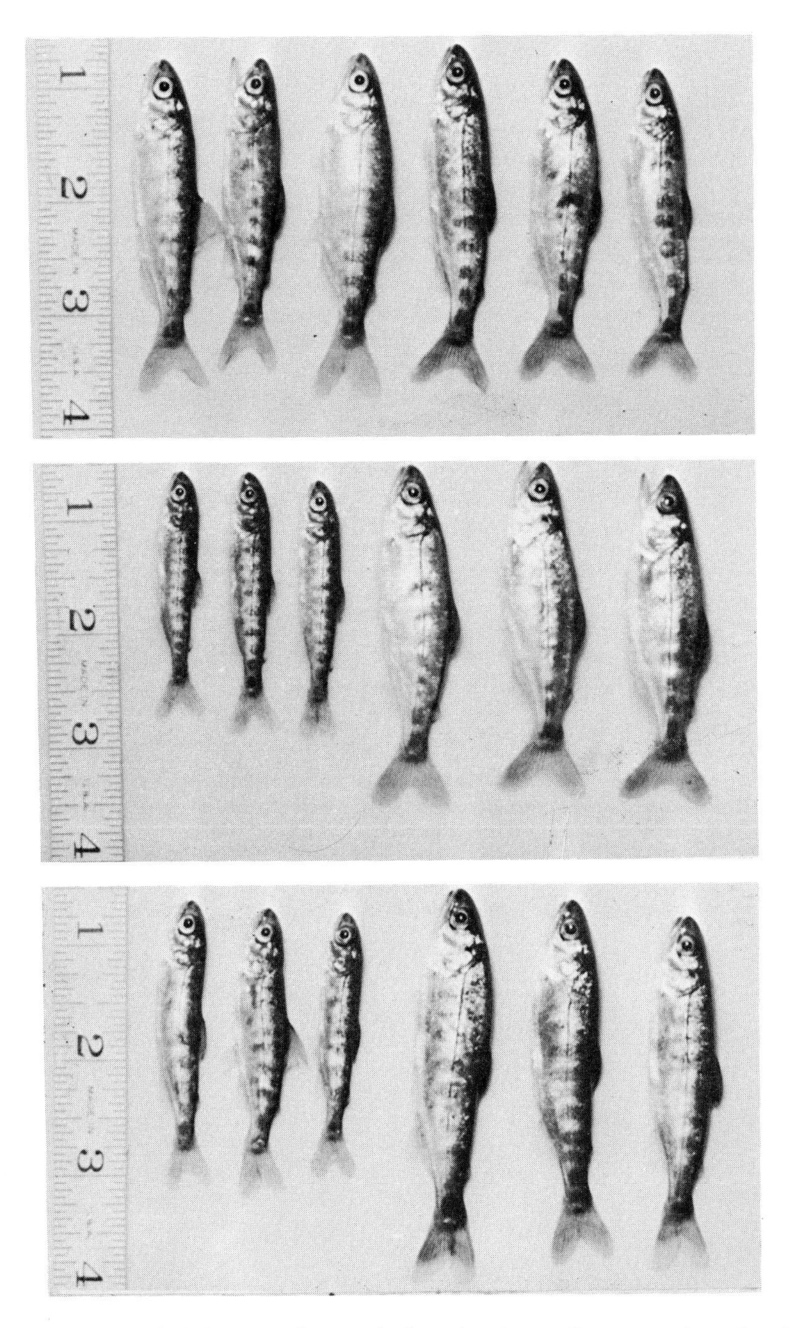

Fig. 3: Amino acid deficient and control chinook salmon. Top row: glutamic acid deficient (first 3) and control fish. Center row: arginine deficient (first 3) and control fish. Bottom row: leucine deficient (first 3) and control fish. (From: Halver, DeLong and Mertz (1)).

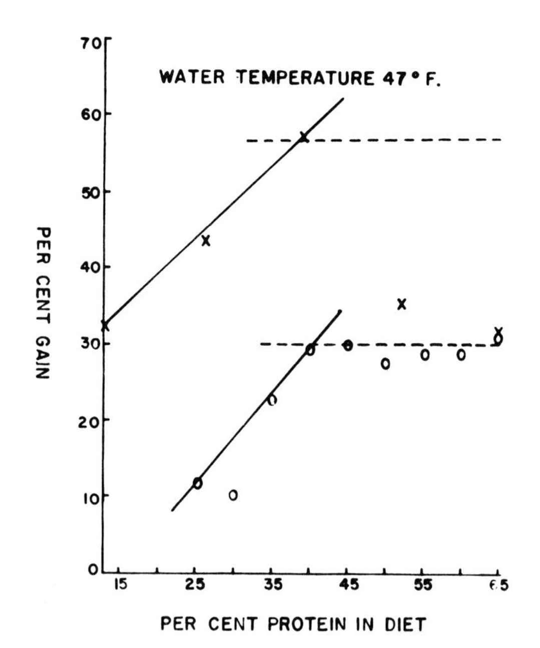

Fig. 4: Protein requirement at 47°F. Top curve: Initial individual average weight of fish, 1.5 gm. Bottom curve: Initial individual average weight of fish, 5.6 gm. (From DeLong, Halver and Mertz (9)).

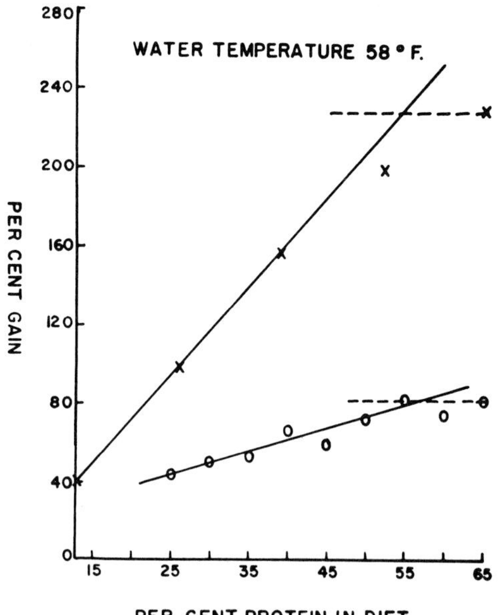

Fig. 5: Protein requirement at 58°F. Top curve: Initial individual average weight of fish. 2.6 gm. Bottom curve: Initial individual average weight of fish, 5.8 gm. (From: DeLong, Halver and Mertz (9)).

THE ROLE OF FATS

R. O. Sinnhuber

INTRODUCTION

The use of fish as an animal in basic research would presume that all the nutritional and dietary needs have been met or may be controlled by the experimenter. Research over the past two decades has demonstrated that most of the dietary essentials required by mammals are also important in the diet of fish. With reference to tissue lipids related to the essential fatty acids, there exists a striking difference between fish and mammals. The polyunsaturated fatty acids in fish are predominantly of the linolenic or ω-3 family while in mammals they are the linoleic or ω-6 series. Consideration of this fundamental difference, which is subject to metabolic and dietary control, and its significance when fish are used as an experimental animal for biological research is the subject of this discussion.

COMPOSITION OF FISH OILS

The fats in fish are characterized by their great variety of fatty acids and their high degree of unsaturation, mainly as mixed triglycerides, and with minor amounts of these fatty acids associated with phospholipids, glyceryl ethers and waxy esters. No review of fish lipids would be complete without reference to the early work of Hilditch (1), Lovern (2, 3) and the marine oils of Canada by Bailey *et al.* (4). An excellent contribution on the chemistry and nutritional properties of fish oils by a number of scientists was recently edited by Stansby (5).

The early laborious procedures designed to elucidate the fatty acid composition of fish oils were a real challenge and required exceptional skill and technique. Gas liquid chromatography more than any other analytical tool revealed the complexities of fish oils and permitted the separation, analysis and characterization of the fatty acids comprising fish lipids. The reader is referred to the works of Ackman, Gruger, Klenk, Shima and many others which are cited in a review by Gruger (6).

The most prominent saturated fatty acid in fish oil is palmitic acid, containing 16 carbons. Limited amounts of C_{14} and C_{18} saturated acids also occur. The unsaturated fatty acids are of the *cis* geometric configuration. The polyunsaturated fatty acids of the divinylmethane or methylene interrupted structure may contain as many as six ethylenic bonds (7). Ackman and Burgher (8) quantitatively identified more than 40 fatty acids in cod liver oil whereas vegetable oil may only contain from eight to ten. It is of interest that brain and nervous tissue as well as cell membranes of higher animals and man contain a generous complement of the higher polyunsaturated fatty acids prevalent in fish oils.

Fatty acids are probably one of the few food components that are unchanged by the digestion process and may be recognized in the tissues as being of dietary origin. Lovern (9) found that when eels were fed exclusively on herring, the depot fat of the eels corresponded almost exactly to a mixture of eel and herring fat. It is well recognized that an organism by biogenesis tends to build and maintain a fatty acid picture characteristic of its species and of a particular tissue or organelle. Other factors which effect the make-up of fish body lipids in addition to diet (10) and the amount and composition of fat consumed are periods of starvation, reproductive cycle, season, and very important in poikilotherms, the environmental temperature (11, 12, 13).

BIO-TRANSFORMATIONS OF FATTY ACIDS IN FISH

The more common unsaturated fatty acids found in fish and mammals may be classified into the following three families: oleic, linoleic and linolenic series and designated as ω-9, ω-6 and ω-3 respectively, numbering from the terminal methyl group with the primary number including the length of the carbon chain. Therefore, according to the accepted nomenclature arachidonic acid which is a C_{20} fatty acid with four double bonds, related to the linoleic series, would be designated as 20:4ω6. In Figs. 1, 2, and 3, the three fatty acid families are depicted. Desaturation by removal of two hydrogens is indicated by arrows to the right and chain elongation by arrows to the left. Present also in lipids is the ω-7 or palmitoleic acid family. Both the 16:1ω7, 16:2ω7 have been found in fish oils (14).

A general statement might be made that fish do not appear to differ greatly from mammals in the metabolism of fatty acids. The studies of Klenk and Kremer (15), Mead *et al.* (16), Kelly *et al.* (17, 18), Brenner *et al.* (19), Higashi *et al.* (20), Lee *et al.* (21) provided evidence that many of the transformations shown in Figs. 1, 2, and 3 occur in fish with the elegant work of Kayama *et al.* (22) in Mead's laboratory using methyl linolenate -1-^{14}C as an outstanding example. They showed that the probable pathway of linolenic acid is 18:3 → 18:4 → 20:4 → 20:5 → 22:5 → 22:6 as depicted in Fig. 3. An excellent review of the metabolism of polyunsaturated fatty acids was recently presented by Mead (23) and a review of the lipid metabolism in fish by the same author (24).

Much of the earlier work was conducted on fish in which the previous dietary history was unknown or the animals were held on varying dietary regimens which often made interruptions difficult. Brenner *et al.* (19) fed a fat free diet, carefully prepared by acetone extraction of the major ingredients and found an increase in the oleic acid series, including eicosatrienoic acid, 20:3ω9. When methyl linoleate was added to the diet, an increase in arachidonic acid was observed confirming the earlier studies by Klenk and Kremer (15) who incubated labeled acetate -1-^{14}C with liver slices from five varieties of fish, including trout, and showed that the synthesis of the C_{20} and C_{22} polyenoic acids occurred by addition of exogenous precursors to the carbonyl end of the linoleic and linolenic series.

Additional evidence for the transformations shown in the figures can be found in the work of Lee *et al.* (21) in which rainbow trout were fed a semi-purified diet containing ten percent corn oil as the only source of lipid from the time of hatching. Corn oil has a simple fatty acid composition and the metabolic conversions which occurred were followed by fatty acid analysis of the triglycerides and phospholipids. Gas chromatographic analysis of corn oil showed that it contained 57, 28 and 1.2 percent of 18:2ω6, 18:1ω9 and 18:3ω3, respectively, as unsaturated fatty acids. After four months on this ration an array of fatty acids was found in the trout lipids providing additional evidence that many of the transformations described in Figs. 2 and 3 had occurred. Fatty acids not present in the diet such as 18:3ω6, 20:2ω6, 20:3ω6, 20:4ω6 and 22:5ω6 were found and may be assumed to originate from 18:2ω6 or linoleic acid.

The small amount of 18:3ω3 or linolenic acid present in corn oil (1.2%) gave rise to 18:4ω3 and 22:6ω3. When additional amounts of 18:3ω3 were added to replace part of the corn oil with soybean oil, or methyl linolenate, 20:5ω3 was found. It is of interest to note that in these experiments no appreciable amounts of 20:4ω3 or 22:5ω3 were detected and finally that 18:3ω3 was not retained in fish tissues, but was used for energy or transformed to the higher unsaturated fatty acids (21).

ESSENTIAL FATTY ACIDS

The studies of Mead *et al.* (16) and Klenk and Kremer (15) discussed in the preceding sections demonstrated that fish are unable to biosynthesize fatty acids with unsaturation in the ω-3 or ω-6 position unless a suitable precursor is provided in the diet. In experimental rations, this is usually furnished by the addition of lipids which contain linoleic or linolenic acids such as corn, cottonseed or soybean oils. Cottonseed oil is devoid of ω-3 acids and corn oil is quite low, containing only about 1% linolenic acid. Nicolaides and Woodall (25) observed a marked depigmentation, described as a change from normal greenish black to a light brown, in the skin of chinook salmon held after hatching for 16-24 weeks on a fat-free diet of semi-purified materials. A similar depigmentation was noted when triolein or linolenic acid was present in the diet but was largely prevented by trilinolein. However, a general repigmentation occurred spontaneously during the recovery experiment, with 3% trilinolein providing the greater response. The authors cited this as evidence for the need of linoleic acid by salmon. Attempts to reproduce these findings with chinook salmon which had been fed the regular experimental ration for six weeks were unsuccessful.

Hagashi *et al.* (20) found that rainbow trout fed a fat-free diet for three months exhibited far more severe symptoms than the depigmentation described by Nicolaides *et al.* (25). The symptoms in trout were described as a hardening and discoloration of the posterior region with a severe erosion of the posterior fin and in extreme cases exposing the terminals of the spinal column. Recovery was observed with the addition to the diet of ethyl linoleate prepared from soybean oil, ethyl linolenate from linseed oil or a highly unsaturated fatty acid fraction from squid oil. Unfortunately, fatty acid composition analysis was not reported on these isolates nor were tissue fatty acids determined. From these studies, it seems that a fat-free diet produces skin disorders in salmon and trout, however, whether it is linoleic, linolenic acid or the higher unsaturated fatty acids which can effect a cure is not completely clear at this time.

GROWTH

In their studies on impaired pigmentation with salmon on an essential fatty acid deficient diet, Nicolaides *et al.* (25) found that trilinolein or linolenic, or both provided a positive growth response. Hagashi *et al.* (20) with rainbow trout held on a fat free diet for two months stated that a highly unsaturated squid oil provided the best growth followed by ethyl linolenate and ethyl linoleate.

In our laboratory, Lee *et al.* (21) demonstrated a requirement for linolenic acid or the ω-3 series in the growth of rainbow trout. Fish on a semi-purified diet after hatching, with corn oil as the sole lipid source after four months grew poorly and exhibited high mortality, Table I. Addition to the diet of linolenate or fish oil produced an immediate growth stimulation and a cessation of mortalities within two weeks. The small amount of linolenic acid present in corn oil was apparently insufficient for normal growth and high mortalities occurred. This may be an example of

TABLE I

EFFECT OF DIETARY LIPIDS ON GROWTH RATE, MORTALITY AND FEED UTILIZATION BY RAINBOW TROUT

Group and diet	Av. wt., week no. 0	4	8	12	Mean gain	Unit feed/ unit gain	Accumulative mortality
	g/fish	g/fish	g/fish	g/fish	g		%
A 10% corn oil	3.0	3.3	5.2	7.2	4.2	1.22	25
B 5% salmon oil + 5% corn oil	3.0	4.0	9.5	16.9	13.9**	0.77	5
C 1% salmon oil + 9% corn oil	3.0	4.0	7.5	10.9	7.9**	1.02	6
D 10% soybean oil	3.2	4.3	8.4	12.4	9.2**	0.77	4
E 1% linolenate[1]	3.0	3.9	7.3	11.4	8.4**	0.92	2

[1] 1% linolenic acid supplied by incorporating 2.16% methyl linolenate concentrate (Nutritional Biochemicals Corporation) into basal diet at the expense of corn oil.

**Significantly different from corn oil diet ($P<0.02$).

From Lee, et al. (21).

competitive inhibition which is discussed in two recent reviews (23, 26). Fatty acid analysis by gas chromatography was conducted on the diets, tissue, neutral lipids and phospholipids. A definite shift in the phospholipids favoring the ω-3 fatty acids with a reduction in ω-6 acids was observed as shown in Table II. The fish maintained on the control diet (corn oil) in addition to poor growth and continued high mortality exhibited an abnormal symptom which might be described as a "shock syndrome." An unusual stress such as that produced during netting and weighing or even a sudden stimulus, as turning on the light, when the fish were in a semi-resting state would cause a rapid swimming motion resulting in a comatose state with a number of fish, five to ten percent, floating motionless in the water. After a short time most of them would regain their equilibrium and start swimming again, occasionally a mortality would result. When dietary supplementations of ω-3 fatty acids or fish oils were provided, this "shock syndrome" could not be initiated and completely disappeared in a few days.

Examination of the 22:6ω3 values of the phospholipids (Table II) show a rapid accumulation of this acid in the supplemented lots. One may speculate that the highly unsaturated 22:5ω6 or 20:4ω6 does not function as well or substitute for the 22:6ω3 and that the lower level of the ω-3 acids found in

TABLE II

PERCENT COMPOSITION OF FATTY ACIDS FROM PHOSPHOLIPIDS OF FISH ON THE TEST DIETS

	Diet			
Fatty Acid	A	B	D	E
14:0	1.05	1.30	1.10	1.51
15:0	trace	0.19	---	trace
16:0	24.73	23.82	16.99	18.15
16:1ω7	2.19	3.97	2.81	3.02
17:0	trace	0.19	---	trace
16:2 ?	trace	0.31	---	trace
18:0	5.80	5.26	7.43	6.51
18:1ω9	16.55	19.25	22.14	17.93
18:2ω6	17.76	11.63	20.52	16.05
18:3ω6	1.58	0.56	0.35	trace
18:3ω3	---	---	1.49	2.86
20:1ω9	0.58*	1.44*	0.69	1.15
18:4ω3	0.17	0.34	---	1.29
20:2 ?	0.79	0.81	1.75	1.67
20:3ω6	4.22	2.71	4.50	3.74
20:4ω6	8.97	4.48	4.76	5.69
20:5ω3	---	1.40	1.50	trace
22:4ω6	0.98	1.13	trace	trace
22:5ω6	11.22	3.27	8.72	7.05
22:6ω3	3.14	17.15	5.69	8.04

*Includes 18:3

From Lee, et al. (21).

the phospholipids of the control rainbow trout is indicative of a deficiency state. Additional work on this unusual condition which seems to be related to membrane function and fatty acid composition is in progress.

EXPERIMENTAL PROCEDURES

Throughout this volume there have been many examples of the use of fish as experimental animals for nutritional and biochemical studies. Research has revealed that many, if not most, biological processes and systems function in fish as they do in higher life forms. It is evident that fish can serve as an additional model recognizing that it has a number of unique and special properties to contribute to the solution of biochemical problems.

The rainbow trout, *Salmo gairdneri*, a highly inbred and commercial species is hardy and easily held and reared in artificial or laboratory surroundings. The main requirement is a supply of fresh water at a temperature ranging from 45 to 65°F. It readily accepts prepared foods and may be artificially propagated. The following list gives a few examples of advantages of the use of fish (trout) as a test animal:

1. Their diet can be carefully controlled from time of hatching with no dependence on maternal nutrition.

2. As many as 1,500 to 2,500 young may be obtained from a single mating.

3. Their growth and metabolism may be altered by changing the water temperature.

4. Rainbow trout is highly inbred with a very rapid growth rate and excellent feed conversion permitting a maximum number of observations to be made with limited amounts of test material.

5. Large numbers of animals can be used permitting frequent random sampling of experimental populations.

6. They endure long periods (one to two months and longer) of low food intake or even starvation which permits a study of the mobilization of stored lipids and other compounds.

SEMI-PURIFIED DIETS

Attempts to prepare and feed trout semi-purified diets composed of casein, starch, salts, lard, cod liver oil and vitamins began as early as 1927 by McCay *et al.* (27). Ten years later, McCay (28) reviewed the biochemistry of fish and described the many problems that confront the experimenter. Only moderate growth was achieved with the semi-purified rations of that period and the trout usually failed to survive unless fresh meat was added to the diet. This led to a search for the elusive "factor H" which in turn stimulated research to establish the nutritional requirements of trout and led to the semi-purified rations of today.

Early workers in fish nutrition were concerned about the loss of finely ground materials and water soluble constituents during feeding and resorted to the use of capsules (29, 30) with reasonable success. The real breakthrough was made by McLaren and her co-workers (31, 32) when gelatin was incorporated in a mixture of semi-purified materials, salts and vitamins to yield a bound diet and replace the gelatin capsule used in the earlier studies. These important contributions form the basis for the present semi-purified rations which have changed little except for the addition of some recently discovered vitamins. From 1951 to 1963 in a series of studies (33-36), workers from Oregon State University and the Fish Commission of Oregon used the McLaren diet with modifications with varying success to rear spring chinook and silver salmon. A semi-purified diet formula was reported by Sinnhuber *et al.* (37) which is similar to the McLaren diet with the addition of supplemental vitamins. Corn oil was the lipid source which in retrospect may now be considered a marginal diet based on its fatty acid composition. The semi-purified ration used by Oregon State University (21) now employs ten percent fish oil (salmon, menhaden or herring) as a lipid source.

Wolf in 1951 (38) described a synthetic diet composed of commercial casein, gelatin lard, cooked potato starch, cellulose flour, minerals and vitamins. Preparation of the diet required cooking of the potato starch. Crisco or lard was the lipid source with 2% cod liver oil added as a vitamin A and D supplement.

In 1957, Halver (39) published the formula of a test diet which was used in his laboratory to determine the water soluble vitamins required for chinook salmon. This semi-purified diet was based on the McLaren formula with the addition of supplemental vitamins and amino acids. Corn oil at the level of 9% was the lipid source. In later experiments, two parts of cod liver oil were substituted for two parts of corn oil as a source of vitamin A and D. Good growth is achieved with this diet and several generations have been reared with this ration as their only source of food (40).

Studies in our laboratory on rainbow trout reared on semi-purified diet containing corn oil as the only oil source of lipid (21) were described earlier in this chapter. The data indicated that trout grew poorly and after four months were showing a constant mortality. The addition of ω-3 fatty acids in the form of salmon oil, soybean oil and methyl linolenate resulted in a considerable improvement in feed utilization, growth rate, and a decrease in mortality. It seems evident from these experiments that 1% linolenic acid present in corn oil fed at 10% in the diet is not sufficient to meet the fatty acid requirements of rainbow trout and additional amounts are needed. It is of interest that apparently 10% corn oil in a semi-purified ration will supply the fatty acid requirements of the laboratory rat but is inadequate for rainbow trout.

As indicated in our studies, when salmon or menhaden oil replaced corn oil as a lipid source in the semi-purified diet from the start of feeding it produced excellent growth and virtually no mortality. This supports the growth studies of Hagashi *et al.* (20) who reported that rainbow trout grew well when fed a highly unsaturated fatty acid ethyl ester fraction prepared from squid liver oil in a semi-purified ration. It would appear from these results reviewed here and unpublished observations from our laboratory that rainbow trout probably require both ω-3 and ω-6 fatty acids with the evidence favoring a far greater requirement for the linolenic or ω-3 series. In 1963, Brockerhoff and Hoyle (41) suggested that the ω-3 or linolenic acid series may have the role in marine animals that the linoleic or ω-6 series have in terrestrial animals. Brenner *et al.* (19) reported evidence that fatty acids of the linoleic acid type were of less importance for fish than linolenic acid series and supported the earlier suggestion of Richardson *et al.* (42) that linolenic may replace the linoleic group. Studies in our laboratory at this time bear out this thesis.

PRACTICAL RATIONS

Meat-meal mixtures. Practical diets for rearing fish are composed of a variety of fresh or frozen animal products and by-products; such as horse meat, slaughter house scrap, liver, spleen, heart, fish, and a mixture of dry animal, fish and vegetable meals supplemented with special concentrates, yeast, distillers and fish solubles, vitamins and minerals. Consideration of rations of this type is beyond the scope of this paper and the reader is referred to other chapters in this volume and specifically to the annual publications of the Cortland Hatchery where studies on trout have been underway since 1932 led by Phillips and his co-workers (43). Tests of hatchery foods in formulating practical rations for salmon were the subject of a series of reports by Burrows *et al.* (44, 45) and Fowler *et al.* (46). Meat-meal mixtures vary in moisture content and are usually prepared as required and fed as a wet mush by various means such as a modified potato ricer, by spoon, or the use of a blower or compressed air (47).

Dry pelleted foods. Dry pelleted diets similar to the chow foods used for feeding other animals was a logical development in fish culture and now are widely used in fish hatcheries everywhere. Phillips *et al.* (47) discusses some of the advantages and disadvantages of dry pelleted foods and suggests that meat supplementation is required to make a complete trout food, which has been the procedure with brood stock, to stimulate increased egg production and fertility.

Frozen pelleted food. A few years ago a formula was developed by investigators at Oregon State University and the Fish Commission of Oregon (48) which combines the features of the meat-meal mixture diets and the dry pellets into a preformed soft-moist pellet which is stored in the frozen state. This ration has been found to be very successful and the results obtained by its use have been the key contributing factor in restoring the salmon runs of Oregon and Washington. It also had an important role in establishing the new coho fishery in the Great Lakes.

Fat in practical rations. The type of fat present in practical rations is dictated by the fatty acid composition of the individual components of the diet. Workers engaged in the rearing of trout and salmon sought to reduce the fat level whenever feasible to avoid rancid or oxidized fat, and the solid or high melting point fats such as those associated with beef by-products (47). There have been few attempts to add fat or supplement formulas of the meat-meal mixture type described earlier. Occasionally, cod liver oil has been added usually as a source of vitamins A and D. Oregon investigators recognizing the value of additional fat as an energy source supplemented the Oregon frozen pellet with corn (48), soy, salmon or herring oil (49), particularly now since solvent extraction is frequently used to remove the fat from vegetable and animal meals.

In commercial dry pelleted foods, fat supplementation with oils and feed grade fats is not usually attempted because of the difficulty in protecting the oil from autoxidation which leads to spontaneous heating and the production of possible toxic substances, destruction of vitamins and even to the loss in availability of certain important amino acids. The addition of protective substances such as antioxidants offers some protection against overheating and the results of autoxidation. In addition, most rations contain alpha tocopherol supplementation as a vitamin and *in vivo* antioxidant. The value of alpha tocopherol in the diet of higher animals as a protection against *in vivo* lipid peroxidation is well established (50). Woodall *et al.* (51) found the alpha tocopherol requirement of chinook salmon fed a diet containing five percent herring triglycerides to be about 3 mg/100 g of dry diet.

Sometime ago Phillips and Podoliak (52), in a review of fats in trout nutrition, stated that three percent cod liver oil in the diet of trout resulted in increased growth which was not attributable to vitamin A or D. In the same article, they state cod liver oil must be fed with caution because of the possibility of damage to the kidneys and liver. In light of present knowledge, it appears possible that the trout may have shown a need for the ω-3 fatty acids present in cod liver oil.

Fish used in research are frequently maintained prior to and during experimentation on a variety of dietary regimes, including dry commercial pellets, all assumed to be adequate for all nutritional needs. Although the requirements for the essential fatty acids have not yet been established, our studies (21) discussed earlier in this chapter direct attention to the importance of the ω-3 fatty acids in trout rations. As no reports are available that give fatty acid composition of commercial stock rations, samples of seven dry pelleted diets were obtained and their fat percentage and fatty acid composition determined by the usual procedure (21). The results shown in Table III indicate considerable variation in fat content and lipid composition. It should be recognized that the lipid composition is dependent on the formula being used at that time and will vary with the components. However, the data are believed to be a cross section of stock diets now employed for rearing fish. The results show the fat content to vary from 5.9 to 14.2 percent.

Examination of the fatty acid composition of the rations indicates a considerable range in values for the essential fatty acids and for the higher unsaturated fatty acids commonly present in fish body lipids. For comparative purposes the ratio of the total of ω-6 and ω-3 acids in the various rations are also presented in Table III. Similar ratios were derived for a few semi-purified diets used with varying success for rearing trout and salmon. The values in Table IV indicate that several commercial pelleted diets have fatty acid ratios similar to the semi-purified diets, in all cases with the ω-6 acids predominating. The actual amount of ω-3 fatty acids present in the diet should be carefully considered when evaluating a ration.

It is of interest that the fatty acid composition as naturally found in fish is the reverse of that found in the rations with the fatty acid ratio in favor of the ω-3 acids. For example, using the data presented by Gruger (6) one obtains the following ω-3/ω-6 ratios: chinook salmon - 11.8, chum salmon - 9.7, coho salmon - 14.8; pink salmon - 19.5 and rainbow trout - 4.9. In our laboratory during the past two years, we have used coho salmon body oil with a ω-3/ω-6 ratio of 16.6 as the total lipid source at ten percent level in a semi-purified diet for rearing rainbow trout and have achieved excellent growth and mortalities of less than one percent (53). The use of fish oil as the total lipid source is a radical departure from current hatchery practice and it is evident that more research is needed before the question of the most desirable level of ω-3 and ω-6 fatty acids in fish rations is answered.

TABLE III

FATTY ACID COMPOSITION OF COMMERICAL FISH RATIONS*

	Ration A Fat–9.0%		Ration B Fat–13.5%		Ration C Fat–5.9%		Ration D Fat–14.3%		Ration E Fat–6.7%		Ration F Fat–9.3%		Ration G Fat–10.5%	
Fatty Acid	% Comp	% in Feed	% Comp	% in Feed	% Comp	% in Feed	% Comp	% in Feed	% Comp	% in Feed	% Comp	% in Feed	% Comp	% in Feed
?	--	--	0.16	0.02	--	--	0.18	0.03	0.18	0.01	0.28	0.03	--	--
?	0.15	0.01	0.16	0.02	0.13	0.01	0.18	0.03	0.31	0.02	--	--	--	--
12:0	0.54	0.05	0.55	0.07	0.54	0.03	0.72	0.10	5.55	0.37	0.20	0.02	--	--
14:0	3.44	0.31	3.04	0.41	4.75	0.28	2.91	0.42	3.82	0.26	4.11	0.38	3.21	0.34
14:1	0.57	0.05	0.76	0.10	0.96	0.06	0.62	0.09	0.26	0.02	0.36	0.03	--	--
15:0	0.51	0.05	0.60	0.08	0.74	0.04	0.53	0.08	0.19	0.01	0.39	0.04	0.46	0.05
16:0	24.55	2.21	20.48	2.76	17.95	1.06	19.31	2.75	17.11	1.15	20.41	1.89	24.47	2.58
16:1	4.29	0.39	4.86	0.65	8.57	0.50	3.77	0.54	5.79	0.39	5.80	0.54	2.96	0.31
17:0	1.68	0.15	1.41	0.19	1.52	0.09	0.96	0.14	0.70	0.05	0.73	0.07	0.68	0.07
16:2 ω6	0.88	0.08	1.07	0.14	0.99	0.06	0.80	0.11	0.59	0.04	0.65	0.06	0.92	0.10
18:0	9.64	0.87	7.11	0.96	5.72	0.34	9.05	1.29	16.11	1.08	5.60	0.52	7.85	0.83
18:1 ω9	24.89	2.24	20.90	2.81	20.88	1.23	25.77	3.66	21.23	1.42	24.96	2.32	22.41	2.36
18:2 ω6	10.86	0.97	21.78	2.93	13.15	0.78	20.75	2.95	22.75	1.52	17.41	1.62	27.55	2.90
18:2 ω3	0.63	0.06	--	--	--	--	--	--	--	--	--	--	2.22	0.23
18:3 ω3	1.89	0.17	7.08	0.55	3.83	0.23	2.71	0.39	1.54	0.10	2.12	0.20	2.65	0.28
20:1 ω11	4.06	0.36	1.42	0.33	4.80	0.28	1.81	0.26	1.40	0.09	5.86	0.55	1.01	0.11
18:4 ω3	4.23	0.38	1.27	0.17	3.35	0.20	--	--	--	--	0.89	0.08	--	--
20:2 ω6	--	--	--	--	0.92	0.05	--	--	0.62	0.04	--	--	2.59	0.27
20:4 ω6	--	--	1.10	0.15	1.24	0.07	1.37	0.19	--	--	--	--	1.02	0.11
22:1	3.99	0.36	1.29	0.17	3.14	0.19	1.76	0.25	--	--	6.21	0.58	--	--
20:5 ω3	2.40	0.22	4.30	0.58	4.10	0.24	1.63	0.23	1.87	0.13	1.84	0.17	--	--
22:6 ω3	0.80	0.07	2.66	0.36	2.70	0.16	5.17	0.74	--	--	2.19	0.20	--	--
Total ω-3		0.90		1.66		0.83		1.36		0.23		0.65		0.51
Total ω-6		1.05		3.22		0.96		3.25		1.60		1.68		3.38
ω-3/ω-6		0.86		0.52		0.86		0.42		0.14		0.39		0.15

*The technical assistance of J. N. Roehm and Kathleen E. Oldfield is greatly appreciated.

TABLE IV

TOTAL DIENE AND POLYENE FATTY ACIDS IN EXPERIMENTAL RATIONS

Diet Description	Percent fatty acids in diet			Reference
	ω-3	ω-6	ω-3/ω-6	
Lipid				
1. Corn, 10%	0.12	5.7	0.021	(25)
2. Corn, 5%; Salmon, 5%[1]	0.87	3.13	0.28	(25)
3. Corn, 7%; Cod Liver, 2%*	1.11	4.04	0.27	(40)
4. Salmon, 10%[2]	3.23	0.19	16.6	(53)

[1]Chinook salmon oil
[2]Coho salmon oil
*Calculated from published data

CONCLUSION

It is apparent from this review of the role of lipids in the diet of fish that the present state of knowledge is quite limited. Recent progress reviewed in this volume indicates an awareness of the growing importance of lipids in the diets of fish. This is of special concern in controlled experiments when purified rations are employed over long periods ot time.

Finally, it may be concluded that defining the metabolic role of lipids in fish and their interrelationships with other dietary factors is a fertile field for research and will occupy the careful experimenter for some time to come.

REFERENCES

1. T. P. Hilditch and P. N. Williams, "The Chemical Constitution of Natural Fats," 4th ed., John Wiley and Sons, New York, 1964.

2. J. A. Lovern, Food Investigation Board Special Report No. 51, Dept. Scientific and Industrial Research. London, 1942.

3. J. A. Lovern, *In* "The Biochemistry of Fish" (R. T. Williams, ed.). Cambridge University Press, London, 1951.

4. B. E. Bailey, N. M. Carter, and L. A. Swain, Fisheries Res. Board Can., Bulletin 89 (1952).

5. M. E. Stansby, "Fish Oils, Their Chemistry, Technology, Stability, Nutritional Properties and Uses," Avi Publishing Co., Inc., Westport,

Conn., 1967.

6. E. H. Gruger, Jr., *In* "Fish Oils" (M. E. Stansby, ed.). Avi Publishing Co., Inc., Westport, Conn., 1967.

7. R. G. Ackman, J. Fisheries Res. Board Can. **21**: 247 (1964).

8. R. G. Ackman and R. D. Burgher, J. Fisheries Res. Board Can. **21**: 319 (1964).

9. J. A. Lovern, Biochem. J. **32**: 676 (1938).

10. J. B. Saddler, R. R. Lowry, H. M. Krueger, and I. J. Tinsley, J. Am. Oil Chemists' Soc. **43**: 321 (1966).

11. R. Reiser, B. Stevenson, M. Kayama, R. B. R. Choudhury and D. W. Hood, J. Am. Oil Chemists' Soc. **40**: 507 (1963).

12. W. G. Knipprath and J. F. Mead, Lipids **1**: 113 (1966).

13. Ibid., Lipids **3**: 121 (1968).

14. E. Klenk and H. Steinbach, Z. Physiol. Chem. **316**: 31 (1959).

15. E. Klenk and G. Kremer, Z. Physiol. Chem. **320**: 111 (1960).

16. J. F. Mead, M. Kayama and R. Reiser, J. Am. Oil Chemists' Soc. **37**: 438 (1960).

17. P. B. Kelly, R. Reiser, and D. W. Hood, J. Am. Oil Chemists' Soc. **35**: 189 (1958).

18. Ibid., J. Am. Oil Chemists' Soc. **35**: 503 (1958).

19. R. R. Brenner, D. V. Vazza, and M. E. DeTomas, J. Lipid Res. **4**: 341 (1963).

20. H. Higashi, T. Kaneko, S. Ishii, M. Ushiyama and T. Sugihashi, J. Vitaminol. **12**: 74 (1966).

21. D. J. Lee, J. N. Roehm, T. C. Yu, and R. O. Sinnhuber, J. Nutr. **92**: 93 (1967).

22. M. Kayama, Y. Tsuchiya, J. C. Nevenzel, A. Fulco and J. F. Mead, J. Am. Oil Chemists' Soc. **40**: 499 (1963).

23. J. F. Mead, *In* "Progress in the Chemistry of Fats and Other Lipids," (R. T. Holman, ed.) Vol. IX. Pergamon Press, New York, 1968.

24. J. F. Mead, *In* "Fish Oils" (M. E. Stansby, ed.). Avi Publishing Co., Inc., Westport, Conn., 1967.

25. N. Nicolaides and A. N. Woodall, J. Nutr. **78**: 431 (1962).

26. R. T. Holman, Federation Proc. **23**: 1062 (1964).

27. C. M. McCay, F. C. Bing and W. S. Dilley, Trans. Am. Fisheries Soc. **57**: 240 (1927).

28. C. M. McCay, *In* "Annual Review of Biochemistry" (J. M. Luck, ed.), pp. 445-468. Annual Reviews, Inc., Palo Alto, California, 1937.

29. J. B. Field, E. F. Herman, and C. A. Elvehjem, Copeia **3**: 184 (1944).

30. B. A. McLaren, E. F. Herman, and C. A. Elvehjem, Arch. Biochem. Biophys. **10**: 433 (1946).

31. B. A. McLaren, E. Keller, D. J. O'Donnell and C. A. Elvehjem, Arch. Biochem. Biophys. **15**: 169 (1947).

32. B. A. McLaren, E. Keller, D. J. O'Donnell and C. A. Elvehjem, Arch. Biochem. Biophys. **15**: 179 (1947).

33. T. B. McKee, R. O. Sinnhuber and D. K. Law, Oregon Fish Comm. Res. Briefs **3**: 22 (1951).

34. T. B. McKee, E. R. Jeffries, D. K. McKernan, R. O. Sinnhuber, and D. K. Law, Oregon Fish Comm. Res. Briefs **4**: 25 (1952).

35. E. R. Jeffries, T. B. McKee, R. O. Sinnhuber, D. K. Law, and T. C. Yu, Oregon Fish Comm. Res. Briefs **5**: 32 (1954).

36. W. F. Hublou, T. B. McKee,D. K. Law, R. O. Sinnhuber, and T. C. Yu, Oregon Fish Comm. Res. Briefs **9**: 57 (1963).

37. R. O. Sinnhuber, D. K. Law, T. C. Yu, T. B. McKee, W. F. Hublou, and J. W. Wood, Oregon Fish Comm. Res. Briefs **8**: 53 (1961).

38. L. E. Wolf, Progressive Fish Culturist **13**: 17 (1951).

39. J. E. Halver, J. Nutr. **62**: 225 (1957).

40. J. E. Halver, Personal Communication (1966).

41. H. Brockerhoff and R. J. Hoyle, Arch. Biochem. Biophys. **102**: 452 (1963).

42. T. Richardson, A. L. Tappel, and E. H. Gruger, Jr., Arch. Biochem.

Biophys. **94**: 1 (1961).

43. Fisheries Research Bulletin. The Nutrition of Trout. New York State Conservation Dept. Albany, N. Y. Published Annually Since 1932.

44. R. E. Burrows, L. A. Robinson and D. D. Palmer, U. S. Fish and Wildlife Service, Special Scientific Report (Fisheries) No. 59, p. 39 (1951).

45. R. E. Burrows, D. D. Palmer, H. W. Newman and R. L. Azevedo, U. S. Fish and Wildlife Service, Special Scientific Report (Fisheries) No. 86, p. 24 (1952).

46. L. G. Fowler and J. L. Banks, U. S. Fish and Wildlife Service, Bureau of Sport Fisheries, Technical Paper 13, p. 1 (1967).

47. A. M. Phillips, A. V. Tunison and G. C. Balzer, U. S. Fish and Wildlife Service, Washington, D. C. Circular 159, p. 1 (1963).

48. W. F. Hublou, J. Wallis, T. B. McKee, D. K. Law, R. O. Sinnhuber, and T. C. Yu, Oregon Fish Comm. Res. Briefs **7**: 28 (1959).

49. D. K. Law, Personal Communication (1969)

50. M. C. Horwitt, C. C. Harvey, B. Century, and L. A. Whitting, J. Am. Dietet. Assoc. **38**: 231 (1961).

51. A. N. Woodall, L. M. Ashley, J. E. Halver, H. S. Olcott, J. van der Veen, J. Nutr. **84**: 125 (1964).

52. A. M. Phillips, Jr., and H. A. Podoliak, Progressive Fish Culturist **19**: 68 (1957).

53. R. O. Sinnhuber, Unpublished Results.

COMMENTS

MR. WOODALL: I was curious about the requirement for your ω-3 fatty acids. These are needed primarily for structural fats for the phospholipids which comprise a rather small amount of the total fat stores in the animal. I would guess without referring back to data that about 15-20% of the total fat would be phospholipids. Have you an idea how much of this should be in ω-3 fatty acids?

PROF. SINNHUBER: According to the data on phospholipid analyses (Table II) when the 22:6ω3 dropped to 3.14% (Diet A), the fish were in trouble and mortalities occurred. When the level rose to 5.69% (Diet D), mortalities ceased and growth resumed. One might assume from this that the former value is critical.

MR. WOODALL: I agree and did not intend to slight that work. What I was comparing was phospholipid and storage fat ratios. I want also to mention the need for real care in getting a fatty acid test diet. We also found that dextrin, although it had a very small amount of fat, contained essential fatty acids and we had to remove these before we could develop any fatty acid deficiency syndromes.

DR. STEINER: Did you characterize the lipids of cancer of trout liver? It is known that in many human tumors one can find lipids which contain ether linkages.

PROF. SINNHUBER: Dr. Fred Snyder at Oak Ridge has worked on lipids in cancerous tissue and has found these unusual lipids. We have not characterized trout hepatoma lipids and plan to do this in the near future.

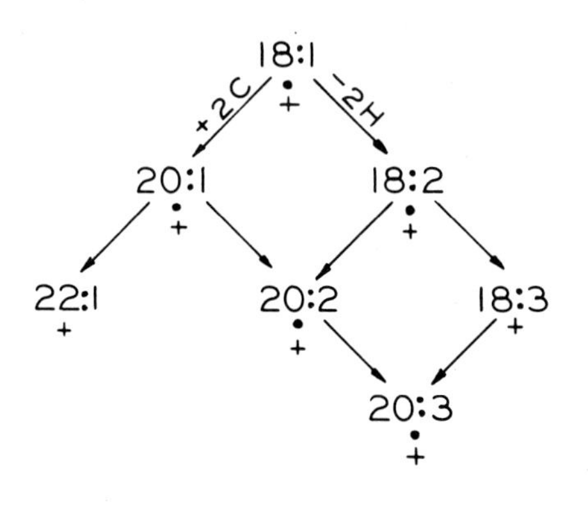

Fig. 1: Oleic acid series ω-9.

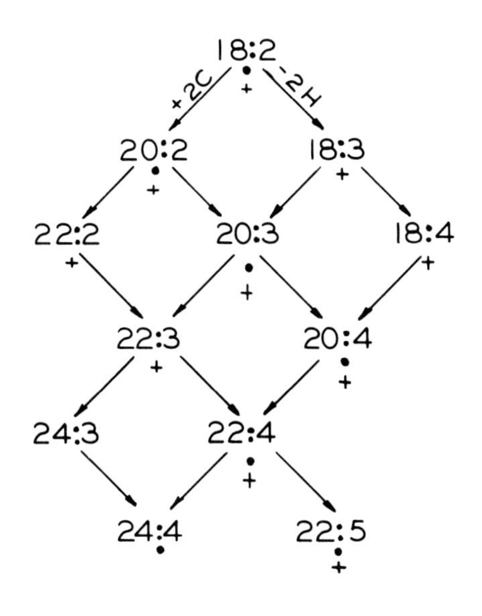

Fig. 2: Linoleic acid series ω-6.

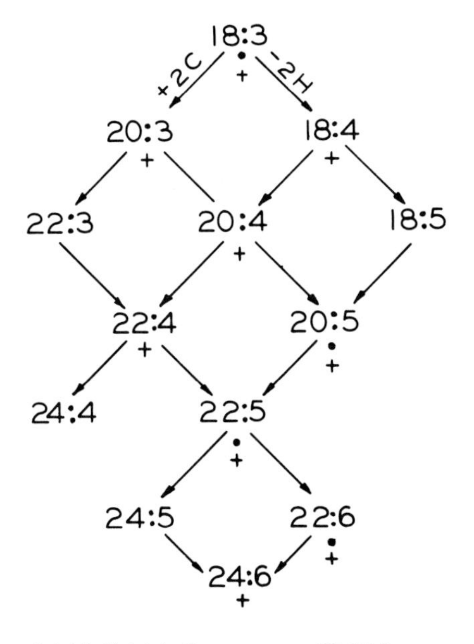

Fig. 3: Linolenic acid series ω-3.

NUTRITIONAL SCORE

W. H. Hastings

INTRODUCTION

"Nutritional Score" measures nutrient content, digestibility and utilization. It may be expressed in chemical, biochemical or biological terms. In a broad sense it may be used to calculate feed formulas, growth curves and the economics of production. In detail, Nutritional Score includes a knowledge of the chemical composition of food, its digestibility, requirements for maintenance and growth, utilization of each nutrient and the disposition of metabolized products in terms of weight gains.

Feeds are formulated and sold on the basis of proximate analyses, a chemical measurement of nutrient classes. A closer guideline to performance is found by measuring digestibility of ingredients, singly or in combination. A still closer guideline is a measurement of utilization. Digestibility and utilization of nutrients by fish will be considered in this paper.

DIGESTIBILITY

Tables of digestibility coefficients (1) and metabolizable energy (2), widely accepted by livestock and poultry nutritionists, are needed for economical fish culture. Digestibility is expressed as percent retention and derived from the equation:

$$\frac{\text{Nutrient intake - Nutrient in feces}}{\text{Nutrient intake}} \times 100 \qquad \text{(A)}$$

or by the indirect method of adding to the nutrient an inert indicator (usually (Cr_2O_3), calculating digestibility of a food component percentages rather than in quantitative amounts.

To date, information on digestion is in terms of nitrogen and calories. These values are, for the most part, one dimensional. Table I lists over one hundred digestion coefficients of ingredients and combinations of ingredients. However, current information is available to question the use of coefficients of digestion unless environmental and management conditions are similar to those in the original test. Digestion coefficients may become two dimensional under the influence of size, age, population density, amount of food, quality of food, frequency of feeding, previous level of nutrition (i.e. starvation), forced versus *ad libitum* feeding, water temperature, nutrient density and presence of toxins.

Initial size and/or age

Gerking (3) and Windell (4) report "there was no difference in the ability of fish of different ages to absorb protein," and Windell added to this "among fish of different ages and body weights." Brown (5) concludes that physiological time to a fish is related to size, not age.

Quite recently Kitamikado *et al.* (6) reported that the level of digestibility of protein for rainbow trout was the same for those fish weighing 10-100 grams, but for smaller fish (<6 grams) digestibility of white fish meal and casein decreased. DeLong *et al.* (7) found that when metabolic activity is at a reduced level as at low water temperatures, young chinook salmon are unable to utilize high levels of protein and a growth depression results, not observed with older fish weighing about 6 grams or more.

TABLE I

DIGESTION COEFFICIENTS OF FISH FEEDS COMPILED FROM DATA IN PUBLICATIONS CITED

Ingredient		Test Fish	Digestion Coefficient		Reference
			Nitrogen	CHO	
Crude protein		Carp	69-92		49
Raw beef heart		Brook trout	95.7		50
Cooked beef heart		Brook trout	95.7		50
Raw beef liver		Brook trout	86.3		50
Pork spleen		Brook trout	92.5		51
Pork spleen		Brook trout	93.7		51
Spleen Skim milk	2 1	Brook trout	79.0		51
Spleen Cottonseed meal	2 1	Brook trout	76.0		51
Spleen Wheat middlings	2 1	Brook trout	76.0		51
Beef liver		Brook trout	95.0		51
Pork spleen Raw corn starch	3 1	Brook trout	87.0		51
Pork spleen Cooked corn starch	3 1	Brook trout	80.0		51
Pork spleen Sucrose	3 1	Brook trout	73.0		51
Pork spleen Beef liver	1 1	Brook trout	96.0		51
Beef liver		Rainbow trout	89.2-94		52
Pork spleen		Rainbow trout	85.3-91.0		52
Spleen Dry meal mixture	1 1	Rainbow trout	91.5-92.4		52
Liver Spleen	1 1	Rainbow trout	86.3-91.7		52

Table I -- continued.

Ingredient		Test Fish	Digestion Coefficient		Reference
			Nitrogen	CHO	
Meal worms		Longear sunfish	97.4		3
Meal worms		Green sunfish	95.7		3
Potato starch Dry chlorells	4 1	Goldfish	25.9±8.7		53
Potato starch Dry chlorells	1 1	Goldfish	47.9±9.6		53
Potato starch Dry chlorells	1 4	Goldfish	54.0-63.0		53
Silk worm pupa Wheat flour Raw starch	5 4 1	Goldfish	80.0		53
Silk worm pupa Wheat flour Raw starch	5 4 1	Rainbow trout	65-70		53
Silk worm pupa		Rainbow trout	80-88		53
Red fish meal		Rainbow trout	62-68		53
Casein		Rainbow trout	96.1		53
Egg albumin		Rainbow trout	83.0		53
Soybean meal		Rainbow trout	87.5		53
White fish meal		Rainbow trout	90.2		53
Frozen fish		Rainbow trout	94		6
Frozen beef liver		Rainbow trout	94		6
White fish meal		Rainbow trout	81		6
Fish meal		Rainbow trout	72		6
Dried silk worm pupa		Rainbow trout	85		6
Dried mysid		Rainbow trout	83		6
Soybean meal		Rainbow trout	71		6

Table I -- continued.

Ingredient		Test Fish	Digestion Coefficient		Reference
			Nitrogen	CHO	
Casein		Rainbow trout	94		6
Casein Gelatin	1 1	Rainbow trout	95		54
Gelatin		Rainbow trout	88.6		54
Casein		Rainbow trout	79		54
Cottonseed meal No. 1		Rainbow trout	84		54
Cottonseed meal No. 2		Rainbow trout	72.7		54
Fish meal No. 1		Rainbow trout	72.7		54
Fish meal No. 2		Rainbow trout	75.6		54
Fish meal No. 3		Rainbow trout	93		54
Salmon meal		Rainbow trout	80.0		54
Feather meal No. 1		Rainbow trout	52.4		54
Feather meal No. 2		Rainbow trout	70.5		54
Brewers' yeast		Rainbow trout	80.5		54
Skimmilk		Rainbow trout	94.7		54
Wheat middlings		Rainbow trout	69.2		54
Whole soybeans		Channel catfish	0		55
Raw ground soybeans		Channel catfish	31		55
Raw ground soybeans Rice hulls	4 1	Channel catfish	30		55
Soybean meal (50% protein)		Channel catfish	72		55
Dehydrated alfalfa meal		Channel catfish	12		55
Poultry biproducts meal		Channel catfish	27		55
Cottonseed meal		Channel catfish	76		55

Table I -- continued

Ingredient		Test Fish	Digestion Coefficient		Reference
			Nitrogen	CHO	
Meat scraps (50% protein)		Channel catfish	42		55
Fish meal (70% herring)		Channel catfish	80		55
Fish meal (60% menhaden)		Channel catfish	73.5		55
Milo gluten meal		Channel catfish	41		55
Corn gluten meal		Channel catfish	80		55
Blood meal		Channel catfish	23		55
Rice bran		Channel catfish	71		55
Distillers' solubles		Channel catfish	67		55
Dehydrated alfalfa meal		Goldfish	66		55
Casein Gelatin	7 3	Rainbow trout	88.7	88.6	14
Casein Gelatin	2 1	Rainbow trout	84.4	85.5	14
Casein Gelatin Glucose	3.5 1.5 5.0	Rainbow trout	87.9	84.1	14
Casein Gelatin Dextrin	3.5 1.5 5.0	Rainbow trout	81.6	80.2	14
Casein Gelatin Raw corn starch	3.5 1.5	Rainbow trout	84.9	60.1	14
Casein Gelatin Cooked corn starch	3.5 1.5	Rainbow trout	85.8	71.0	14
Casein Gelatin Cellulose flour	3.5 1.5 5.0	Rainbow trout	59.8	36.6	14

Table I -- continued

Ingredient		Test Fish	Digestion Coefficient		Reference
			Nitrogen	CHO	
White fish meal Potato starch	9 1	Rainbow trout	81		56
White fish meal Potato starch	8 2	Rainbow trout	32		56
White fish meal Potato starch	6 4	Rainbow trout	78		56
White fish meal Potato starch	4 6	Rainbow trout	74		56
Casein Potato starch	9 1	Rainbow trout	96		56
Casein Potato starch	8 2	Rainbow trout	94		56
Casein Potato starch	6 4	Rainbow trout	89		56
Casein Potato starch	4 6	Rainbow trout	82		56
Casein Potato starch	2 8	Rainbow trout	79		56
Silk worm pupa Potato starch	8.5 1.5	Rainbow trout	86		56
Silk worm pupa Potato starch Worm oil	7.65 1.35 1.00	Rainbow trout	86		56
Silk worm pupa Potato starch Worm oil	6.8 1.2 2.0	Rainbow trout	86		56
Silk worm pupa Potato starch Worm oil	5.95 1.05 3.00	Rainbow trout	83		56
Casein Potato starch Olive oil	8.5 1.5 0	Rainbow trout	97		56
Casein Potato starch	7.65 1.35	Rainbow trout	97		56

Table I -- continued

Ingredient		Test Fish	Digestion Coefficient		Reference
			Nitrogen	CHO	
Olive oil	1.00				
Casein Potato starch Olive oil	6.8 1.2 2.0	Rainbow trout	96		56
Casein Potato starch Olive oil	5.95 1.05 3.00	Rainbow trout	96		56
White fish meal Bread crumbs	1 2	Rainbow trout	85.3	48.2	15
White fish meal Wheat flour	1 2	Rainbow trout	88.0	22.0	15
White fish meal Bread crumbs	1 1	Rainbow trout	91.4	81.7	15
White fish meal Bread crumbs	2 1	Rainbow trout	94.0	83.9	15
White fish meal Bread crumbs	3 1	Rainbow trout	92.8	90.0	15
Casein Gelatin Glucose	5 1.5 2	Rainbow trout		99.3	16
Casein Gelatin Glucose	4 1.5 3	Rainbow trout		99.0	16
Casein Gelatin Glucose	3 1.5 4	Rainbow trout		99	16
Casein Gelatin Glucose	2.0 1.5 5	Rainbow trout		99.6	16
Casein Gelatin Glucose	1.0 1.5 6.0	Rainbow trout		99.5	16
Casein Gelatin Sucrose	5.0 1.5 2.0	Rainbow trout		99.5	16

Table I -- continued

Ingredient		Test Fish	Digestion Coefficient		Reference
			Nitrogen	CHO	
Casein Gelatin Sucrose	4.0 1.5 3.0	Rainbow trout		98.8	16
Casein Gelatin Sucrose	3.0 1.5 4.0	Rainbow trout		99.1	16
Casein Gelatin Sucrose	2.0 1.5 5.0	Rainbow trout		99.2	16
Casein Gelatin Sucrose	1.0 1.5 6.0	Rainbow trout		98.8	16
Casein Gelatin Lactose	5.0 1.5 2.0	Rainbow trout		94.4	16
Casein Gelatin Lactose	4.0 1.5 3.0	Rainbow trout		95.3	16
Casein Gelatin Lactose	3.0 1.5 4.0	Rainbow trout		97.4	16
Casein Gelatin Lactose	2.0 1.5 5.00	Rainbow trout		97.2	16
Casein Gelatin Lactose	1.0 1.5 6.0	Rainbow trout		96.4	16
Casein Gelatin Dextrin	5.0 1.5 2.0	Rainbow trout		77.2	16
Casein Gelatin Dextrin	4.0 1.5 3.0	Rainbow trout		74.8	16
Casein Gelatin Dextrin	3.0 1.5 4.0	Rainbow trout		60.0	16

Table I -- continued

Ingredient		Test Fish	Digestion Coefficient		Reference
			Nitrogen	CHO	
Casein Gelatin Dextrin	2.0 1.5 5.0	Rainbow trout		50.1	16
Casein Gelatin Dextrin	1.0 1.5 6.0	Rainbow trout		45.5	16
Casein Gelatin Potato cooked starch	5.0 1.5 2.0	Rainbow trout		69.2	16
Casein Gelatin Potato cooked starch	4.0 1.5 3.0	Rainbow trout		65.3	16
Casein Gelatin Potato cooked starch	3 1.5 4	Rainbow trout		52.7	16
Casein Gelatin Potato cooked starch	2 1.5 5	Rainbow trout		38.2	16
Casein Gelatin Potato cooked starch	1 1.5 6.0	Rainbow trout		26.1	16

Feeding level

Nose (8) testing nitrogen and Davies (9) testing total energy, reported that digestion efficiency increased as food intake increased. Davies demonstrated lower "energy extraction efficiency" (his term for digestibility) at low intake levels and postulated an optimum range beyond which efficiency would drop. Kinne (10) observed the passage of undigested food in the feces of fish on unrestricted intake and related this to digestibility.

Testing single and multiple meals of natural organisms, Windell (11) found that although gastric evacuation increased with the level of food intake, the rate of digestion remained about the same. Currently Windell (4) feels that "digestibility is considered a function of rate of . . . gastric evacuation."

For a wide range of food intake, the coefficient of digestion as measured by several authors seems to remain constant. At low levels, which would be found for small percentages of nutrients in a food or for small percentages of food used per body weight of fish, digestion efficiency as found by the formula (A) is consistently low. This may be due to the additive effect of metabolic nitrogen, so that the formula:

$$\frac{\text{Intake N - (Fecal N - Metabolic N)}}{\text{Intake N}} \times 100 \qquad \text{(B)}$$

truly reflects the absorption of nitrogen from a test material independent of metabolic nitrogen and results in a higher coefficient at low food intake than "apparent digestibility" as found by equation (A). Nose (8) concludes that "true digestibility of protein is constant and independent of protein level of the diet." This affirms a summary by Gerking (12): "in the present calculations (N absorbed/N consumed) the protein nitrogen consumed is considered to be equivalent to that absorbed."

For natural foods and test diets simulating the texture and moisture content of natural foods, we find that as food intake increases, apparent digestibility increases, and then decreases for abnormal intake. Recently Windell (4) has found that feed pellets behave differently. Following a high gastric evacuation of digestible food at low levels of consumption, digestion leveled off to 50% for pellet intake beyond 1.2% of body weight and for fish fed three times daily. Excessive intake (2.5% of body weight) was associated with finding hard pellets in the intestinal tract beyond the range of gastric activity. With both coldwater and warmwater commercial fish culture dependent upon dry feedstuffs in multiple combinations, intensive research is needed to compare coefficients of digestibility as reported for common wet hatchery foods.

Feeding technique

In order to have known and uniform food intake, various techniques of forced feeding are often used. Dodd *et al.* (13) reports a suitable mechanical device and data to show that dogfish sustain good growth fed in this way.

Windell (11) reports that one of the initial effects of manual manipulation of forced-feeding is either immediate or delayed regurgitation. He measured gastric evacuation 18 hours after *ad libitum* feeding and found that the average decrease was 63.1% with a range of 48.4 to 78.3%. For the forced-feeding group, the average decrease was 82.1% with a range of 52.8 to 100%. He concluded that forced-feeding not only increased the rate of digestion but also caused increased variation among individual fish. Therefore, if a fixed time in which digestion less than 100% were selected for measuring fecal nitrogen as a ratio to intake nitrogen, the effect of feeding method should be examined.

Feeding history

Another factor to be considered in establishing reliable data for digestibility is the effect of starvation or the previous level of feeding. Workers who report a fasting period before test feeding to assure complete removal of all food from the alimentary tract often do not state the length of that time. Windell (11) found that holding fish for test feeding may influence digestibility. Bluegill sunfish starved for 7, 14 and 25 days showed a progressive decrease in rate of digestion. Examination of the digestive tract of starved fish showed morphological changes (shrunken and atrophied) in the pyloric caeca. It appears that the pyloric caeca have a dual function in fish, being a site of enzyme secretion and nutrient absorption, since normal digestive function was not restored during the 22-hour period of sampling. Slow passage of food from the stomach under conditions of previous starvation may be an efficiency mechanism to allow enzymes and digestive juices to attain a level of excretion essential for the amount of substrate present. For comparison with other tests, history of starvation or other deprived levels of nutrient intake may be important for accurately measuring digestibility as total input minus fecal material at given time intervals.

Feeding frequency

Kinne (10), Davies (9) and Windell (11) have suggested that frequent feeding may influence digestibility via forcing food through the digestive tract for fast maximum absorption. There is not sufficient evidence on the use of natural foods (insects, larvae, worms) to prove that a maximum stomach capacity can be exceeded for sunfish and channel catfish so that a fish fed *ad libitum* will mechanically hasten undigested food beyond the point of digestion. However, Kinne (10) watched the desert pupfish (extremely competitive) on an unrestricted diet excrete undigested natural food. Since references by Windell (11, 4) relate that gastric absorption and immediate intestinal absorption take place in fish, frequency of feeding may be important for measuring the coefficient of digestion by the historical ratio.

Interaction of other nutrient classes

So far, two dimensional responses have been shown for variations in digestion due to feeding management and fish size. The digestibility of protein in foods containing various levels of other nutrients (starch, fat) indicates a third dimensional response due to nutrient density. Kitamikado *et al* (6) found a drop in protein digestion as white fish meal was replaced by potato starch in the diet of rainbow trout. Increased oil in the diet had no effect on protein digestion even though protein percentage decreased. (They also observed that digestibility of fish meal was improved appreciably by using the portion passing through a No. 50 sieve (297 microns opening). Whether this portion contained less bone and more tissue protein or was representative of the entire material but in smaller particles, was not determined, but it presents another dimension in measuring digestibility.) Again in rainbow trout, Smith (14) observed that protein nitrogen was equally digested in diets containing 50% glucose, dextrin, raw and cooked corn starch, but in those containing 50% cellulose flour both nitrogen and total digestibility were reduced. Carbohydrate digestion measured by difference (total energy minus nitrogen energy) appeared lower for corn starch than for glucose and dextrin.

Inaba *et al.* (15) reported decreased starch digestibility in rainbow trout (tested by the Cr_2O_3 method) as this nutrient increased in diets containing fish meal and/or dried mysids. Singh and Nose (16) showed a constant high digestibility of glucose, sucrose and lactose regardless of their concentrations in the diet. Digestibility of dextrin and potato starch was lower than for the di- or monosaccharides and decreased further with increasing levels in the diet. This suggested that at higher concentrations trout possess insufficient amylase to hydrolyze polysaccharides to soluble form.

A "maximum ingestion level" of digestible carbohydrate has been suggested by Phillips *et al.* (17) and Strand (18), beyond which high-glycogen livers result. The simultaneous addition of epinephrine to high-carbohydrate food prior to feeding accelerated weight gains of salmon (*Salmo salar*) and restored values of hepatic glycogen to normal.

CHEMICAL EVALUATION

As more is published on fish requirements and digestibility of nutrients, other than major classes (protein, total energy), other expressions of evaluation may be used. Several foods have been measured in terms of "Chemical Score" (19), which is a comparison of the essential amino acids in a single protein or in a mixed feed with those of whole egg protein, the "score" being defined as "100 minus the greatest percentage of deficit." Another measure is the "Essential Amino Acid Index," developed by Oser (20) and Mitchell (21). This evaluates a protein on the basis of the relative quantities of all the essential amino acids, again using whole egg protein as a standard. The E.A.A. Index shows greater correlation with biological values of formula feeds than does the chemical score (22). Tables II and III show common ingredients evaluated in terms of chemical score and essential amino acid index.

TABLE II

PROTEIN EFFICIENCY RATIO (GRAMS FISH GAIN/GRAM PROTEIN INTAKE) FOR TEST DIETS AND DIETARY INGREDIENTS FED TO CHINOOK SALMON
(Courtesy, W. E. Shanks, BSFW, Portland, Oregon)

Diet Ingredient	Weight Gain	Protein Fed	P.E.R.	%Protein Utilization
Casein	63.2	26.5	2.38	31.44
Liver	114.6	37.7	3.04	39.4
Cottonseed meal	13.7	15.4	0.91	10.00
Fish meal	96.2	36.3	2.65	35.43
Brewers yeast	21.1	25.0	0.85	10.03
Protein 27: Dextrin 10	71.0	89.5	0.80	10.3
Protein 27: Dextrin 22	104	98.1	1.17	13.8
Protein 27: Dextrin 34	169	120.5	1.40	18.3
Protein 32.5: Dextrin 10	153	122.6	1.25	15.8
Protein 32.5: Dextrin 22	237	150.8	1.57	20.4
Protein 32.5: Dextrin 34	340	191.3	1.78	23.1
Protein 40: Dextrin 10	294	197.7	1.49	19.3
Protein 40: Dextrin 22	353	220.8	1.60	20.8
Protein 40: Dextrin 34	359	253.7	1.42	18.4
Protein 47.5: Dextrin 10	382	235.6	1.62	21.1
Protein 47.5: Dextrin 22	496	312.7	1.59	20.4
Protein 47.5: Dextrin 34	433	313.4	1.38	18.0

* As Casein

TABLE III

EVALUATION BY CHEMICAL SCORE AND ESSENTIAL AMINO ACID INDEX OF COMMON FEED INGREDIENTS AND A FEW DIETS

Dietary Ingredient	Chemical Score	Ist Limiting	Essential A.A. Index*
Whole egg - Orr and Watts	100	none	99.2
Salmon eggs	85	tryptophan	91.0
Casein	82	arginine	88.4
Past. Salmon Carcass	78	tryptophan	85.8
Autolyzed Salmon Carcass	78	tryptophan	85.5
Herring meal	76	tryptophan	84.0
Commercial Feed No. 1	76	tryptophan	84.1
Commercial Feed No. 2	76	tryptophan	83.4
Commercial Feed No. 3	71	tryptophan	78.5
Commercial Feed No. 4	71	tryptophan	75.6
Commercial Feed No. 5	74	tryptophan	81.6
Commercial Feed No. 6	68	tryptophan	74.4
Dried Skimmilk	76	tryptophan	82.9
Drackett Soybean	75	methionine	82.2
Salmon Viscera	74	tryptophan	82.0
Turbot	72	tryptophan	79.1
Soybean meal	71	methionine	77.5
Oregon Moist Pellet	69	tryptophan	75.6
McNenny Formula No. 31	68	tryptophan	75.3
Salmon meal	67	isoleucine	74.4
Tuna Viscera	66	tryptophan	72.9
Sesame meal	63	lysine	72.5
Brewers Yeast	63	methionine	70.1
Wheat germ meal	62	tryptophan	67.6
Distillers solubles	59	tryptophan	65.0
Cottonseed meal	59	tryptophan	64.6
Wheat middlings	56	isoleucine	61.8
Shrimp meal	50	tryptophan	55.6
Crab solubles	44	tryptophan	48.3

* Ratio (%) of essential amino acids in test material to standard (egg protein).

UTILIZATION

Ideally, utilization of nutrients is expressed in comparable units, i.e. efficiency of energy intake is evaluated by measuring energy used and stored during a time period. Nutrients are commonly reported today as dry weights, calories, protein or a combination of these. The relative merits of using calories or protein are discussed by Gerking (23) and by Mann (24), either reference being appropriate to the understanding of fish response to feed. Ostapenya and Sergeev (25) found that a number of natural foods reflect a simple and regular relationship between protein, fat, carbohydrate and utilization, enabling the investigator to interpret results of experiments where equivalents were not reported. In other experiments using food mixtures or artificial foods, extensive chemical analyses are required.

Indices of Utilization

Given reliably determined and appropriately reported measures of growth and food intake, indices of utilization are generally derived as simple ratios. Two such are commonly employed. The first, familiar in fish culture as the ratio of food intake to weight gain, is called the "Conversion Factor." This term is variously known throughout the world as "Absolute food quotient" (26), "Nutrient Quotient" (27), and "Utilization Coefficient" (28). Since the percentage of food intake appearing as equivalent units of growth varies from zero at maintenance level to a maximum value of about 50 percent, values of conversion will rarely be less than 2.0 and values of 1.0 or less are a theoretical absurdity. In practice it has become conventional to express food consumed in dry weight and fish gain in wet weight. While such values may be useful comparative indices for a series of tests conducted at one time and place, they reflect a lack of accuracy unless exact equivalent units of dry and wet weights are also reported.

For pond culture, Swingle (29) proposed two calculations for food conversion. The "S" value = pounds of food used divided by total pounds of fish produced by natural plus added feeds, and the "C" value, food used divided by total pounds of fish produced minus that produced in a control pond without added food.

A second ratio used in measuring utilization is simply the inverse of "Conversion Factor," i.e. weight gain/weight of food consumed. This is the "Conversion Efficiency" of Kinne (10) and as used by him is calculated by "dry weight of fish gain x 100/dry weight of food," an index comparable to any research data recalculated to a dry weight, calorie or protein intake.

Utilization calculated in this form has been used to report the broad trophic relations of fish in a controlled environment. Davis and Warren (30) studied the energy relations of a laboratory system containing known levels of fish biomass and food biomass. Tabulations were made of: fish biomass (cal.), food consumed (cal.), food not assimilated (cal.), food assimilated (cal.), growth (cal.), respiration (cal.), growth as percentage of food consumed (cal.), and growth as percentage of food assimilated (cal.). Reference will be made later to the effect of level of food intake on efficiency of energy utilization, but sufficient now is the observation that at optimum intake nearly 37% of ingested and 50% of assimilated biomass was used for growth. Approximately one-half of the food energy intake is likely to be used for metabolic activities (31); therefore, values of conversion efficiency greater than 50% are high.

Factors Affecting Food Utilization

A variety of biotic and abiotic factors affect food utilization (32). A few of these have been analyzed by research workers and are possible guidelines for the economic usefulness of food conversion as a physiological tool.

Level of feeding. A certain quantity of energy is needed for basal, routine or active metabolism. This energy must come from body tissues or from ingested food. If fish eat an amount of food just equal to the metabolic expense at a certain plane of activity, no production results and food conversion calculated as ingested food divided by weight gain is infinity. Increments of food beyond that needed for maintenance will be metabolized into weight gain and calculated as finite conversion.

Under conditions of pond equilibrium where natural foods support a static carrying capacity, any additions to the food supply may cause a spectacular increase in fish production. Warren *et al.* (33) reported that most of the natural food produced in a test stream was required for trout maintenance. Added nutrients (in the form of cannery waste) increased the biomass in terms of kilocalories per square meter, allowing a two-fold increase in food consumption and a resulting seven-fold increase in trout growth.

Just as there is a minimum level of food required for maintenance, there is a minimum level required for optimum growth. In a balanced food, the chemical nutrients which satisfy the inherent capacity for growth at a certain level of metabolic activity, are contained in a minimum weight. DeLong *et al.* (7) express this in searching for the minimum protein (or amino acid level) for optimum growth. Swingle (34) appraises the change in food conversion at different levels of feeding: "the relationship between conversion rate and percentage fed is parabolic in nature with the parabola opening upward." Fig. 1 shows that for channel catfish, conversion varies with the amount of food used, being best at an intake slightly more than 2 percent of body weight. Similar data are available for Tilapia (35).

Other factors contributing to conversion must be standardized in order to determine accurately the relationship of the independent variable "amount of food." Conversion shown in Fig. 1 is the ratio of dry food in pounds and net yield of wet fish in pounds. Ingestion of some natural food by pond fish contributes to a lower conversion value than may apply to the same food used in raceways, troughs and aquariums. In the above reference (34) the nutrient requirements of 730 half-pound fish stocked per acre so greatly exceeded that available from natural food that it may be concluded that weight gains were caused only by added food.

Pentelow (36) found that brown trout (*Salmo trutta*) responded to increasing amounts of fresh water shrimp (*Gammarus pulex*) with increasing values for food conversion. A low plateau (maximum conversion) was observed for natural food intake 4 to 7 percent of mean body weight. Fish fed at one percent of body weight did not grow. This level, amounting to 10 milligrams of food per gram of fish weight, was considered a maintenance requirement. Pentelow proposed a "Net Efficiency" figure as more suitable than gross conversion, calculated by subtracting the maintenance amount of food from the gross amount and dividing this by fish weight gain. Net Efficiency was approximately 3 milligrams per gram, dry weight per fish weight at all levels of food intake.

Maintenance amounts of food for routine metabolism have been reported by experimental observations (9), and by calculations (28). Calculated values based on oxygen consumption have been worked out for several species. Winberg's equation for salmon is:

$$Q = 0.498\ w^{0.176} \qquad \text{(C)}$$

where Q = oxygen consumption in ml per hour
w = fish weight in grams

The equation may be used to show that a 100-gram fish requires 16.48 ml O_2 per hour or 395.5 ml O_2 per day. Since 1000 ml O_2 are equivalent to 4.69 Calories at a R.Q. of 0.70, the maintenance requirement for a 100-gram salmon is 1.8 Calories. This value may be related to food requirements by dividing it by the net energy per gram of food. If a dry food contains 4.0 Calories digestible energy per gram, then 0.450 grams will maintain a 100-gram fish for one day. Converted to milligrams to correspond with units given by Davies (9), 4.5 mg/g. fish/day is similar to observed data. (See Table IV for comparative values of maintenance by observed and calculated methods.)

TABLE IV

A COMPARISON OF THE MAINTENANCE REQUIREMENTS OF FISH BY OBSERVED AND CALCULATED METHODS

Species	Type of food	Average Weight Fish, gms.	Approximate mg dry feed per gm fish/day	Reference
Trout	Gammarus	17.5	4.7	Pentelow (36)
Trout	Minced meat	50.0	4.0	Brown (60)
Plaice	Mytilus	33.0	3.0	Dawes (61)
Goldfish	Enchytraeus	21.5	3.0	Davies (9)
*Salmon	Std. Metabolism	-----	4.5	Winberg (28)
*Ave. all fish	Std. Metabolism	----	3.0	Winberg (62)

*Calculated values for salmon and "all fish" from Winberg's formulas (28).

Food requirements for a wide variety of conditions of fish activity, water temperature and stress conditions may be calculated from oxygen consumption using tables and graphs published by Beamish (37), Wohlschlag and Juliano (38), and Dollar (39).

An expression of intake versus growth in terms of total energy was shown by Davis and Warren (30) in a comprehensive study of the bioenergetics of the sculpin. As food biomass (expressed as Calories) increased, growth increased but at an increasing cost for maintenance. Efficiency of energy utilization decreased as food intake increased beyond 68.5% of maximum. Respiration at the highest level of food intake was found to be four times that at the lowest level. Over-all assimilation efficiency was determined by calculation to be 82% (See Table V. Energy utilization by yearling sculpin for a period of 55 days.)

TABLE V

ENERGY UTILIZATION BY YEARLING SCULPINS FOR A PERIOD OF 55 DAYS (From: Davis and Warren (30))

Fish Biomass Calories	Food Consumed Calories	Food Assimilated Calories	Growth Calories	Growth as % of Assimilated
1346	467	383	60	15.7
1246	505	414	106	25.6
1451	511	419	77	18.4
1440	993	814	327	40.2
1352	1011	829	225	27.1
1372	1024	840	307	36.5
1587	1754	1438	607	42.2
1490	1795	1472	692	47.0
1591	1800	1476	674	45.7
1554	2005	1644	568	34.5
1507	2078	1704	656	38.5
1559	2618	2147	822	38.3

In a previous reference Gerking (12) showed that the efficiency of protein absorption was not affected by different rates of feeding; he also found that above the level of maintenance, nitrogen utilization as weight gain was in direct proportion to the amount of nitrogen consumed. The efficiency of protein utilization for growth, calculated as:

$$\frac{\text{Protein N retained}}{\text{Protein N absorbed}} \times 100$$

increased asymptotically to about 32%. He concluded, "the maximum efficiency of protein utilization for growth is apparently related only to the maximum rate of protein consumption . . ."

Population density. Several research workers have considered that fish population density was not a decisive factor in food utilization and growth, that high stocking could be corrected by feeding, and improved water chemistry, that crowding reduced food consumption because of competition and that efficiency of food utilization was of more importance. Huq (40) concluded that competition for food contributes only in a limited way to growth and that weight gains decreased over time in proportion to stocking rate. Kinne (10) worked with the desert pupfish which require a minimum territory and continually fight to establish such a space. Decreased growth in aquariums with increased stocking was due partly to a reduction in food intake even when offered unrestricted supply and partly to increased maintenance required due to nervous excitation. Kinne postulated that the accumulation of metabolic products and low dissolved oxygen depressed metabolism. Burrows (41) observed reduced growth rates and reduced physical stamina with pre-disposition to bacterial gill disease as a result of changes in water chemistry induced by crowding.

An explanation for reduced stamina and growth under conditions of stress from crowding in mammals was found by Christian (42) to be due to an exhaustion of the adreno-pituitary system inherent in high populations. Here no accumulations of metabolic products forced a change in the environment so that an adaptation was made to a mechanical or psychological force not a chemical one.

Davies (43) found the effect of crowding to be lowered heat production corresponding to depressed metabolic rate. This occurs with decreased level or quality of food, low dissolved oxygen, presence of stress chemicals, temperatures above or below optimum range, presence of disease organisms or exhaustion of the endocrine-hormonal system.

Temperature. The question of temperature dependence is a complex one because:

a. Optimum conversion, maximum growth and optimum protein/calorie ratio are all temperature dependent functions.
b. The parameters of these functions have not been satisfactorily worked out.
c. While a number of studies have been made on effect of temperature on growth and food efficiency, few have been concerned with maximum growth or optimum conversion.

Optimum growth and food conversion as a function of temperature have recently been analyzed by Allen and Strawn (44). At 31°C channel catfish fingerlings showed the fastest growth and at 29°C the best food conversion (See Table VI). Values for conversion ranged from 2.57 at 21°C to a low of 1.35 at 29°C, increasing to 5.92 at 36°C. It was difficult to acclimate fish to 36°C water, 100% mortality occurring in one replicate tank.

TABLE VI

70-DAY TEST GROWTH OF CHANNEL CATFISH AT VARIOUS TEMPERATURES (From: Allen and Strawn (44))

Temperature	MM. Mean Fork Length*	Food Conversion
21° C	38.1	2.57
23°C	50.3	2.36
25° C	53.1	2.09
27° C	64.3	1.40
29° C	66.8	1.35
31° C	65.9	1.43
32°C	67.6	1.65
33° C	65.1	1.88
34° C	61.4	1.92
35° C	56.7	2.32
36° C	38.1	5.92

* L_0 = 15.3 mm.

Primarily studying optimum protein density for growth and food conversion in channel catfish, Dupree and Sneed (45) found that at 21°C conversion was not as good as at 24°C. Table VII shows that for casein a slight difference in conversion was found, lower for fish in 24.4°C water than at 20.6°C, but for vegetable proteins (wheat gluten and soybean) a decrease in temperature resulted in much less conversion efficiency.

For silver salmon stocked in production raceways, food conversion appears to be related more to season than to water temperature, size of fish or amount of food used. Cairns (46) presented hatchery statistics for the 1960-1961 seasons as shown in Table VIII. These data analyzed statistically indicate that food conversion may be correlated with time (April being the point of best conversion) but not with water temperature.

TABLE VII

DATA SHOWING THE INTER-RELATIONSHIPS BETWEEN FOOD CONVERSION AND WATER TEMPERATURE FOR SEVERAL LEVELS OF PROTEIN IN TEST DIETS FOR CHANNEL CATFISH FINGERLINGS (From: Dupree and Sneed (45))

Percent Protein		Food Conversion	
		76°F (24.4°C)	69°F (20.6°C)
Casein -	11.7	2.2	2.8
	17.1	1.6	1.7
	22.5	1.1	1.3
	28.3	0.9	1.1
	33.6	0.9	1.1
	39.0	0.9	1.0
	50.4	0.8	1.3
Wheat gluten -	11.3	1.8	2.6
	16.1	1.7	2.9
	21.1	1.4	2.9
	26.1	1.2	2.6
	30.2	1.1	4.3
	34.6	1.0	4.3
	43.5	1.1	-----
Soybean protein -	11.6	2.4	3.6
	16.8	1.8	3.4
	22.1	1.3	2.9
	27.7	1.2	2.2
	32.6	1.1	1.5
	37.7	1.2	2.3
	47.9	1.6	2.7

For the desert pupfish (10) pre-adult growth was maximum at a water temperature of 30°C, with optimum food conversion at 20°C. Older fish (indicated by length) made better use of food at lower temperatures than did young fish. Kinne concluded "Conversion Efficiency is not maximal within the range of the temperature optimum for growth." Such a coincidence could, of course, hardly be expected, since the general activity (spawning, locomotion, fighting, territorial behavior, etc.) is much higher at 30° than at 20°C where no spawning and very little territorial activity is observed.

Small speckled trout fed minnows (47) exhibited optimum growth at 13°C with the ratio of food consumed to weight gain showing an increase above that temperature.

The interaction of metabolic plane of activity with growth and food conversion was reported by Dupree and Sneed (45) (Table VII). DeLong *et al.* (7) stated "It is possible that the metabolic activity at low water temperature is at such a reduced level that it does not permit rapid elimination of excess amino acid pool of very young fish receiving high levels

TABLE VIII

RELATIONSHIP BETWEEN FOOD CONVERSION AND SEASON FOR YOUNG SILVER SALMON (Courtesy, D. Cairns, Carson NFH, BSFW, Carson, Washington)

	1960		1961	
Month	Water Temperature	Food Conversion	Water Temperature	Food Conversion
March	42	4.2	43.1	2.0
April	44	1.9	44.9	2.2
May	45	2.9	44.7	2.6
June	45.4	2.8	46	2.5
July	46	2.6	46.1	3.1
August	46	3.9	46.1	3.3
September	45	5.4	43.9	3.4
October	44	4.3	42.8	3.8
November	43.3	4.4	41.2	5.7
December	41.8	4.9	40.0	6.3
January	41.9	3.8	39.8	3.1
February	42.8	2.9	40.1	2.8

of peptide-bound and free amino acids. At higher water temperatures, however, metabolism and enzymatic activity in the fish would be at a sufficiently high level to permit adequate utilization of protein."

Food quality. Utilization of individual proteins and mixtures of primarily animal or vegetable sources were determined by Dupree (48) in terms of channel catfish growth and conversion. The results at two temperatures are summarized in Table IX. The best animal proteins were whole egg, casein and fish meal (conversions of 1.4, 1.6 and 1.6 respectively in the warmer water) and the best plant protein was soybean meal (conversion 3.0 in warmer water). A mixture of the animal proteins gave better weight gain and food conversion than any protein tested alone or a mixture of all plant proteins.

Age, size. The relationship between protein absorbed (digested) and that used for growth decreases as fish become larger. Gerking (3) found a linear

TABLE IX

UTILIZATION OF PROTEINS AND MIXTURES OF PROTEIN AT TWO WATER TEMPERATURES AS MEASURED BY PERCENTAGE WEIGHT GAIN AND CONVERSION (From: Dupree (48))

Protein	75-88°F		65°F	
	% Gain	Conversion	% Gain	Conversion
Fish meal	165	1.6	86	3.1
Zein	7	31.6	none	
Casein	176	1.6	91	2.9
Whole egg	202	1.4	125	2.1
Wheat gluten	57	5.0	18	14.6
Soybean	92	3.0	61	4.3
Gelatin	56	4.9	30	8.8
Blood plasma			63	4.2
Whole Blood			5	49.2
Blood albumin			10	26.3
All-animal	235	1.2	135	1.9
All vegetable	91	3.0	39	6.8

decrease in protein utilized by longear sunfish and a curvilinear decrease in protein utilized by green sunfish as beginning weight increased. Kinne (10) reported essentially the same response, but at five different temperatures. At optimum temperature his conversion efficiency decreased with size increase but at temperatures lower than optimum efficiency improved as fish became larger and acclimated to a higher plane of metabolism. For his 25°C, 30°C and 35°C tests, efficiency dropped rapidly with age as indicated by length. At his highest temperature, the largest fish (22 weeks old) required 32.1 mg of food (wet natural worms) to yield one mg weight gain.

Our work on age-group I, II and III channel catfish show a decreasing conversion with age. In spite of careful attention to feeding management (the use of feeding trays as an indication of daily requirements, monthly sampling and re-calculation of amount fed), older fish gained significantly less on the same food than younger fish. (See Table X. Feed conversion data for catfish fed a basic food).

TABLE X

FEED CONVERSION DATA FOR CATFISH FED A BASIC FEED[1]

Fish Age and Species	Stocking Rate	Conversion	Year
White catfish 1st year	12,000/Acre	0.9	1963
Channel catfish 1st year	12,000	0.9	1963
Blue catfish 1st year	12,000	1.0	1963
White catfish 2nd year	1,500	2.4	1964
Channel catfish 2nd year	1,500	2.2	1964
Blue catfish 2nd year	1,500	2.7	1964
Channel catfish 2nd year	1,500	1.3	1965
White catfish 3rd year	1,000	2.0	1965
Channel catfish 3rd year	1,000	1.8	1965
Blue catfish 3rd year	1,000	2.5	1965
Channel catfish 3rd year	1,500	3.2	1965
Channel catfish 2nd year	1,500	1.3	1966

[1]Feeds made 1965 and following contained added vitamins.

DISCUSSION

Fish growth is not a simple response to ingested nutrients, but a complex function of several dependent variables. Production in a water impoundment may be expressed by the equation:

$$W_t = f\,(T + S_r + W_o + H + F_a + F_q + F_m + F_f + D + W_p + \ldots\ldots) \qquad \text{(D)}$$

Where:
T = time
S_r = stocking rate
W_o = initial size
H = water temperature
F_a = amount of feed
F_q = quality of feed
F_m = method of feeding
F_f = frequency of feeding
D = disease and parasitism
W_p = water parameters

Under standard conditions of production, basic qualitative nutrient requirements may be found reliably by careful attention to purified dietary components and physical properties of the test diet, and from these growth may be predicted in replicate lots, e.g. Halver *et al.* (58). Quantitative requirements are extremely difficult to determine, since imbalance, interrelations, and individual variations generate special interpretative procedures (57). Transferring quantitative fundamental data to commercial fish culture involves feeding for expected production of a pond population, rather than for individual requirements.

The growth potential of separate genetic stocks has not been proposed in formula (D) since factors therein are basically outside the scope of "nutritional score." Wohlfarth and Moav (59) are currently engaged in testing genetic variations in carp production. In discussing the regression of gained weight on initial body weight, they say "when different stocks of fish are tested for genetic differences of their growth rates, the regression of gained weight on initial weight should be considered and corrected for. Environmental factors may affect the magnitude of this regression, so that a "correction" term appropriate to one set of conditions is *not* necessarily applicable to another." The regression of gained weight (over a given time) on initial weight was found for carp to be:

$$y = 97.27 + 4.934\ x \qquad \text{(E)}$$

Where:
x = initial weight in grams
y = final weight gain in grams

Equations for growth of channel catfish as influenced by stocking rate have been determined at the Fish Farming Experimental Station, Stuttgart, Arkansas. For initial weights of 21 grams and stocking rates from 1,000 to 20,000 per acre, weight at the end of a six-months' season is shown by the equation:

$$W = 698\ S^{-0.679} \qquad \text{(F)}$$

Where:
W = average weight in grams
S = stocking rate in thousands

and for an initial weight of 7.5 grams, the equation is:

$$W = 305.14\ S^{-0.467} \qquad \text{(G)}$$

Numerical coefficients for all the factors in formula (D) will permit a theoretical maximum production. Actually, there is a phase response around a maximum due to inherent physical variables within each factor. The sensitivity of fish in a determinable environment makes it suitable as a biological tool for studying these variables.

REFERENCES

1. F. B. Morrison, "Feeds and Feeding," Appendix Table I. 22 Ed.,

Morrison Publishing Co., Clinton, Iowa, 1959.

2. C. H. Hubbell, "Feedstuffs," Feedstuffs Analysis Table, June 8, 1968.

3. D. Gerking, Physiol. Zool. **25**: 358 (1952).

4. J. T. Windell, Personal Communication.

5. M. E. Brown, *In* "The Physiology of Fishes," (M. E. Brown, ed.), Vol. I, Academic Press, N. Y., 1957.

6. M. Kitamikado, T. Morishita, and S. Tachino, Bull. Japan. Soc. Sci. Fisheries **30**: 46 (1964).

7. D. C. DeLong, J. E. Halver, and E. T. Mertz, J. Nutr. **65**: 589 (1958).

8. T. Nose, Bull. Freshwater Fisheries Res. Lab. **17**: 97 (1967).

9. P. M. C. Davies, Comp. Biochem. Physiol. **17**: 67 (1964).

10. O. Kinne, Physiol. Zool. **33**: 288 (1960).

11. J. T. Windell, Invest. Indiana Lakes & Streams **7**: 185 (1966).

12. S. D. Gerking, Physiol. Zool. **33**: 288 (1955).

13. J. M. Dodd, M. H. I. Dodd, and C. K. Goddard, Nature **184**: 1660 (1959).

14. R. R. Smith, J. Fisheries Res. Board Can. (In Press).

15. D. Inaba, C. Ogino, C. Takamatsu, T. Ueda, and K. Kurokawa, Bull. Japan. Soc. Sci. Fisheries **29**: 242 (1963).

16. R. P. Singh and T. Nose, Contrib. Freshwater Fish. Res. Lab. No. 210, U. of Toyko, 1967).

17. A. M. Phillips Jr., A. V. Tunison, and D. R. Brockway, Fish. Res. Bull. No. 11, N. Y. State Cons. Dept., Albany, 1948.

18. F. L. Strand, N. Y. Fish and Game J., January, 1958.

19. H. H. Mitchell and R. J. Block, J. Biol. Chem. **163**: 599 (1946).

20. B. L. Oser, J. Am. Diet. Assoc. **27**: 396 (1951).

21. H. H. Mitchell, Biological value of protein and amino acid interrelationships *In* "Methods for evaluation of nutritional adequacy and status." Chicago Quartermaster Food and Container Institute, 1954.

22. R. J. Block and K. W. Weiss, Amino Acid Handbook. Charles C. Thomas, Publisher, Springfield, Ill. 1956, pp. 138.

23. S. D. Gerking, "Introduction to the technical meeting." The Biological Basis of Freshwater Fish Production. Blackwell Sci. Publ. pp xi-xlv, 1967.

24. K. H. Mann, *In* "The Biological Basis of Freshwater Fish Production." (S. D. Gerking, ed.) Blackwell Sci. Pub., 1967, pp. 243.

25. A. P. Ostapenya and A. I. Sergeev, Voprosy Ikhtiologii **3**: 177 (1963).

26. W. Schaeperclaus, Textbook on pond culture, 1933.

27. Marcel Huet, Traite de Pisciculture, 1960, pp. 369.

28. G. G. Winberg, Translation Series 194. Fisheries Res. Bd. Canada. pp. 202, Table 32, 1956.

29. H. S. Swingle, Proc. 12th Ann. Conf. S. E. Assoc. Game and Fish Comm. 63, 1958.

30. G. E. Davis and C. E. Warren, J. Wildlife Management **29**: 846 (1965).

31. J. E. Paloheimo and L. M. Dickie, J. Fisheries Res. Board Can. **23**: 869 (1966).

32. N. I. Chugunova, "Age and Growth Studies in Fish." Transl. Nat. Sci. Found., Washington, 1963, pp. 132.

33. C. E. Warren, J. H. Wales, G. E. Davis and P. Doudoroff, J. Wildlife Management **28**: 617 (1964).

34. H. S. Swingle, Fisheries Res. Repts. to the Faculty, Auburn University, Auburn, Ala.

35. Auburn University, Department of Fisheries Report to the Faculty, Auburn, Alabama, 1960.

36. F. K. T. Pentelow, J. Exp. Biol. **16**: 446 (1939).

37. F. W. H. Beamish, Can. J. Zool. **42**: 189 (1964).

38. D. E. Wohlschlag and R. O. Juliano, Limnol. Oceanog. **4**: 195 (1959).

39. A. M. Dollar, U. S. Trout Farmers' Convention, Las Vegas, Nevada, October 8, 1964.

40. M. F. Huq, Scientific Researches. Vol. II (3): 112, 1965. The East

Regional Laboratories, Dacca.

41. R. E. Burrows, "Effects of Accumulated Excretory Products on Hatchery-Reared Salmonoids." U. S. Dept. Int. FWS, Res. Rept. 66, 1964.

42. J. J. Christian, J. Mammalogy **31**: 247 (1950).

43. P. M. C. Davies, Comp. Biochem. Physiol **17**: 983 (1966).

44. K. O. Allen and K. Strawn, Proc. 21st Ann. Conf. S. E. Assoc. Game and Fish Comm. 63, 1967.

45. H. K. Dupree and K. E. Sneed, U. S. Dept. Int., FWS Tech. Papers No. 9: p. 21, 1966.

46. D. Cairns, Personal Communication.

47. N. S. Baldwin, Trans. Am. Fisheries. Soc. **86**: 323 (1956).

48. H. K. Dupree, Administrative Report, Progress in Sport Fishery Research for 1966. Bureau of Sport Fisheries and Wildlife Resource Publ. 39. p. 130, 1967.

49. M. Knauthe, Zeitschrift Fischerei **6**: 317 (1898).

50. S. M. Morgulis, J. Biol. Chem. **36**: 391 (1918).

51. A. V. Tunison, D. R. Brockway, J. M. Maxwell, A. L. Dorr, and C. M. McCay, Fish. Res. Bull. No. 4, N. Y. State Conserv. Dept. Albany, N. Y., 1942.

52. E. M. Wood, "Methods for Protein Studies with Trout with Applications to Four Selected Diets," Thesis, Cornell Uni., 1952.

53. T. Nose, Bull. Freshwater Fisheries Res. Lab. **10**: 1 (1960).

54. W. E. Shanks, Personal Communication.

55. W. H. Hastings, Administrative Report, Progress in Sport Fishery Research for 1963. U. S. Dept. Int. Circ. **210**: 48, 1964.

56. M. Kitamikado, T. Morishita, and S. Tachino, Bull. Japan. Soc. Sci. Fisheries **30**: 50 (1964).

57. R. E. Chance, E. T. Mertz, and J. E. Halver, J. Nutr. **83**: 177 (1964).

58. J. E. Halver, D. C. DeLong, and E. T. Mertz, J. Nutr. **63**: 95 (1957).

59. W. G. Wohlfarth and R. Moav, Genetic investigation and breeding methods of carp in Israel. Fish Culture Research Station, Dor, Israel, (in press).

60. M. E. Brown, J. Exp. Biol. **22**: 118 (1946).

61. B. Dawes, J. Marine Biol. Assoc., U.K. **17**: 103 (1931).

62. G. G. Winberg, Translation Series 362. Fisheries Res. Board Can., pp 11, 1961.

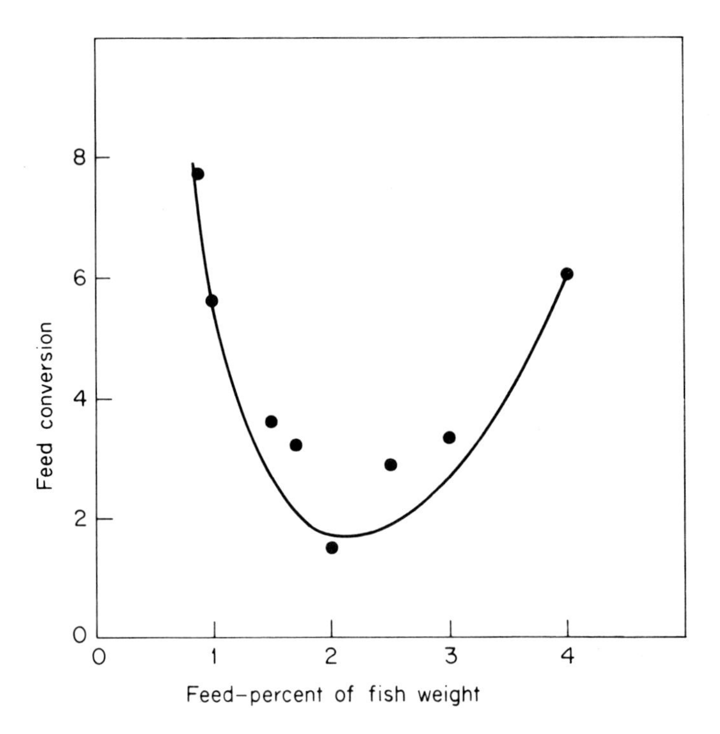

Fig. 1: Pounds dry feed per pound wet fish weight gain versus feeding rate. (Courtesy H. S. Swingle, Auburn University, Auburn, Alabama.)

INORGANIC SALT EFFECTS ON GROWTH, SALT WATER ADAPTION, AND GILL ATPase OF PACIFIC SALMON

W. S. Zaugg and L. R. McLain

INTRODUCTION

Studies concerned with the influences of inorganic elements on the life cycle of fish can be listed, for the most part, under either of two categories: toxicity or osmoregulation. References containing results of experiments designed primarily to determine inorganic nutritional requirements of fishes are indeed scarce, not because of lack of desire on the part of the investigators, but rather, because of the difficulty in experimentally controlling the uptake and excretion of inorganic elements by these animals. The ability of fish to exchange salts across the gill membranes, and even the skin, greatly complicates attempts to determine quantitatively those inorganic elements needed by the fish for maximal growth. Roeder and Roeder (1) reported that addition of ferrous sulfate to water in aquariums containing swordtail (*Xiphophorus helleri*) and platyfish (*X. maculatus*) caused increased growth rates, but the amount actually utilized by the fish for the accelerated growth was not determined. This is an example, however, of how fish can utilize mineral components of the water for nutritional purposes. There is no question that studies of ion absorption and excretion through the gill have nutritional as well as osmoregulatory implications.

On the other hand, the absorption of inorganic ions accompanying food consumption and digestion, a process which might ordinarily be considered fundamentally nutritional, may affect the osmoregulatory system. This is indicated by the report of Phillips (2) that the amount of calcium absorbed from the water varies with the quantity of calcium present in the diet. Low dietary calcium caused an increase in the amount absorbed from the environmental water, and the opposite effect occurred with high dietary calcium. The interdependence of osmoregulation and mineral nutrition, at least with respect to certain inorganic ions, is widely recognized.

Smith (3) showed that teleosts residing in a salt water environment maintain water balance by drinking the medium and excreting much of the accumulated monovalent salts extrarenally. Since this initial work, there have been a number of publications in this area (see review by Parry, (4) and Potts (5)). Hickman (6) determined that 99% of the sodium, 98% of the potassium, 93% of the chloride, and 60% of the calcium swallowed by the southern flounder was removed extrarenally. In some species drinking accounts for only a small part of the total influx of ions such as sodium and chloride, with the body surface, in particular the gill, playing a major role (4,5). As a consequence of the large influx of salt, whether from passive diffusion through the body surface or ingestion by drinking, fish living in a marine environment must be able to excrete large quantities of inorganic ions in order to maintain a suitable electrolyte balance. Since there is very little urine production and the excretory processes are principally extrarenal involving the gill, this organ must possess a highly efficient sodium pump or Na^+, K^+-stimulated adenosine triphosphatase (ATPase) activity. This has been experimentally verified by Epstein, *et al.* (7) and Kamiya and Utida (8) who have shown dramatic increases in the gill Na^+, K^+-stimulated ATPase activity of killifish and Japanese eel upon salt water adaption. Comparable changes in the ATPase activity of gills from salmon would be expected upon transition to a salt water environment.

This paper reports the effects of salt enriched diets on growth and some of the osmoregulatory processes associated with salt water adaption of young Pacific salmon. Results show that excess dietary salt caused reduced growth rates and decreased efficiency of diet utilization in young coho salmon (*Oncorhynchus kisutch*). However, fish fed supplemental salt adapted to full strength sea water with fewer mortalities than fish receiving a control diet. Migrating size coho salmon fed a sodium chloride enriched diet showed moderately elevated gill microsomal Na^+, K^+-stimulated ATPase activity. Animals fully adapted to salt water had a much more elevated activity, however. This high Na^+, K^+-stimulated ATPase decreased when the salt water adapted fish were returned to fresh water.

EXPERIMENTAL METHODS

Test diet. Ingredients of the complete test diet, fed to all control fish, are listed in Table I. For salt enriched diets the desired amounts of salt were

TABLE I

INGREDIENTS OF BASIC DIET

	g/300 g diet
Vitamin free casein	38
White dextrin	28
Gelatin	12
Mineral supplement[1]	4
Vitamin supplement[2]	9
Oil mixture[3]	9
Water	200

[1]Salt mixture No. 2, U.S.P. XIII (Nutritional Biochemicals Corp.), plus (in mg.): $AlCl_3 \cdot 6\ H_2O$, 0.6; KI, 0.6; CuCl, 0.4, $MnSO_4 \cdot H_2O$, 3.2; $CoCl_2 \cdot 6\ H_2O$, 4.0; $ZnSO_4 \cdot 7\ H_2O$, 12.
[2]Containing in mg: choline chloride, 500; inositol, 200; ascorbic acid, 100; niacin, 75; calcium pantothenate, 50; riboflavin, 20; menadione, 4; pyridoxine·HCl, 5; thiamine, 5; folic acid, 1.5; biotin, 0.5; vitamin B_{12}, 0.01; plus alpha cellulose flour, 8 g.
[3]Containing in grams: corn oil, 7; cod liver oil, 2; alpha tocopherol acetate, 0.04.

added dissolved in the water component. Diets were formulated primarily as described previously (9). Fish were fed twice daily, five days a week, and weighed every two weeks as described by Woodall and LaRoche (10).

The "Instant Ocean" salt mixture used in the feeding trials and in the salt water adaption studies was obtained from Aquarium Systems, Inc., Wycliff, Ohio. This mixture was composed principally of (%): NaCl, 65.2, $MgSO_4 \cdot 7H_2O$, 16.3; $MgCl_2 \cdot 6H_2O$, 12.8; $CaCl_2$, 3.3; KCl, 1.7; $NaHCO_3$, 0.5; and small amounts of other salts.

Salt water adaption trials. Salt water adaption trials were carried out by placing 20 fish (fasted for 2 days) from each diet in 6-liter aquariums containing 3.5 liters of "Instant Ocean" at 100% sea water strength (39.5 g Instant Ocean salt/1). On analysis, this salt solution was found to contain 35 g salts/kg water. The water temperature was maintained at 9.5 - 10.0°C. A constant stream of air was introduced through air stones and the water was changed daily. Fish were not fed during salt water residence. Mortalities were removed every one to four hours, depending on the rate of death.

Adenosinetriphosphatase. ATPase activities were studied on preparations of gill microsomes. Gill filaments (0.05 - 0.20 g wet wt.-blotted) were trimmed from the arches of sacrificed or anesthetized fish (0.01% tricaine methanesulfonate-Sandoz) and placed in ice cold 0.3 M sucrose for rinsing.

Filaments were removed from the sucrose solution, blotted, weighed if desired, and placed in 2.0 ml cold sucrose (0.3M)-EDTA* (5mM)-deoxycholate (0.1%) solution, pH 7.3, then homogenized thoroughly in a small glass homogenizer. The homogenate was placed in a 4 ml centrifuge tube, and the homogenizer was rinsed with 1.5 ml of the sucrose-EDTA-deoxycholate solution which was added to the homogenate. The resultant mixture was centrifuged (Spinco) for 30 minutes at 9,000 X g (max.) to sediment debris and mitochondria. The supernatant liquid from this centrifugation was decanted and centrifuged at 60,000 X g (max.) for 60 minutes. The supernatant liquid was discarded and the pellet and tube rinsed with 0.3 M sucrose-5 mM EDTA, pH 7.3 (no deoxycholate). The pellet was then suspended with a glass homogenizer in 0.3 to 0.6 ml of the sucrose-EDTA solution (volume depending on weight of gill filaments). The suspension was placed in a conical glass tube and centrifuged for 15 minutes at 700 X g (max.) in a swinging bucket rotor (International No. 269). This removed larger, clumped particles. The supernatant suspension was decanted and used fresh or after being frozen no more than one week.

Reaction mixtures contained (mM): $MgCl_2$, 2.0; Tris-HCl (Sigma), 13; $(Tris)_4ATP$, 2.7; and when employed NaCl, 50; KCl, 10; in a final volume of 1.5 ml; final pH 7.0. $(Tris)_4ATP$ was prepared by passing a solution of 1.0 g Na_2ATP (Sigma) in 15.0 ml glass distilled water through a column containing 25 g Dowex 50W resin, 200-400 mesh, acid form, into 2.5 ml of 2.5 M Trizma Base (Sigma). The column was rinsed with 60 ml of water in 15 ml aliquots. The pH of the combined eluates was adjusted to 7.0 with a small amount of 2.5 M Trizma Base. The concentration of ATP in this solution was determined spectrophotometrically at 259 mμ using a molar absorptivity of 15,400 (11). Reactions were initiated by introduction of the microsomal suspension containing from 0.03 to 0.1 mg protein and were allowed proceed for 10 minutes at 30°C. The reaction was stopped by addition of 0.1 ml cold 35% $HClO_4$. Samples were placed on ice after $HClO_4$ addition, then centrifuged for 10 minutes at 600 X g in a refrigerated centrifuge. Aliquots (0.5 ml) were used for inorganic phosphate determinations by atomic absorption spectrophotometry (12). The aliquot was added to 4.5 ml water and 3.0 ml 2-octanol followed by 0.5 ml of the ammonium molybdate reagent. Immediate extraction by shaking avoided excess acid hydrolysis of ATP. Prompt addition of citrate (1.0 ml) and a second shaking prepared the samples for reading. Standards contained all components of the reaction mixture with the exception of ATP and microsomes. A normal reaction mixture incubated 10 minutes at 30° without microsomes (microsomes were added after $HClO_4$) served as the blank.

EFFECT OF DIETARY SALT ON GROWTH

Table II shows the average gain in young coho salmon during 16 weeks of feeding on NaCl supplemented diets. All diets containing added NaCl caused less weight gain than in controls. The addition of only 1.5% salt resulted in 7% less weight gain than the controls fed no added salt. Moreover, salt

*Ethylenediaminetetraacetic acid, disodium salt.

TABLE II

EFFECT OF DIETARY NaCl ON GROWTH OF COHO SALMON (Oncorhynchus Kisutch)

Added NaCl (% dry wt.)	Average initial wt. (g)	Average final wt. (g)[1]	Average gain (g)	% of control
0	0.56	3.27	2.71	-----
1.5	0.55	3.07	2.52	93.0
3.0	0.58	2.87	2.29	84.5
6.0	0.56	2.64	2.08	76.7
12.0	0.57	2.46	1.89	69.6

[1]After 16 weeks of feeding, 500 fish/diet.

interfered with diet utilization, resulting in decreased conversion efficiency. As seen in Table III, up to 20% more diet (excluding the weight of NaCl) was required to produce a gram of gain when 6% salt was present in the diet. With the weight of salt included, 35% more diet per gram gain was required when 12% NaCl was present. There were only small differences in the total quantities of diet consumed (weight of salt included), indicating that the presence of salt up to 12% did not greatly affect the appetite. Obviously, the total amount of nutrients per gram of diet consumed was less in diets containing added salt.

TABLE III

EFFECT OF DIETARY NaCl ON DIET EFFICIENCY IN COHO SALMON (Oncorhynchus Kisutch)

Added NaCl (% dry wt.)	Diet Consumed[1,2] (g dry wt.)	Wt. gained[2] (g)	g diet per g gain	% more than control
0	902 -----	1266	0.71	--
1.5	871 (884)	1180	0.74 (.75)	4(6)
3.0	849 (875)	1092	0.78 (.80)	8(13)
6.0	854 (908)	1005	0.85 (.90)	20(27)
12.0	763 (867)	907	0.84 (.96)	18(35)

[1]Minus added NaCl. Numbers in parenthesis include NaCl.
[2]After 16 weeks of feeding, 500 fish/diet.

Table IV shows that decreased growth also occurred when a mixture of salts (Instant Ocean) was added to the diet. Although not tabulated here,

TABLE IV

EFFECT OF DIETARY INSTANT OCEAN SALTS
ON GROWTH OF COHO SALMON
(Oncorhynchus Kisutch)

Added Salts (% dry wt.)	Average initial wt. (g)	Average[1] final wt. (g)	Average gain (g)	% of control
0	1.13	3.90	2.77	----
2	1.12	3.69	2.57	93
4	1.08	3.58	2.50	90
8	1.03	3.37	2.34	84

[1]After 16 weeks of feeding, 1000 fish/diet.

determinations of diet conversions also showed reduced efficiencies with added salt.

DIETARY SALT AND ADAPTION TO SEA WATER

One of the primary objectives of the salt feeding experiments was to measure effects of increased salt consumption on the ability of young salmon to adapt to sea water. Fig. 1 shows the results of one salt water adaption trial. In this experiment feeding salt resulted in fewer mortalities when the animals were subjected to salt water of 100% sea water salinity. Although, in the many experiments conducted, there was not always a direct correlation between the level of salt fed and the mortality rate, we consistently observed greater survival with fish that had received supplementary dietary salt. This can be seen also in the data of Tables V and VI. The results in Table V indicate that at some time between 7 and 12 weeks on diets supplemented with Instant Ocean salts, young coho began to show increased survival times in sea water.

Table VI shows data collected from two separate experiments, comparing the TM_{50} values (time (hrs.) for 50% mortality) and numbers of surviving fish (after the stated exposure time) fed the control diet and a diet containing 4% NaCl - 2% KCl. From the second set of results it can be seen that sea water was tolerated with less difficulty after only 2 weeks on the salt enriched diet. Of the data reported in this table, the only experiment that failed to show a definite beneficial effect of NaCl-KCl-supplemented diets during sea water adaption was the one dated 8/19.

DIETARY SALT AND ADENOSINETRIPHOSPHATASE ACTIVITY

In May 1968 a small number of yearling juvenile coho salmon, which had been released from the local hatcheries for downstream migration, were retrieved from the river. One half of this group of fish was started on the complete test diet while the remainder received this same diet to which NaCl

TABLE V

EFFECT OF DIETARY INSTANT OCEAN SALTS ON THE LENGTH OF TIME (HRS) FOR 50% MORTALITY (TM_{50}) OF COHO SALMON IN SYNTHETIC SEA WATER

		Instant Ocean Salts Added to Diet (% dry wt.)							
		0		2		4		8	
Date (1968)	Weeks on Diet	Av. length[1]	TM_{50}[2]	Av. length	TM_{50}	Av. length	TM_{50}	Av. length	TM_{50}
4-30	5	6.3	32	6.1	32	6.2	41	6.2	39
5-13	7	6.6	36	6.9	35	6.6	33	6.5	38
6-17	12	7.2	39	7.3	44	7.2	66	7.3	66
6-25	13	7.3	42	7.4	64	7.1	55	7.4	55
7-1	14	8.0	44	---	---	---	---	8.1	71
7-8	15	7.6	38	---	---	---	---	7.7	56

[1]20 fish per aquarium having this average length (cm).
[2]Hours.

TABLE VI

EFFECT OF DIETARY SODIUM AND POTASSIUM CHLORIDE ON SURVIVAL OF COHO SALMON IN SYNTHETIC SEA WATER

			Control			+ NaCl-KCl (4 - 2% dry wt)		
Date (1968)	Weeks on Diet	Hours Exposure to salt water	Number of survivors[1]	TM_{50}	Av. length[2]	Number of survivors	TM_{50}	Av. length
4-30	5	72	0	32	6.3	4	42	6.4
5-13	7	72	1	36	6.6	2	38	6.5
6-17	12	162	0	39	7.2	8	72	7.4
6-25	13	143	0	42	7.3	8	99	7.6
7-1	14	163	1	44	8.0	6	82	8.0
7-8	15	167	1	38	7.6	2	57	7.6
7-15	1	167	3	52	8.0	6	54	8.0
7-22	2	336	4	47	7.9	6	71	8.1
8-5	4	168	0	50	8.2	7	100	8.4
8-19	6	168	2	40	8.5	0	50	8.4
8-26	7	192	8	180	9.2	14	---	9.1

[1]Number of fish surviving after exposure to salt water for the number of hours shown in column 3.
[2]20 fish per aquarium having this average length (cm).

(12%) had been added. Periodically, samples were taken from these two groups and gills were used to prepare microsomes for ATPase studies. The results are summarized in Table VII. From July 1 to October 2 the collected data indicated a moderate elevation in the Na^+, K^+-stimulated ATPase activity of the salt-fed fish. Since the fish did not eat well until after June 1, the lack of a difference in this activity between salt-fed fish and controls on June 13 was probably a consequence of limited diet consumption.

TABLE VII

EFFECT OF DIETARY SODIUM CHLORIDE ON COHO GILL MICROSOMAL ATPase ACTIVITY

Date (1968)	ATPase Activity (μmoles/mg/hr)					
	Control Diet			Control Diet + NaCl (12%)		
	Mg^{++}	$+Na^+, K^+$	Na^+, K^+ Stimulated	Mg^{++}	Na^+, K^+	Na^+, K^+ Stimulated
5-26*	32	40	8	---	---	---
6-13*	34	44	10	23	35	12
7-1*	---	---	---	26	58	32
7-8*	37	50	13	34	65	31
8-8	23	34	11	16	40	24
8-15*	27	35	8	24	41	17
8-22	---	---	---	21	40	19
10-2	41	55	14	33	61	28
Average	32	43	11	26	51	25

*Frozen preparations.

The Na^+, K^+-stimulated ATPase activity of gill microsomes is increased in young coho salmon which have become adapted to sea water (Table VIII). There was a 3-fold increase in this activity after 4 weeks of sea water residence. Exposure to fresh water after four weeks in sea water resulted in reduced Na^+, K^+-stimulated ATPase activity. Four weeks of fresh water residence following the period of sea water exposure resulted in the return of this activity to the value found in controls. It is significant that the Mg^{++}-dependent ATPase activity is lower in salt water adapted fish (28 compared to 53 for controls), and that this activity rose again when the fish were placed in fresh water following salt water residence. This is also evident from the data in Table IX.

Table IX reports ATPase activities of gill microsomes prepared from 2

TABLE VIII

EFFECT OF SALT WATER ADAPTION AND DEADAPTION ON COHO GILL MICROSOMAL ATPase ACTIVITY[1]

Treatment (No. of fish)	ATPase activity (μ moles/mg/hr)		
	Mg^{++}	$+Na^{+}, K^{+}$	Na^{+}, K^{+} - Stimulated
Fresh water controls (4)	53	75	22
Salt water 4 wks (4)	28	89	61
Salt water 4 wks - fresh water 2 wks (3)	45	84	39
Salt water 4 wks - fresh water 4 wks (3)	44	64	20

[1]Length of all fish 9.5 to 10.2 cm. fork length.

TABLE IX

SALT WATER AND FRESH WATER EFFECTS ON CHINOOK GILL MICROSOMAL ATPase ACTIVITY

Initial Treatment	ATPase activity (μ moles/mg/hr)						
	Mg^{++}	$+Na^{+}, K^{+}$	Na^{+}, K^{+} - stimulated	Second treatment	Mg^{++}	$+Na^{+}, K^{+}$	Na^{+}, K^{+} - stimulated
14 days SW[2] - 13 days FW	52	113	61	21 days SW	47	73	26
14 days SW - 13 days FW	38	123	65	21 days FW	35	114	79
21 days SW	25	98	73	28 days FW	43	92	49
Fresh water only	51	71	20	-----	---	---	---
Spawning chinook[3]	19	25	6	-----	---	---	---

[1]30-40 g fish from 12/66 hatch.
[2]SW = Salt Water; FW = Fresh Water.
[3]30-lb. male chinook obtained from holding pond of Little White National Fish Hatchery.

year old chinook salmon that had been exposed alternately to salt and fresh water. Gill filaments were trimmed from two arches of each anesthetized fish and used for microsome preparation following the period of exposure to a particular environment. Lines one and two show that a relatively high Na^+, K^+-stimulated ATPase was present after 14 days in salt water, then 13 days in fresh water. However, this activity was decreased after an additional 21 days in fresh water (line 1) and resembled that observed in the fresh water fish (line 4). When the 21 days were spent in salt water, however, there was an increase in the Na^+, K^+-stimulated activity (line 2) and a drop in the Mg^{++}-dependent activity. The fish exposed to salt water for 21 days (line 3) had a Mg^{++}-dependent ATPase activity lower than the control and a high Na^+, K^+-stimulated ATPase activity. After 28 days in fresh water, these activities were returning toward control values.

An adult, spawning male (line 5) having been in fresh water for about 2 months, had lost all Na^+, K^+-stimulated activity characteristic of salt water adapted fish. In fact, all ATPase activities were much lower than those of young, fresh water fish.

DISCUSSION

Impaired growth resulting from increasing the NaCl content of diets has been reported in rats (13). In further studies it was shown that excessive amounts of either sodium chloride or sea salts were much more toxic to rats fed diets deficient in B vitamins (14). It is not likely that a deficiency existed in the above fish diets since the B vitamins had been added in excess (15). It is of interest that a commercially prepared pellet fed to young salmon at a local production hatchery contains nearly 15% (dry wt.) salts, of which 6% is sodium chloride. Whether such levels of salts in these diets causes some growth inhibition is not known, but in view of the results reported here, it seems likely. Our original objective was to discover if feeding a salt supplemented diet to young coho salmon could condition them to make the transition from fresh water to salt water with less difficulty. We hoped that ingestion of elevated quantities of salts would activate the excretory processes involved in the elimination of these salts when swallowed by drinking sea water in a marine environment.

Using a survival as an indication of adaptability, several studies were made by introducing fish on test diets directly into full strength synthetic sea water. The results of such experiments (Fig. 1, Tables V and VI) showed that inclusion of salt in the diet, whether Instant Ocean salts or a NaCl-KCl combination, resulted in greater survival. Survival rates were increased after only two weeks of feeding the NaCl-KCl supplemented diet (Table VI). In other experiments NaCl-enriched diets produced similar results in salt water adaption studies with young coho salmon. Under the conditions used in these experiments mortalities were observed for up to two weeks in salt water. Fish surviving this two week period seemed to be completely adapted and very few mortalities were observed thereafter. Mortality rates in these salt water adaption studies are higher than those reported by Conte (16) for juvenile coho salmon, probably because of the greater salinity (35 compared to 30 g/kg).

The Na^+, K^+-stimulated ATPase activity in the nasal salt gland of ducks has been increased by feeding salt (17, 18). This gland provides an extrarenal route for electrolyte excretion, and in this sense, functions much as does the gill of a salt water adapted salmon which also has an elevated Na^+, K^+-stimulated ATPase activity (Tables VIII and IX). Attempts to increase this activity in the gill microsomes of young coho salmon by dietary salt were partially successful. The results in Table VII show a moderate elevation of activity, but still far short of the total capacity as judged by the activities measured in gills of fully adapted fish. Greater activation of the extrarenal excretory system by dietary salt may be difficult to achieve because normal renal functions in fresh water provide large amounts of dilute urine into which the excess salt load can be discharged. In addition, electrolytes would tend to diffuse passively through the gill membrane into the dilute environmental water with no need for an active sodium pump. However, it is significant that this system can be activated during fresh water residence.

The use of fish provides a unique opportunity for studying the development of a Na^+, K^+-stimulated ATPase activity. In fact, this opportunity is available for the study of any biochemical change in the gill which accompanies the salt water adaption process. The gill is a readily available organ from which samples can be removed under anesthesia without complicated surgical procedures or undue stress to the subject. Gill filaments can be removed from the arches with essentially no blood loss if care is taken to avoid injury to the afferent and efferent branchial arteries which lie next to the gill arch. Even when these are accidently cut, blood loss is minimized by rapid clotting and the fish appears to suffer no ill effects.

The gill-clipping technique makes it possible to follow the ATPase activities at two or three time intervals or to study the effects of varying treatments of salt and fresh water in a single fish. This method was used to obtain most of the data in Table IX.

Tables VIII and IX show changes in both the Mg^{++}-dependent and the Na^+, K^+-stimulated ATPase activities which accompany acclimation to either a salt or fresh water environment. The drop in Mg^{++}-dependent activity which seems to occur simultaneously with adaption to salt water and the increased activity associated with re-adaption to fresh water account for much of the changes observed in the Na^+, K^+-stimulated ATPase activity. It is tempting to speculate that upon exposure to salt water there is a conversion of the Mg^{++}-dependent form of the enzyme to a form requiring Na^+ and K^+ for full activation. Such a conversion has apparently been produced *in vitro* by treatment of brain microsomes with deoxycholate and NaCl or KCl (19). Matsui and Schwartz (20) also suggested that the conversion of heart microsomal Mg^{++}-dependent ATPase to the Na^+, K^+-dependent form occurred on treatment with deoxycholate and NaI.

Although we have consistently observed the decrease in Mg^{++}-dependent ATPase activity on adaption of coho or chinook salmon to salt water, this phenomenon may not be universal. Kamiya and Utida (8) failed to observe a significant change in the Mg^{++}-dependent activity in the gills of salt water adapted eels while there was a dramatic increase in the Na^+, K^+-dependent activity. Further work is necessary to resolve these differences and to determine the real significance of these changes.

There can be no doubt that salt water adaption results in an increase in the Na^+, K^+-ATPase in gill microsomes. From our experiments with migrating size coho salmon, this elevated activity is not found unless the fish is exposed to salt water, albeit some increase is obtained by feeding salt-laden diets. Therefore, the development of this particular phase of the adaption process must wait until the downstream migration is completed and the fish becomes exposed to a salt water environment. Preliminary work with coho salmon indicates that residence in 100% sea water for about one week is required to produce the first observable changes in Na^+, K^+-stimulated ATPase activity. Experiments now in progress should show relationships between development of this ATPase activity and other indications of adaption to salt water, such as serum electrolyte levels.

CONCLUSIONS

Young coho salmon (*Oncorhynchus kisutch*) gained less weight and demonstrated lower efficiencies of diet utilization than controls when fed salt-enriched diets (1.5 to 12%). However, the salt-fed fish suffered fewer mortalities when exposed to salt water of 100% sea water strength (35 ppt). The Na^+, K^+-stimulated ATPase activity of gill microsomes was elevated moderately by inclusion of 12% NaCl in the diet. Salt water adaption, however, resulted in a three to four-fold increase in this activity, which decreased again to normal fresh water levels when the fish was placed in fresh water. A decrease in the Mg^{++}-dependent ATPase activity was associated with salt water adaption. Re-adaption to fresh water caused this activity to rise again to the normal fresh water level. It is suggested that these results indicate that, in the presence of salt water, the Mg^{++}-dependent enzyme may be converted to a form requiring Na^+ and K^+ for full activation.

These experiments and those of other investigators have only begun to probe the complex biochemistry and physiology of osmoregulatory processes involved in the maintenance of electrolyte balance during salt water adaption of fishes. It is very probable that the basic mechanisms of ion transport activated in the gill under these conditions are similar to those present in tissues of warm blooded animals. Therefore, studies of the biochemistry and physiology of ion transport in the gill will not only produce important data directly related to fish, but should also provide valuable information for a more complete understanding of similar mechanisms in tissues of other animals. Thus, the use of fish as experimental animals can become very important to basic research in ion transport mechanisms.

REFERENCES

1. M. Roeder and R. H. Roeder, J. Nutr. **90**: 86 (1966).

2. A. M. Phillips, Jr., Trans. Am. Fisheries Soc. **88**: 133 (1959).

3. H. W. Smith, Am. J. Physiol. **93**: 480 (1930).

4. G. Parry, Biol. Review **41**: 392 (1966).

5. W. T. W. Potts, Ann. Rev. Physiol. **30**: 73 (1968).

6. C. P. Hickman, Jr., Can. J. Zool. **46**: 457 (1968).

7. F. H. Epstein, A. I. Katz, and G. E. Pickford, Science **156**: 1245 (1967).

8. M. Kamiya and S. Utida, Comp. Biochem. Physiol. **26**: 675 (1968).

9. J. E. Halver, J. Nutr. **62**: 225 (1957).

10. A. N. Woodall and G. LaRoche, J. Nutr. **82**: 475 (1964).

11. R. M. Bock and Nan-Sing Ling, Arch. Biochem. Biophys. **62**: 253 (1956).

12. W. S. Zaugg and R. J. Knox, Anal. Biochem. **20**: 282 (1967).

13. C. R. Meneely, R. G. Tucker and W. J. Darby, J. Nutr. **48**: 489 (1952).

14. A. B. Morrison and H. P. Sarett, J. Nutr. **68**: 231 (1958).

15. J. E. Halver, Proc. Vth International Congress of Nutrition, 191 (1960).

16. F. P. Conte, H. H. Wagner, J. Fessler, and C. Gnose, Comp. Biochem. Physiol. **18**: 1 (1966).

17. S. A. Ernst, C. C. Goertemiller, Jr., and R. A. Ellis, Biochim. Biophys. Acta **135**: 682 (1967).

18. G. L. Fletcher, I. M. Stainer and W. N. Holmes, J. Exp. Biol. **47**: 375 (1967).

19. J. Järnefelt, Biochem. Biophys. Res. Commun. **17**: 330 (1964).

20. H. Matsui and A. Schwartz, Biochim. Biophys. Acta **128**: 380 (1966).

COMMENTS

DR. CONTE: First, I would like to congratulate Dr. Zaugg on an ingenious approach. Your findings certainly are analogous to physiological acclimation; going through lower salinities gradually seems to precondition the animal better. This fits; if you give a little bit of salt you stimulate the system a little bit, then consequently, you allow for better survival. You have clearly explained that mechanism of acclimation versus adaptation. The second point is, as discussed before, are we dealing with two enzymes or are we dealing with one enzyme? It certainly looks from this data that we are dealing with one enzyme with two sites that may be inter-related on the molecule.

PROF. SINNHUBER: What would you recommend as a salt level for the Oregon Moist Pellet?

DR. ZAUGG: The salt level in the Oregon Moist Pellet, that is the variation four fed at Willard National Fish Hatchery, has about 15% salt content by dry weight according to our analysis. This seems quite high. We do not really know what effect it had because these fish are raised, of course, under different circumstances than our laboratory fish. We do at the present time have some troughs on Oregon Moist Pellet comparing it with our diet so maybe we can tell later.

DR. CONTE: Not being a nutritionist but trying to get a good diet, neither one of the diets works well when you put these fish into the marine environment. We have tried now for several years to carry these animals on into maturity at Newport in ocean waters, giving them the artificial diets. It has not worked. I do not have an explanation for these failures, but I think that Dr. Zaugg has a very good point that I thought was made in Dr. Mertz's talk. I pose the question again, in the test diet what is the percent protein, and then what is the minimum protein required when one adds salt? You are changing the entire condition for assimilation of those products in the intestine, and I think you have a good test system to measure conversion of protein.

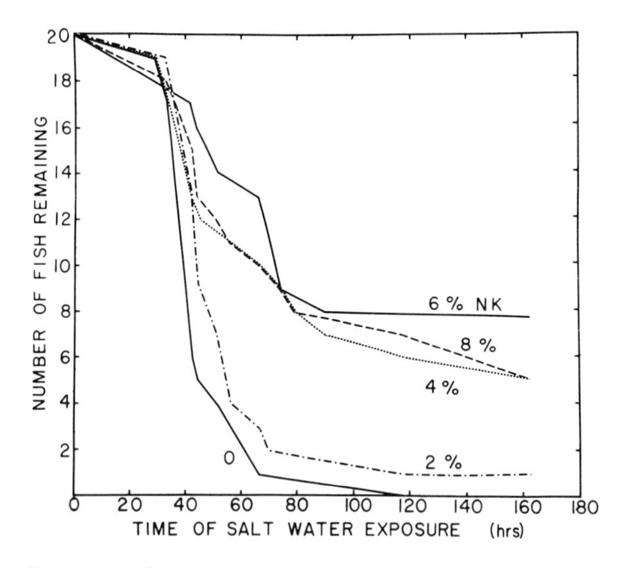

Fig. 1: Mortality rates of young coho salmon exposed to artificial sea water after receiving salt supplemented diets. Diets contained Instant Oceans salts in the quantities indicated (% dry wt.). The curve labeled 6% NK resulted from fish fed a diet containing 4% NaCl and 2% KCl. Other details are given in the Experimental section.

PARTICIPANTS

Ghaleb Abu-Erreish
University of South Dakota

Bill Allen
Westmar College
LeMars, Iowa

Sister Anthony
Heelan High School
Sioux City, Iowa

Richard L. Applegate
South Dakota State University
Brookings, South Dakota

Hector F. Balegno
University of South Dakota

L. Beckman
N. C. Reservoir Investigations
Pierre, South Dakota

Norman Benson
N. C. Reservoir Investigations
Yankton, South Dakota

Iftikhar H. Bhatti
Buena Vista College
Storm Lake, Iowa

Gerald R. Bouck
Pacific N. W. Water Laboratory
Corvallis, Oregon

W. R. Bridges
Fish Genetics Laboratory
Beulah, Wyoming

Thomas Briggs
Oklahoma Medical Center
Oklahoma City, Oklahoma

Donald R. Buhler
Oregon State University
Corvallis, Oregon

Ross V. Bulkley
Iowa State University
Ames, Iowa

Robert W. Burwell
Bureau of Sport Fisheries and Wildlife
Washington, D. C.

Sister Mary Odile Cahoon
College of St. Scholastica
Duluth, Minnesota

Ronald Chandler
Buena Vista College
Storm Lake, Iowa

K. V. R. Choudari
University of Nebraska
Lincoln, Nebraska

James B. Christiansen
Buena Vista College
Storm Lake, Iowa

T. Clifford
South Dakota State University
Brookings, South Dakota

Elgie B. Coacher
Black Hills State College
Spearfish, South Dakota

Joe B. Coacher
Black Hills State College
Spearfish, South Dakota

Ole B. Conn
University of Nebraska
Omaha, Nebraska

D. Crossley
Buena Vista College
Storm Lake, Iowa

Gerald J. Crowley
Sierra Nevada Aquatic Research Lab.
Bishop, California

Richard A. De Long
Graceland College
Camoni, Iowa

Ray Dillon
University of South Dakota

PARTICIPANTS

James Divelbiss
Westmar College
LeMars, Iowa

Robert H. Dolan
Sioux Falls Washington High School
Sioux Falls, South Dakota

Dennis Doupnik
Westmar College
LeMars, Iowa

Virgil E. Dowell
University of Northern Iowa
Cedar Falls, Iowa

Ed Dunbar
Southern State College
Springfield, South Dakota

Janet Egger
Mount Marty College
Yankton, South Dakota

J. Elrod
N. C. Reservoir Investigations
Pierre, South Dakota

Sister Veronica Fasbender
Mount Marty College
Yankton, South Dakota

R. Ford
Department of Game, Fish & Parks
Pierre, South Dakota

Charles Gasaway
Yankton, South Dakota

S. Gloss
South Dakota State University
Brookings, South Dakota

M. Goldman
University of South Dakota

John S. Gottschalk
Bureau of Sport Fisheries and Wildlife
Washington, D. C.

Melvin Greenblatt
University of Nebraska
Omaha, Nebraska

Yvonne Greichus
South Dakota State University
Brookings, South Dakota

M. Hannon
South Dakota State University
Brookings, South Dakota

Bryon Harrell
University of South Dakota

John Hartung
Dakota Wesleyan University
Mitchell, South Dakota

John Hasting
Dakota Wesleyan University
Mitchell, South Dakota

T. Hassler
N. C. Reservoir Investigations
Pierre, South Dakota

C. L. Hills
Dakota Wesleyan University
Mitchell, South Dakota

Robert C. Hiltibran
Illinois Natural History Survey
Urbana, Illinois

James F. Heisinger
University of South Dakota

William Houk
Dakota Wesleyan University
Mitchell, South Dakota

Ron Howard
South Dakota State University
Brookings, South Dakota

Joseph B. Hunn
Fish Control Laboratory
La Crosse, Wisconsin

Clarence Johnson
Western Fish Nutrition Laboratory
Cook, Washington

Howard E. Johnson
Michigan State University
East Lansing, Michigan

PARTICIPANTS

Fred C. June
N. C. Reservoir Investigations
Pierre, South Dakota

Jerry Kaiser
Yankton, South Dakota

K. B. Kerr
Salsbury Laboratories
Charles City, Iowa

John A. Kounas
University of South Dakota

D. Lamb
South Dakota State University
Brookings, South Dakota

Mary Ann Lambert
Kansas State University
Manhatten, Kansas

Stanley E. Lane
University of South Dakota

Geoffrey Lawrence
Cornell University
Ithaca, New York

J. C. MacLeod
Freshwater Institute
Winnipeg, Canada

Donald C. Malins
U. S. Bureau of Commercial Fisheries
Food Science Pioneer Research Lab.
Seattle, Washington

Helmut K. Mangold
Hormel Institute
University of Minnesota
Austin, Minnesota

S. V. Manohar
Freshwater Institute
Winnipeg, Canada

Finley D. Marshall
University of South Dakota

Danny Martin
Yankton, South Dakota

Wayne Marty
Westmar College
LeMars, Iowa

William Maunz
Buena Vista College
Storm Lake, Iowa

John McGuire
Kansas State Unviersity
Manhatten, Kansas

Helen Melaragno
Kansas State University
Manhatten, Kansas

K. Mirabella
Buena Vista College
Storm Lake, Iowa

Vivian Moody
Westmar College
LeMars, Iowa

Robert J. Muncy
Iowa State University
Ames, Iowa

S. Nading
Buena Vista College
Storm Lake, Iowa

Otto W. Neuhaus
University of South Dakota

R. Nichols
Department of Game, Fish & Parks
Pierre, South Dakota

John G. Nickum
South Dakota State University
Brookings, South Dakota

Ross F. Nigrelli
New York Aquarium; Osborn Labs.
of Marine Science
Brooklyn, New York

John H. Olwin
30 North Michigan Avenue
Chicago, Illinois

PARTICIPANTS

John Paulsrud
Hormel Institute
University of Minnesota
Austin, Minnesota

Robert J. Peanasky
University of South Dakota

Margaret H. Peaslee
University of South Dakota

D. R. Progulske
South Dakota State University
Brookings, South Dakota

LeRoy Ross
Westmar College
LeMars, Iowa

Richard Ruelle
Yankton, South Dakota

George D. Ruggieri
New York Aquarium; Osborn Labs. of Marine Science
Brooklyn, New York

Donald M. Sand
Hormel Institute
University of Minnesota
Austin, Minnesota

Arthur Scheier
Academy of Natural Sciences
Philadelphia, Pennsylvania

Hermann Schlenk
Hormel Institute
University of Minnesota
Austin, Minnesota

James Schmulbach
University of South Dakota

Betty Schneider
Westmar College
LeMars, Iowa

Jewell Schock
Wayne State College
Wayne, Nebraska

William R. Shawler
Westmar College
LeMars, Iowa

Edwin H. Shaw
University of South Dakota

Zakariya Shihabi
University of South Dakota

Raymond C. Simon
Oregon State University
Corvallis, Oregon

John Sinclair
Buena Vista College
Storm Lake, Iowa

Gary D. Small
University of South Dakota

Richard D. Spall
Oklahoma State University
Oklahoma City, Oklahoma

Robert Sparks
University of South Dakota

Friedrich Spener
Hormel Institute
University of Minnesota
Austin, Minnesota

V. Starostka
South Dakota State University
Brookings, South Dakota

W. Thorn
South Dakota State University
Brookings, South Dakota

Dale Tigges
Buena Vista College
Storm Lake, Iowa

M. Gene Ulrich
Westmar College
LeMars, Iowa

E. Urban
Buena Vista College
Storm Lake, Iowa

PARTICIPANTS

Wendell Utech
Westmar College
LeMars, Iowa

Gary Vance
Westmar College
LeMars, Iowa

L. Van Ray
Department of Game, Fish & Parks
Pierre, South Dakota

Bruno von Limbach
Fish Genetics Laboratory
Beulah, Wyoming

Charles Walburg
N. C. Reservoir Investigations
Yankton, South Dakota

Karl H. Wegner
University of South Dakota

Patricia White
University of South Dakota

Duane Winter
Buena Vista College
Storm Lake, Iowa

A. N. Woodall
Sierra Nevada Aquatic Research Lab.
Bishop, California

M. Yurkowski
Freshwater Institute
Winnipeg, Canada